THE METABOLISM OF ARSENITE

Arsenic in the Environment

Series Editors

Jochen Bundschuh

University of Southern Queensland (USQ), Toowoomba, Australia
KTH Royal Institute of Technology, Stockholm, Sweden

Prosun Bhattacharya

KTH-International Groundwater Arsenic Research Group, Department of Land and Water Resources Engineering, KTH Royal Institute of Technology, Stockholm, Sweden

ISSN: 1876-6218

Volume 5

ISGSD

International Society of Groundwater for Sustainable Development

Credit for Front Cover Photo:

Photograph by Heather Jamieson, Queen's University, Kingston, Ontario, Canada

Microbial biofilm from Giant Mine, Yellowknife, Northwest Territories, Canada. The biofilm was located growing in an abandoned stope below seepage from a diamond drill hole approximately 152 m below chambers where arsenic trioxide from roasting operations has been stored (230 m below land surface). The biofilm, which is about 2 m in length, has high levels of arsenic (i.e. $> 1\,\mathrm{g\,L^{-1}}$) and contains members of all three domains of Life: Bacteria, Archaea and Eukarya.

The Back cover structure of arsenite was made by Wolgang Nitschke, CNRS, Marseille, France

The Metabolism of Arsenite

Editors

Joanne M. Santini & Seamus A. Ward

Institute of Structural and Molecular Biology,
University College London, UK

CRC Press

Taylor & Francis Group

Boca Raton London New York

CRC Press is an imprint of the
Taylor & Francis Group, an **informa** business

A BALKEMA BOOK

First issued in paperback 2018

CRC Press/Balkema is an imprint of the Taylor & Francis Group, an informa business

© 2012 Taylor & Francis Group, London, UK

Typeset by MPS Limited, Chennai, India

Published by: CRC Press/Balkema
P.O. Box 447, 2300 AK Leiden, The Netherlands
e-mail: Pub.NL@taylorandfrancis.com
www.crcpress.com – www.taylorandfrancis.com

Library of Congress Cataloging-in-Publication Data

The metabolism of arsenite / [edited by] Joanne M. Santini, Seamus A. Ward. — 1st ed.
  p. cm. — (Arsenic in the environment)
 Includes bibliographical references and index.
 ISBN 978-0-415-69719-4 (hardback)
 1. Arsenic—Metabolism.   2. Arsenic—Environmental aspects.   3. Drinking water—Arsenic content.   4. Arsenic in the body.   5. Arsenic cycle (Biogeochemistry)   I. Santini, Joanne M.   II. Ward, Seamus A.
 TD427.A77M48   2012
 612.3'926—dc23
                                                                        2012003955

ISBN 13: 978-1-138-07782-9 (pbk)
ISBN 13: 978-0-415-69719-4 (hbk)

About the book series

Although arsenic has been known as a 'silent toxin' since ancient times, and the contamination of drinking water resources by geogenic arsenic was described in different locations around the world long ago – e.g., in Argentina in 1917 – it was only two decades ago that it received overwhelming worldwide public attention. As a consequence of the biggest arsenic calamity in the world, which was detected more than twenty years back in Bangladesh, West Bengal, India and other parts of Southeast Asia, there has been an exponential rise in scientific interest that has triggered high quality research. Since then, arsenic contamination (predominantly of geogenic origin) of drinking water resources, soils, plants and air, the propagation of arsenic in the food chain, the chronic effects of arsenic ingestion by humans, and their toxicological and related public health consequences, have been described in many parts of the world, and every year, even more new countries or regions are discovered to have arsenic problems.

Arsenic is found as a drinking water contaminant in many regions all around the world, in both developing as well as industrialized countries. However, addressing the problem requires different approaches which take into account the different economic and social conditions in both country groups. It has been estimated that 200 million people worldwide are at risk from drinking water containing high concentrations of arsenic, a number which is expected to further increase due to the recent lowering of the limits of arsenic concentration in drinking water to $10\,\mu g\,L^{-1}$, which has already been adopted by many countries, and some authorities are even considering decreasing this value further.

The book series 'Arsenic in the Environment' is an inter- and multidisciplinary source of information, making an effort to link the occurrence of geogenic arsenic in different environments and the potential contamination of ground and surface water, soil and air and their effect on the human society. The series fulfills the growing interest in the worldwide arsenic issue, which is being accompanied by stronger regulations on the permissible Maximum Contaminant Levels (MCL) of arsenic in drinking water and food, which are being adopted not only by the industrialized countries, but increasingly by developing countries.

The book series covers all fields of research concerning arsenic in the environment and aims to present an integrated approach from its occurrence in rocks and mobilization into the ground- and surface water, soil and air, its transport therein, and the pathways of arsenic introduction into the food chain including uptake by humans. Human arsenic exposure, arsenic bioavailability, metabolism and toxicology are treated together with related public health effects and risk assessments in order to better manage the contaminated land and aquatic environments and to reduce human arsenic exposure. Arsenic removal technologies and other methodologies to mitigate the arsenic problem are addressed not only from the technological perspective, but also from an economic and social point of view. Only such inter- and multidisciplinary approaches, will allow case-specific selection of optimal mitigation measures for each specific arsenic problem and provide the local population with safe drinking water, food, and air.

We have the ambition to make this book series an international, multi- and interdisciplinary source of knowledge and a platform for arsenic research oriented to the direct solution of problems with considerable social impact and relevance rather than simply focusing on cutting edge and breakthrough research in physical, chemical, toxicological and medical sciences. The book series will also form a consolidated source of information on the worldwide occurrences of arsenic, which otherwise is dispersed and often hard to access. It will also have role in increasing the awareness and knowledge of the arsenic problem among administrators, policy makers and company executives and in improving international and bilateral cooperation on arsenic contamination and its effects.

Consequently, we see this book series as a comprehensive information base, which includes authored or edited books from world-leading scientists on their specific field of arsenic research, but also contains volumes with selected papers from international or regional congresses or other scientific events. Further, the abstracts presented during the homonymous biannual international congress series, which we organize in different parts of the world is being compiled in a stand-alone book series 'Arsenic in the Environment – Proceedings' that would give short and crisp state of the art periodic updates of the contemporary trends in arsenic-related research. Both series are open for any person, scientific association, society or scientific network, for the submission of new book projects. Supported by a strong multi-disciplinary editorial board, book proposals and manuscripts are peer reviewed and evaluated.

Jochen Bundschuh
Prosun Bhattacharya
(*Series Editors*)

Editorial board

Table of contents

Preface

Seamus A. Ward & Joanne M. Santini

1 ARSENIC: ENVIRONMENTAL AND MEDICAL

Arsenic (As) is toxic to most living cells in most organisms. Hundreds of millions of people in over 70 countries are at risk from contamination in drinking water – contamination that can cause a range of debilitating illnesses and several fatal cancers (for review see Chapter 3: Marcos and Hernández). Arsenic can enter the water supply in rainwater or as wind-blown dust from volcanoes, but especially through industrial pollution. It may also be mobilized from solid minerals to aqueous solution, at rates depending on such factors as mineral chemistry and grain size, pH, and microbe-driven redox processes. In Chapter 1 Kossoff & Hudson-Edwards review the chemistry and mineralogy of As and the processes involved in As cycling. They also summarize data on the concentrations of dissolved As in rain, rivers *etc.* The numbers are striking: in rainwater and rivers, concentrations of As in industrially polluted areas were three orders of magnitude higher than baseline levels. Groundwaters in the UK (i.e. baseline) contained $<0.5–10\,\mu g\,L^{-1}$; whereas natural As-rich aquifers contained up to $5{,}000\,\mu g\,L^{-1}$; geothermal groundwaters up to $50{,}000\,\mu g\,L^{-1}$ and most mining-affected groundwaters $50–1{,}000\,\mu g\,L^{-1}$. Close to a Texan As herbicide plant the level reached $408{,}000\,\mu g\,L^{-1}$; and the concentration in seepage from Giant mine, in Canada's Northwest Territories, has been recorded at $1{,}058{,}000\,\mu g\,L^{-1}$ – five to six orders of magnitude above baseline. Chapter 2 (Bromstad and Jamieson) focuses on this mine, one of the largest concentrations of arsenite at the Earth's surface.

At Giant mine, soluble As was produced when sulfides were oxidized during processing for gold; crushing of the rock created small grains and dust, from which As can escape into solution. As the authors point out, the extreme level of contamination in the seepage water is not the result of poor management. Throughout its lifetime (1949–1999) the mine was run according to best practice. Its complex history, however, has left a complex set of problems to be solved. The chapter explains these, and discusses the remediation program, which may not be able to eliminate As release, but will significantly reduce it and will prevent it from increasing as a result of flooding.

While remediation may be possible in the case of specific industrial sites, however large, the problem of naturally As-rich aquifers or geothermal groundwaters is harder to solve. The consequence is major and widespread sickness and mortality caused by chronic exposure to contaminated drinking water. The mechanisms by which As causes diseases such as cancers of bladder, skin and lung are not yet fully understood, but are unlikely to be simple. One thing that does seem clear is that susceptibility to As varies. In Chapter 3, Marcos and Hernández examine the role of genetics in the variation in As susceptibility. If we know which genes confer resistance to As we may be able to determine which mechanisms are most important in causing cancers. The authors review the evidence concerning several hypothesized damage mechanisms, e.g., As may act as a co-carcinogen by inhibiting DNA repair enzymes, or it may cause oxidative stress, with the resulting DNA damage. They go on to examine whether As-susceptibility is associated with specific alleles at loci relevant to As-damage mechanisms (e.g. DNA repair; resistance to oxidative stress, etc.). A number of intriguing results emerge, implicating loci involved in DNA repair, in As metabolism and in regulating the cell cycle. As the authors point out, there are still many groups of genes to be examined, and it is not yet clear whether the same loci are involved in susceptibility to the different cancers. This is a young and rapidly developing field, likely to yield many novel insights in the years ahead.

2   RESISTANCE, METABOLISM, OXIDATION, EVOLUTION

As Kossoff & Hudson-Edwards point out (Chapter 1), 'bacteria are probably directly or indirectly responsible for most of the redox transformation of As in the environment.' Chapters 4–7 are concerned with why this should be so and with how the transformations are achieved. They focus on direct effects: i.e. redox transformations catalyzed by As-specific enzymes. In Chapter 4, Stolz introduces As resistance and metabolism. 1) Resistance involves the reduction by ArsC of arsenate [As(V)] to the (more toxic) arsenite [As(III)], which can then be removed by an As(III)-specific exporter, ArsB. In some organisms the As(III) is methylated, producing a range of more toxic organic As compounds, some of which are highly volatile. 2) The second group of processes involve metabolism of As to gain energy. A number of bacteria have been shown to acquire energy by oxidizing As(III) to As(V), either aerobically (reviewed in Chapter 5) or anaerobically using alternative electron acceptors such as nitrate (reviewed in Chapter 6). Finally, in a wide range of Bacteria and Archaea the respiratory arsenate reductase (Arr) catalyzes As(V) respiration: i.e. the use of As(V) as terminal acceptor of electrons from organic or inorganic donors (summarized in Chapter 4).

The focus of this book is prokaryotic metabolism of As(III), which is catalyzed by two distantly related As(III) oxidase enzymes, Aio and Arx, the former of which is the most well studied. It is a heterodimeric bioenergetics enzyme with complex redox centers involved in electron transfer (see Chapter 7 by Heath *et al.*). Homologues of Aio have been found in both Archaea and Bacteria. This, together with other geochemical, phylogenetic and biochemical data, suggests that it is an ancient bioenergetic enzyme which was present in the last universal common ancestor, i.e. before the split between the Bacteria and Archaea c. 3.5 billion years ago (see Chapter 10). Arx, however, has been found in only a few organisms, all of which respire anaerobically with As(III) as the electron donor (see Chapters 6 and 10). It is closely related to the respiratory As(V) reductase (Arr), which van Lis *et al.* in Chapter 10 suggest evolved from polysulfide reductase.

3   GENOMICS AND REGULATION

Strains of prokaryotes rarely act alone. Instead, they influence As cycling as part of a community or a consortium. To understand how such an entity functions we need to know which As-related genes are present and under what conditions they are expressed. Since only a small fraction of prokaryotes can be cultured, community function is now often analyzed using culture-independent methods: genomics (examining what genes are present in the community), transcriptomics (mRNA), proteomics (proteins) and, potentially, metabolomics (products of catalysis by enzymes). Bertin *et al.* in Chapter 8, review the results of such culture-independent studies. They have revealed widespread induction of detoxification genes by As, and more restricted induction of the *aio* and *ars* genes involved in As(III) oxidation and As(V) detoxification. They also show that detoxification genes commonly appear in (horizontally transferred) genetic islands, which they share with genes for oxidative stress resistance, DNA repair or biofilm formation – all traits thought to be important in As resistance.

In Chapter 9 Wojnowska and Djordjevic examine the regulation of As(III)-oxidation gene expression. [It is arsenite oxidation that is the most important process for bioremediation (Chapter 11) and biosensing (Chapter 12).] Expression of the As(III) oxidase (*aio*) genes is induced by a two-component signal transduction system induced by As in the periplasm. This is not a 'typical' two-component signal transduction system, i.e. with a sensor histidine kinase that detects the signal and phosphorylates the transcriptional regulator which binds to DNA, switching on transcription. Instead, another protein in the periplasm, designated AioX, detects the signal and may itself act as the ligand for the sensor in place of As(III). An understanding of the regulation of As(III) oxidation in a wider range of organisms will be a useful aid in interpreting the results of community transcriptomics.

4 APPLICATIONS: BIOREMEDIATION AND BIOSENSORS

Remediation of As contamination has generally involved oxidation of the highly toxic As(III) to As(V), which is readily adsorbed to various solids that can be easily removed by filtration etc. But oxidation of As(III) is very slow without the addition of environmentally undesirable oxidants. Consequently, work is being carried out on the use of biological As(III) oxidizers. This is reviewed in Chapter 11 by Delavat *et al.* They examine a range of methods of 'active' remediation. These high-input systems use pure cultures or consortia that are either inoculated or spontaneously developed; the As(III) may be oxidized aerobically or, under anoxic conditions, anaerobically with (e.g.) nitrate as electron acceptor. Bench- or pilot-scale studies have found that both pure cultures and consortia are effective at high As concentrations. The authors also highlight several important areas requiring more work. It is not known whether any bioremediation systems can bring As concentrations down to levels below the World Health Organization recommended Maximum Contaminant Level of $10\,\mu g\,L^{-1}$. And almost nothing is known of any biological processes involved in zero-input 'passive' systems relying on natural attenuation of As. Clearly bioremediation is a promising set of methods meriting, and needing, substantial further work.

It is difficult to manage As levels unless they can also be measured reliably. Most available chemical methods suffer from one or more disadvantages: there is a high cost per sample; samples must be analyzed by trained personnel in the laboratory; methods for field use fail to detect As at the low levels required, etc. As French *et al.* (Chapter 12) point out, biosensors may provide a solution – an inexpensive field-based method of measuring As at low concentrations. These may be based on whole-cell responses, in which (perhaps transgenic) cells react measurably to the solute of interest. Alternatively, an enzyme such as As(III) oxidase may be immobilized on, e.g., an electrode, such that when it oxidizes As(III) the electrons are passed to the electrode, yielding a measurable current. The authors review types of output (luminescent; amperometric etc.), sensitivity to As, whole-cell and enzyme-based sensing, and a remarkable range of options for future development. They also point out that hardly any As-biosensing systems have been tested, so there is scope for a great deal more important research.

About the authors

James Ajioka is a Senior Lecturer in the Department of Pathology, University of Cambridge, with an interest in synthetic biology. His core research interests are host-parasite interactions in the apicomplexan parasite *Toxoplasma gondii*. He was co-leader of the University of Cambridge entries in the International Genetically Engineered Machine competition (iGEM).

Florence Arsène-Ploetze is a microbiologist with a PhD from the University Paris VI, France. She completed her post-doctoral work at the University of Freiburg (Germany) studying the heat shock response in *E. coli*. She is currently an Assistant Professor at the University of Strasbourg, France and her research interests include environmental genomics, microbial adaptation and stress response.

Philippe N. Bertin is a biologist. He obtained his Master's Degree in Ecology and PhD in Microbiology from the University of Louvain (Belgium). He was a researcher at the Pasteur Institute in Paris and is currently Professor of Microbiology and Genomics at the University of Strasbourg (France). He leads a research team working on adaptive responses in microorganisms and heads a network of laboratories on arsenic metabolism.

Mackenzie Bromstad obtained her BSc in Earth and Planetary Science from McGill University in Montreal (Canada) and MSc in Geological Sciences from Queen's University, Kingston (Canada). Her MSc research focused on arsenic geochemistry at Giant Mine, located in the Canadian Arctic. She is currently a geological contractor in the mineral exploration industry with a continued interest in environmental geochemistry.

François Delavat is a Microbiologist with a Masters degree from the University of Strasbourg (France). He is currently completing his PhD at the Génétique Moléculaire, Génomique, Microbiologie Unit from the University of Strasbourg. He is interested in understanding the structure and function of a microbial community from an Acid Mine Drainage (Carnoulès, France), focusing on the cultured organisms and their involvement in polymer degradation.

Kim de Mora is an engineer and synthetic biologist. During his MEng degree at the University of Edinburgh, he was a member of the iGEM (International Genetically Engineered Machine competition) team which developed a novel biosensor for arsenic. Subsequently he undertook a PhD in synthetic biology at the University of Edinburgh. During his PhD, Kim spent a year as a visiting researcher at Harvard Medical School investigating peroxisome engineering in *Saccharomyces cerevisiae* and promoter standardization in *E. coli*. His current research involves heart regeneration in adult zebrafish at Queen's Medical Research Institute, University of Edinburgh.

Snezana Djordjevic is a structural biologist with a PhD from Medical College of Wisconsin. She entered the world of bacterial signaling as an HHMI research associate in the laboratory of Professor Ann Stock at the CABM, New Jersey. She investigated structural and mechanistic aspects of methylation regulation of chemotaxis receptors by methyltransferase CheR and methylesterase CheB. She is currently a Senior Lecturer at University College London where she studies two-component signal transduction systems in *Rhizobium* sp. str. NT-26 and *Mycobacterium tuberculosis*.

Simon Duval completed his Masters and PhD in Bioinformatics-Structural Biochemistry and Genomics at the Aix-Marseille University (France). He undertook his research in biochemistry and biophysics at the Laboratoire de Bioénergétique et Ingénierie des Protéines at the CNRS (Marseille). Since then he has been a post-doctoral fellow at the Centre de l'Energie Atomique, Grenoble and Cadarache and is currently completing his second post-doc at Utah State University (Salt Lake City).

Alistair Elfick is a bioengineer with degrees in Mechanical Engineering and Biomedical Engineering (PhD) from the University of Durham (UK). He has won a Fulbright Commission Distinguished Scholar's Award and a Royal Academy of Engineering, Global Research Award enabling him to experience biomedical engineering research at the University of California, Berkeley. He returned to Scotland in 2004 to take an EPSRC Advanced Research Fellowship at the University of Edinburgh. His core research interest is the development of novel instrumentation for the measurement of biological samples. He was founding Director of the Edinburgh Centre for Synthetic Biology and its recent merger with the Centre for Systems Biology has created SynThSys Edinburgh, of which he is now co-Director. He has also co-authored the Royal Academy of Engineering report on Synthetic Biology, 2009.

Christopher French is a microbiologist and biotechnologist, with a degree in Biotechnology and Bioprocess Engineering (Massey University, New Zealand), and PhD in microbial biotechnology (Institute of Biotechnology, University of Cambridge). He is currently a lecturer in the School of Biological Sciences, University of Edinburgh, and member of the interdisciplinary Centre for Synthetic and Systems Biology. His current research is centered around synthetic biology, and includes improved methods for DNA assembly, as well as applications in biosensors, including a novel biosensor for detection of arsenic in groundwater, and processes for the conversion of biomass to useful products.

Lucie Geist graduated with a MSc in Biochemistry and Microbiology from the University of Strasbourg, France. She is currently a PhD student in microbiology at the University of Strasbourg. Her research involves understanding the impact of environmental conditions, including arsenic, on biofilm formation.

David Halter is a microbiologist with a PhD in Microbiology from the University of Strasbourg (France). He is currently working at the Department of Microbiology and Molecular Medicine, University of Geneva (Switzerland) and his research interests are in descriptive and functional genomics (transcriptomics, proteomics, metabolomics).

James Haseloff is a plant biologist working in the Department of Plant Sciences, University of Cambridge. His scientific interests are focused on the engineering of plant morphogenesis, using microscopy, molecular genetics, computational and synthetic biology techniques (www.haseloff-lab.org). Prior to joining the Department of Plant Sciences he served as group leader at the Medical Research Council Laboratory of Molecular Biology in Cambridge where his group developed advanced imaging techniques and modified fluorescent proteins for efficient use in plants. Before this, Jim was a research fellow at Harvard Medical School, working on trans-splicing ribozymes. He has also worked at the CSIRO Division of Plant Industry, Canberra, and developed methods for the design of the first synthetic RNA enzymes with novel substrate specificities.

Matthew D. Heath graduated with a BSc in Biochemistry and Microbiology from The University of Sheffield. He is currently pursuing an interdisciplinary PhD, funded by the Wellcome Trust, in the Institute of Structural and Molecular Biology at University College London. His current project is aimed at using a variety of techniques to characterize and compare wild-type and recombinant arsenite oxidases.

Alba Hernández is a Biologist and obtained a PhD in Genetics from the Autonomous University of Barcelona (UAB) (Spain). She is currently an Associate Professor in the Department of Genetics at the UAB. Her research is focused on the effects of arsenic in exposed populations, concentrating on the influence of genetic variability in arsenic-related risk. She is also involved in *in vivo* and *in vitro* projects on the mechanisms of arsenic carcinogenesis, with special interest in understanding the role that arsenic-induced oxidative DNA damage and deregulation of cell differentiation processes have on the acquired cancer phenotype.

Karen A. Hudson-Edwards is a geologist with a BSc from Queen's University, Kingston, (Canada), MSc from Memorial University of Canada and PhD from the University of Manchester (UK). Currently, she is a Reader in environmental geochemistry and mineralogy at Birkbeck, University of London. Her research focuses on understanding contaminant, nutrient and water cycling in Earth surface environments. Specific topics of interest include mechanisms of arsenic and manganese pollution of groundwater, controls on the cycling of contaminants in mining-affected environments, and the formation and stability of environmental minerals.

Heather E. Jamieson is a Professor in the Department of Geological Sciences and Geological Engineering at Queen's University, Kingston (Canada). She also holds an appointment and teaches courses in the School of Environmental Studies at Queen's. Her expertise is in the area of environmental geochemistry, particularly the mineralogical controls on the mobility of metals and metalloids (notably arsenic) in mine waste and the application of synchrotron-based X-ray experiments and other microanalytical methods to metal speciation in mine tailings, soils, sediments and household dust. Much of her fieldwork is in the Canadian Arctic but she has also conducted research in Nova Scotia, California, Montana, Spain and Australia.

Nimisha Joshi has a BSc in Botany and Chemistry, a MSc in Environmental Science from Lucknow University, India and a MRes in microbial biotechnology from the University of Edinburgh. She was involved in the development of a novel arsenic biosensor and biosensors to detect heavy metals such as zinc and copper using synthetic biology techniques. She is currently undertaking a PhD at the School of Geosciences, University of Edinburgh where she is studying the bactericidal impact of nanoparticles and the possible role of extracellular polysaccharides as a defence mechanism.

Sandrine Koechler is an assistant engineer in biology at the French National Centre for Scientific Research. She is currently working at the University of Strasbourg (France) on descriptive, functional and comparative analyses of various arsenic-metabolizing microorganisms and microbial communities from contaminated environments.

David Kossoff is a geochemist with a B.Sc. degree in chemistry and Earth science from the Open University and a Ph.D. in environmental geochemistry from Birkbeck, University of London. Currently, he is a post-doctoral researcher in environmental geochemistry at Birkbeck, investigating the bioleaching of copper-bearing ores and the significance of the mineral scorodite as a sink for arsenic and antimony in mining-contaminated environments.

Marie-Claire Lett is a microbiologist with a Masters degree and PhD from the University Louis-Pasteur, Strasbourg (France). She is currently Professor of Microbiology and Microbial Genetics at the Faculty of Life Sciences, University of Strasbourg. Her main research interests are in gene transfer of mercury resistance, gene transfer in aquatic environments, microbial metabolism of arsenic, survival mechanisms of stressed bacteria, culture of "uncultivable" bacteria and function of an arsenic-contaminated environment.

Didier Lièvremont is a biologist with a Masters degree from the University of Franche-Comté (France) and a PhD from the University Joseph Fourier, Grenoble (France). He is currently an Associate Professor in Microbiology at the Department of Biological Engineering, Institute of Technology, University of Strasbourg (France). His areas of interest include biogeochemical cycling of arsenic, bacterial diversity, bacterial microenvironment and bacterial biotechnology.

Marie Marchal is an environmental microbiologist and has a Master's degree and PhD from the University of Strasbourg (France). She is currently working as a postdoctoral researcher at the Swiss Federal Institute of Aquatic Science and Technology (EAWAG, Zürich) and has research interests in microbial communities assembly and resistance to environmental stresses.

Ricard Marcos is a Biologist and Professor in Genetics at the Universitat Autònoma de Barcelona (Spain). He has worked for a long time in Genetic Toxicology and Human Biomonitoring evaluating the effects of several environmental contaminants, mainly arsenic. *In vitro* and *in vivo* approaches have been used to determine the genotoxic risk associated to arsenic exposure, as well as the underlying mechanisms involved in arsenic genotoxicity. Specific studies have been conducted to determine the existence and origin of individual differences in arsenic susceptibility. He coordinates the Mutagenesis Group and is the President of the Spanish Society of Environmental Mutagenesis.

Wolfgang Nitschke completed his Masters in Physics and a PhD in Biochemistry from the University of Regensburg (Germany). He is currently at the Laboratoire de Bioénergétique et Ingénierie des Protéines (BIP/CNRS, Marseille/France) and his research interests are bioenergetics, metalloproteins and evolution. http://bip.cnrs-mrs.fr/bip09/

Ronald S. Oremland is a geomicrobiologist specializing in physiology/biochemistry of anaerobes isolated from extreme environments, and conducting field studies in soda lakes and hot-springs. He received a BSc in Biology from Rensselaer Polytechnic Institute, and his PhD from the Rosenstiel School of Marine and Atmospheric Sciences of the University of Miami. He did a postdoctoral associateship at the NASA Ames Research Center, after which he joined the US Geological Survey in Menlo Park, California where he has been for the past 35 years. His current title is Senior Scientist.

Thomas H. Osborne is an environmental microbiologist. He graduated with a BSc in Biology (with honours in Biotechnology) from the University of Edinburgh and was awarded a PhD in Microbiology from University College London. During his PhD he investigated the microbial oxidation of arsenite at low temperatures. He is currently undertaking post-doctoral research in the Institute of Structural and Molecular Biology, University College London and is investigating the influence of microorganisms on natural arsenic pollution.

Chad W. Saltikov is an environmental microbiologist specializing in molecular-genetics of microbial anaerobic respiratory processes that influence the biotransformation of pollutants in the environment. He received a BSc in Molecular Biology/Biochemistry and Aquatic Biology from the University of California, Santa Barbara; PhD in environmental toxicology from the University California, Irvine; and Postdoctoral Scholar in geomicrobiology at California Institute of Technology. In 2004, he joined the Department of Microbiology and Environmental Toxicology at University of California, Santa Cruz and is currently an Associate Professor.

Joanne M. Santini is an environmental microbiologist with interests in how microorganisms metabolize metals and metalloids such as arsenic and how this information can be used for biotechnological purposes (e.g. the development of biosensors). She was awarded a PhD in microbiology from La Trobe University, Melbourne (Australia) and is currently a Lecturer in the Institute of Structural and Molecular Biology, University College London. http://www.ucl.ac.uk/santini-lab

Barbara Schoepp-Cothenet has a PhD in biophysics and biochemistry from the Institut National Agronomique, Paris-Grignon. She is currently at the Laboratoire de Bioénergétique et Ingénierie des Protéines (CNRS, Marseille, France) and her research interests include bioenergetics, protein structure/function relationships, metalloenzymes and evolution.

John F. Stolz is Professor in the Department of Biological Sciences and Director of the Center for Environmental Research and Education at Duquesne University. He received a PhD from Boston University in microbial ecology and evolution. His main research interests are the microbial metabolism of metals and metalloids (e.g., arsenic, selenium, nitrate, and chromate) and microbial ecology and evolution studying community structure in stratified microbial mats and the mouse colon.

Robert van Lis is a biologist who completed his Masters degree in environmental science at Wageningen University, The Netherlands and a PhD in biochemistry at the National Autonomous University of Mexico. He returned to Europe, *via* the USA, to Germany and then France, and is now working in the field of bioenergetics in the Laboratory of Bioenergetics and Protein Engineering (BIP), CNRS Marseille (France).

Seamus Ward is a biologist-ecologist with a BA from Oxford University and a PhD from the University of East Anglia (UK). He has held fellowships at the University of East Anglia and the Agricultural University of Wageningen (Netherlands), and a lectureship and senior lectureship at La Trobe University in Melbourne Australia. He has since lectured at Birkbeck, University of London, and at University College London, where he is an honorary senior lecturer. His research has focused mainly on mathematical modeling in agriculture and on population and community ecology; in 2006 he joined the UCL team working on arsenite oxidation.

Marta Wojnowska graduated from University College London with a BSc in Genetics. She is currently involved in studying bacterial signal transduction as a part of her Wellcome Trust PhD project at UCL. The primary focus of her project is on the complexity of two-component signalling systems in *Rhizobium* sp. str. NT-26.

CHAPTER 1

Arsenic in the environment

David Kossoff & Karen A. Hudson-Edwards

1.1 INTRODUCTION

Arsenic (As) is toxic to most living cells. As such, it has been used throughout history as a mild poison hair removal for treatment of skin diseases and bubonic plague, and as a deadly poison in pesticides and for human murder (Cullen, 2008). In the mid-1990s chronic As pollution of groundwater was recognized in Asia (Chatterjee *et al.*, 1995; DPHE/BGS/MML, 1999), and since that time, many more As-polluted aquifers, soils and materials have been recognized. Many groundwaters that are used as drinking water sources are polluted with As (Smedley and Kinniburgh, 2002), and consumption of these waters can lead to melanosis, gangrene, cancer and, ultimately, death (Hopenhayn, 2006). Sun *et al.* (2004) have estimated that the health of 200 million people in more than 70 countries worldwide is at risk from drinking As-polluted groundwaters. Predictions are that long-term exposure to As will result in 1.2 million cases of hyperpigmentation, 600,000 cases of keratosis, 125,000 cases of skin cancer and 3,000 fatalities per year from internal cancers in Bangladesh alone (Yu *et al.*, 2003).

As a result of this serious global health issue, the literature on As in the environment is large and growing. This chapter gives a brief overview of this work, covering the chemistry, mineralogy and distribution of As in the Earth's surface environment. For further information readers are referred to the reviews of Matschullat (2000), Smedley and Kinniburgh (2002, 2004), Yudovich and Ketris (2005), Vaughan (2006), Plant *et al.* (2007), Cullen (2008), Drahota and Filippi (2009), Lièvremont *et al.* (2009), Ravenscroft *et al.* (2009), Sharma and Sohn (2009), and the literature cited therein and in this chapter.

1.2 CHEMISTRY AND MINERALOGY OF ARSENIC

1.2.1 *Chemical characteristics of arsenic*

Arsenic, with an atomic number of 33, has only one stable isotope (^{75}As) and an average atomic weight of 74.92. It is the 47th most abundant of the 88 naturally-occurring elements, with an average crustal abundance of 1.5 mg kg^{-1}. Occurring in group 15 of the Periodic Table, with an electronic configuration of [Ar] $3d^{10}4s^24p^3$, As is classified as a metalloid. It can exist as As(−III), As(−I), As(0), As(III) and As(V) (Table 1.1).

1.2.2 *Chemical forms of arsenic in gas*

The major gaseous As-bearing compound is the highly toxic arsine (Chauhan *et al.*, 2008; Table 1.1). Industrially, arsine is used in doping gas mixtures used for the production of particular semiconducting materials, principally gallium arsenide (Feng *et al.*, 2008). In the 19th century As was employed as a green dyeing agent (particularly for wall papers), and it has been speculated that the death of Napoleon I was the result of mold growing on such wallpaper and causing the release of toxic gaseous trimethyl arsine (Kintz *et al.*, 2007). In the wider environment arsine will only form under the most reducing of conditions (Sharma and Sohn, 2009). As pH falls arsine formation becomes more favored; indeed, arsine, together with mono-, di- and tri-methylarsine, are released from surfaces of marshy soil or swamps under anoxic conditions (Duker *et al.*,

Table 1.1.   Major chemical forms of arsenic.

Oxidation state of As	Gaseous forms	Aqueous forms	Solid (mineral) forms	Organic forms
As(–III)	Arsine [H_3As] Trimethylarsine [$As(CH_3)_3$]			
As(–I)			Arsenopyrite [FeAsS] Arsenical pyrite [$Fe(S,As)_2$] Loellingite [$FeAs_2$] Niccolite [NiAs] Orpiment [As_2S_3] Realgar [AsS] Enargite [Cu_3AsS_4] Tennantite [$(Cu,Ag,Fe,Zn)_{12}As_4S_{13}$] Cobaltite [CoAsS] Gersdorffite [NiAsS]	
As(0)			Elemental arsenic [As]	
As(III)		Arsenite [AsO_3^{3-}] Monohydrogen arsenite [$HAsO_3^-$] Dihydrogen arsenite [$H_2AsO_3^-$] Arsenous acid [H_3AsO_3]		
As(V)		Arsenate [AsO_4^{3-}] Monohydrogen arsenate [$HAsO_4^{2-}$] Dihydrogen arsenate [$H_2AsO_4^-$] Arsenic acid [H_3AsO_4]	Scorodite [$FeAsO_4$] (see Table 1.2 for other examples)	
Organic As				Monomethylarseonic acid (MMAA) $CH_3AsO(OH)_2$ Dimethylarsinic acid (DMAA) $(CH_3)_2As(O)OH$ Dimethylarsine $(CH_3)_3As$

2005). These gases can then be quickly converted by oxidation to As(V)-bearing aqueous phases (Turpeinen *et al.*, 2002; Table 1.1).

Arsine has been studied as a possible chemical weapon agent (CWA), but was rejected due to its instability (Chauhan *et al.*, 2008). However, it has been used as a precursor for other, more stable, gaseous As compounds used as CWAs (e.g., adamsite (10-chloro-5,10-dihydrophenarsazine) and lewisite (2-chlorovinyldichloroarsine); Graeme and Pollack Jr., 1998).

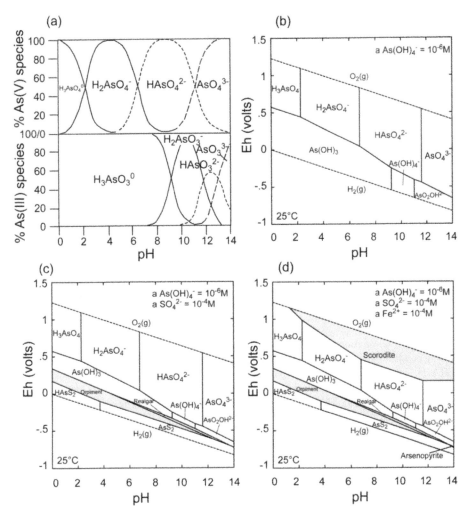

Figure 1.1. Examples of inorganic As(III) and As(V) species in water. (a) Percentage arsenite [As(III)] and arsenate [As(V)] species in water as a function of pH at an ionic strength of 0.04 M. Adapted from Meng *et al.* (2000); (b–d) Eh-pH stability diagrams for aqueous species of As at 25°C, 1 bar total pressure and an activity of As(OH)$_4^-$ of 10^{-6} M. (c) additionally shows the effect of SO$_4^{2-}$ on the stability fields, and (d) shows the effects of both SO$_4^{2-}$ and Fe^{2+} on the stability fields. The upper and lower limits of stability of water are shown by the dashed lines. Diagrams b to d were constructed using the Act2 program in The Geochemist's Workbench® package, version 8.0 Standard.

1.2.3   *Chemical forms of arsenic in waters*

The major factors controlling the speciation of As in natural waters are redox potential (Eh), pH, temperature and the presence of other chemicals, as shown in Figure 1.1. The major forms of As in waters are the inorganic species As(III) (arsenite) and As(V) (arsenate). The organic forms shown in Table 1.1 occur rarely in waters. Figure 1.1 and Table 1.1 show that the major aqueous forms of As(III) are H$_3$AsO$_3^0$, H$_2$AsO$_3^-$, HAsO$_3^{2-}$ and AsO$_3^{3-}$, and those of As(V) are H$_3$AsO$_4^0$, H$_2$AsO$_4^-$, HAsO$_4^{2-}$ and AsO$_4^{3-}$. In Figures 1.1b, 1.1c and 1.1d, the downwardly sloping boundary that begins at c. 0.5 volts and pH 0 and extends to c. 0 volts and pH 7 separates the As(V) (oxidized) species from the As(III) (reduced) species. This shows that acidic conditions tend to stabilize

Table 1.2.   Examples of secondary arsenic minerals and compounds.

Mineral group	Mineral Name	Chemical formula
Oxide	Arsenolite	As_2O_3
Oxide	Claudetite	As_2O_3
Oxide	ArsenicPentoxide	As_2O_5
Fe-arsenate	Arseniosiderite	$Ca_2Fe_3O_2(AsO_4)_3 \cdot 3H_2O$
Fe-arsenate	Parasymplesite	$Fe_3(AsO_4)_2 \cdot 8H_2O$
Fe-arsenate	Pharmacosiderite	$K[Fe_4(OH)_4(AsO_4)_3] \cdot 6.5H_2O$
Fe-arsenate	Scorodite	$FeAsO_4 \cdot 2H_2O$
Fe-arsenate	Symplesite	$Fe_3(AsO_4)_2 \cdot 8H_2O$
Fe-arsenate	Yukonite	$Ca_7Fe_{12}(AsO_4)_{10}(OH)_{20} \cdot 15H_2O$
Fe sulfoarsenates	Beudantite	$PbFe_3(AsO_4)(SO_4)(OH)_6$
Fe sulfoarsenates	Tooeleite	$Fe_6(AsO_4)_4(SO_4)(OH)_4 \cdot 4H_2O$
Fe sulfoarsenates	Zýkaite	$Fe_4(AsO_4)_3(SO_4)(OH) \cdot 15H_2O$
Ca-Mg-arsenates	Hörnesite	$Mg_3(AsO_4)_2 \cdot 8H_2O$
Ca-Mg-arsenates	Pharmacolite	$Ca(HAsO_4) \cdot 2H_2O$
Other metal arsenates	Annabergite	$Ni_3(AsO_4)_2 \cdot 8H_2O$
Other metal arsenates	Erythrite	$Co_3(AsO_4)_2 \cdot 8H_2O$
Other metal arsenates	Köttigite	$Zn_3(AsO_4)_2 \cdot 8H_2O$
Other metal arsenates	Mimetite	$Pb_5(AsO_4)_3Cl$

As(III) species, while more basic conditions tend to stabilize their As(V) equivalents. Thus, in oxidizing conditions (i.e., Eh > 0), the As(V) species $H_2AsO_4^-$ dominates in pH < c. 6.9 solutions, and $HAsO_4^{2-}$ in pH > c. 6.9 solutions H_3AsO_4 and AsO_4^{3-} occur in strongly acidic and alkaline solutions, respectively. In reducing conditions (i.e., Eh < −0.3), the As(III) species $As(OH)_3$ dominates at pH < c. 9, and $As(OH)_4^-$ and AsO_2OH^{2-} at pH > c. 9. The addition of SO_4^{2-} to the system suppresses the fields of $As(OH)_3$, $As(OH)_4^-$ and AsO_2OH^{2-} in favor of sulfur-bearing aqueous As species and sulfide minerals that incorporate As (Fig. 1.1c). The addition of SO_4^{2-} and Fe^{2+} to the system suppresses the fields of $As(OH)_3$, $As(OH)_4^-$ and AsO_2OH^{2-}, $H_2AsO_4^-$, $HAsO_4^{2-}$, AsO_4^{3-}, and also favors solid scorodite formation. If sulfur activities are higher than those shown in Figure 1.1b and 1.1c, then species such as $As_3S_4(SH)_2^-$ and $AsO(SH)_2^-$ can occur (Schwedt and Rieckhoff, 1996; Nordstrom and Archer, 2003).

1.2.4   Mineralogy of arsenic

1.2.4.1   Major As-bearing minerals

Arsenic is a component of more than 300 minerals, including arsenates (c. 60% of total As-bearing minerals), sulfides and sulfosalts (c. 20%), oxides (c. 10%) and arsenites, arsenides, native elements and metal alloys (c. 10%) (Bowell and Parshley, 2001; Drahota and Filippi, 2009). Some of the major As-bearing minerals are summarized in Table 1.1, and others most commonly found as secondary minerals formed by weathering are in Table 1.2. As a solid, As occurs in the native element form (As), and as sulfide (e.g., arsenopyrite, pyrite, loellingite, realgar) and arsenate minerals (e.g., scorodite, beudantite, yukonite). In Earth surface environments, the native As and sulfides are regarded as primary forms, since it is their weathering that leads to the formation of secondary oxides and arsenates (Table 1.2; section 1.4.3.3). Although arsenopyrite contains proportionately more As in its structure, the most commonly occurring As-bearing sulfide mineral is arsenical pyrite. Clark (1960) deduced from theoretical phase-equilibria studies that the maximum equilibrium solubility of As in pyrite at 600°C is 0.53 wt.%, but field-based studies have suggested that pyrite can contain up to 2–3 wt.% As (e.g., Savage et al., 2000).

Compared with arsenical pyrite and arsenopyrite, realgar and orpiment are rare primary sources of As. Their presence has, however, been reported in sulfate-reducing sediments, where realgar is predicted to be more stable under reducing conditions than orpiment, particularly if pyrite is present (Root *et al.*, 2009). In Cu-rich deposits enargite (Lattanzi *et al.*, 2008) and tennantite (Bruckard *et al.*, 2010) may be significant As-bearing phases, while in Co- and Ni-rich deposits, cobaltite (Kwong *et al.*, 2007) and gersdorffite (Senior *et al.*, 2009), respectively, may be present as primary As-bearing minerals.

The native and sulfide forms of As mostly form in moderate to high-temperature, hydrothermal environments (>350°C), but they can also form authigenically at lower temperatures (<200°C; e.g., Simon *et al.*, 1999). Arsenical pyrite, for example, can form under low temperature reducing conditions where there is associated decomposing organic matter and high concentrations of S. This authigenic pyrite will incorporate much of the soluble As available in pore waters (e.g., Lowers *et al.*, 2007). Coal measures can contain significant amounts of As owing to their contents of such pyrite (Yudovich and Ketris, 2005).

1.2.4.2 *As in other minerals*
Arsenic can also be incorporated in trace amounts in a variety of minerals such as Fe, Mn or Al oxides, hydroxides and sulfates, other hydrous and anhydrous sulfates, phosphates and hydrous silicates (Foster *et al.*, 1998; Hochella Jr. *et al.*, 1999; Roussel *et al.*, 2000; Savage *et al.*, 2000; Hudson-Edwards *et al.*, 2005; Walker *et al.*, 2009; Mao *et al.*, 2010; Fawcett and Jamieson, 2011). The As is taken up by sorption onto the mineral surface (Mohan and Pittman, 2007), or by incorporation through co-precipitation (see section 1.4.1., below, for more details). Table 1.3 gives examples of the As concentrations of a selection of these minerals. These data show that As is strongly sorbed by secondary Fe minerals, particularly Fe(III) oxyhydroxides, whose surface areas and thus density of sorption sites, is correspondingly large (Raven *et al.*, 1998). Arsenate ions can substitute for sulfate and phosphate in the jarosite solid solution series of minerals (e.g., hydronium jarosite, plumbojarosite) and in phosphate minerals (e.g., apatite), respectively.

1.3 DISTRIBUTION OF ARSENIC IN THE ENVIRONMENT

1.3.1 *Arsenic in water*

Examples of As concentrations in rain, river, lake, estuarine, ocean and groundwaters are shown in Table 1.4. They range widely from very low (<0.005 µg L^{-1}) to very high (1,058,000 µg L^{-1}), as discussed below.

1.3.1.1 *Rainwater*
The baseline range of concentrations of As in rainwater is 0.02–0.42 µg L^{-1} (Smedley and Kinniburgh, 2002). Arsenic in the atmosphere is mostly in the particulate form, and is sourced from aeolian mobilization, volcanoes, marine aerosols and, most significantly, industrial processes. It is deposited on the ground by wet or dry deposition from smelter and fossil fuel-burning operations (Plant *et al.*, 2007). 'Wet' deposition refers to that proportion of the atmospheric As that is dissolved by rainwater. For example, elevated As concentrations in snowfall are reported as being associated with industrial emissions in the Orlické Hory Mountains near the Czech–Polish border. Snow concentrations fell from 15 µg L^{-1} in the winters of 1984–1986 to <2 µg L^{-1} in those of 2003–2005, owing to reductions in regional industrial emissions (Dousová *et al.*, 2007). Similarly, in the Rio Grande, Brazil, rainwater downwind of a superphosphate manufacturing plant has As concentrations up to 35.6 µg L^{-1} (Mirlean and Roisenberg, 2006).

1.3.1.2 *River waters*
The baseline concentration of As in river water is 0.83 µg L^{-1} (range 0.13–2.1 µg L^{-1}) (Table 1.4). River water concentrations above these values are due to volcanic inputs of As, or from leaching of As from underlying bedrock or unconsolidated sediment, or from contamination. For example, concentrations as low as 0.13 µg L^{-1} were found in rivers draining the pristine karst limestone of former Yugoslavia (Seyler and Martin, 1991). In contrast, As concentrations of up to 160 µg L^{-1}

Table 1.3.   Arsenic concentrations in selected minerals. Adapted from in Smedley and Kinniburgh (2002), with updating and additions.

Mineral	As concentration range ($mg\,kg^{-1}$)	Reference
Sulfide minerals		
Pyrite	Up to 50,000	Savage *et al.* (2000)
Pyrrhotite	Minor	Peters *et al.* (2006)
Marcasite	Up to 10,000	Craw *et al.* (2000)
Mackinawite	Up to 5,000	Craw *et al.* (2000)
Galena	<5–60	Boyle and Jonasson (1973)
Sphalerite	Minor, as a solid solution, but up to 578 in inclusions	Ye *et al.* (2011)
Chalcopyrite	10–1,000	Boyle and Jonasson (1973)
Oxide minerals		
Haematite	Up to 160	Baur and Onishi (1969)
Fe(III) oxyhydroxide	Up to 2,000	Pichler *et al.* (1999)
Magnetite	2.7–41	Bauer and Onishi (1969)
Ilmenite	<1	Baur and Onishi (1969)
Silicate minerals and rocks		
Quartz	0.4–1.3	Baur and Onishi (1969)
Feldspar	<0.1–2.1	Baur and Onishi (1969)
Biotite	1.4	Baur and Onishi (1969)
Amphibole	1.1–2.3	Baur and Onishi (1969)
Olivine	0.08–0.17	Baur and Onishi (1969)
Pyroxene	0.05–0.8	Baur and Onishi (1969)
Garnet	150–1,200	Charnock *et al.* (2007)
Smectite from geothermal area	1,500–4,000	Pascua *et al.* (2005)
Carbonate minerals and rocks		
Calcite	1–8	Boyle and Jonasson (1973)
Dolomite	<3	Boyle and Jonasson (1973)
Siderite	<3	Boyle and Jonasson (1973)
Sulfate minerals		
Gypsum/anhydrite	<1–6	Smedley and Kinniburgh (2002)
Plumbojarosite	Up to 3,570	Kocourková *et al.* (2011)
Jarosite	Up to 710	Kocourková *et al.* (2011)
Hydronium jarosite	Up to 890	Kocourková *et al.* (2011)
Other minerals and rocks		
Apatite	<1–1,000	Bauer and Onishi (1969)
Ba-phosphate	Up to 5,000	Craw *et al.* (2000)
Halite	<3–30	Stewart (1963)

have been reported in the Gibbon River, which drains the Norris Geyser Basin (Yellowstone, USA) (McCleskey *et al.*, 2010).

Mining and industrial pollution of river courses is a widespread global problem. For example, the Ríos Odiel and Tinto drain most of the Iberian pyrite belt (western Spain and eastern Portugal), which is one of the largest massive sulfide deposits in the world with at least seven deposits being larger than 100 Mt (Rosa *et al.*, 2010). The sulfide mineralogy is variable across the belt, but is principally composed of pyrite with significant local concentrations of sphalerite, galena, chalcopyrite, tetrahedrite and arsenopyrite (e.g., Ruiz and Arribas, 2002). In a study that encompassed two years of sampling the Odiel and Tinto rivers exhibited mean As concentrations of 441 and 1,975 $\mu g\,L^{-1}$, respectively (Sarmiento *et al.*, 2009), owing primarily to the weathering of these minerals.

Table 1.4. Arsenic concentrations in waters. Adapted from Smedley and Kinniburgh (2002), with updating and additions.

Water type, body and location	Average or range of As ($\mu g\,L^{-1}$)	Reference
Rainwater		
Maritime	0.02	Andreae (1980)
Terrestrial (USA)	0.013–0.032	Andreae (1980)
Coastal (mid-Atlantic)	0.1 (<0.005–1.1)	Scudlark and Church (1988)
Snow (Arizona)	0.14 (0.02–0.42)	Barbaris and Betterton (1996)
Seattle rain impacted by Cu smelter	16	Crecelius (1975)
Rio Grande, Brazil	Up to 35.6	Mirlean and Roisenberg (2006)
River water		
Baseline (various)	0.83 (0.13–2.1)	Andreae *et al.* (1983); Froelich *et al.* (1985); Seyler and Martin (1991)
Rio San Antonio de los Cobres, Argentina	13–9,310	Hudson-Edwards and Archer (2012)
Rio Tinto, Spain	200–14,000	Hudson-Edwards *et al.* (1999)
Polluted European rivers	4.5–45	Seyler and Martin (1990)
Dordogne, France	0.7	Seyler and Martin (1990)
Lake water		
Baseline British Columbia	0.28 (<0.2–0.42)	Azcue *et al.* (1994, 1995)
Baseline France	0.73–9.2	Seyler and Martin (1989)
Baseline Japan	0.38–1.9	Bauer and Onishi (1969)
Western USA	0.38–1,000	Benson and Spencer (1983)
Ontario	35–100	Azcue and Nriagu (1995)
Estuarine water		
Baseline Norway	0.7–2.0	Abdullah *et al.* (1995)
Baseline British Columbia	1.2–2.5	Peterson and Carpenter (1983)
Loire Estuary, France	Up to 16	Seyler and Martin (1990)
Tamar Estuary, UK	2.7–8.8	Howard *et al.* (1988)
Ocean water		
Deep Pacific and Atlantic	1.0–1.8	Cullen and Reimer (1989)
Coastal Spain	1.5 (0.5–3.7)	Navarro *et al.* (1993)
Coastal Australia	1.3 (1.1–1.6)	Maher (1985)
Groundwaters		
Baseline UK	<0.5–10	Edmunds *et al.* (1989)
Arsenic-rich aquifers shown in Figure 1.2	10–5,000	Das *et al.* (1995); BGS and DPHE (2001); Nicolli *et al.* (1989); Del Razo *et al.* (1990); Luo *et al.* (1997); Hsu *et al.* (1997); Varsányi *et al.* (1991)
Mining-affected groundwaters	50–1,000	Wilson and Hawkins (1978); White *et al.* (1963); Ellis and Mahon (1977)
Giant Mine, Yellowknife	Up to 1,058,000	Osborne *et al.* (2010)
Geothermal groundwaters	<10–50,000	Bauer and Onishi (1969); White *et al.* (1963); Ellis and Mahon (1977); Romero *et al.* (2003)
As herbicide plant, Texas	408,000	Kuhlmeier (1997a, b)

1.3.1.3   *Lake waters*

Reported baseline concentrations of As in lake waters range from 0.06–1.9 µgL^{-1} (Table 1.4), but the heterogeneous distribution of As in the crust, coupled with climatological factors, can result in particular lakes having very high As loadings. Lakes in arid areas, for example, often exhibit extreme elevated As concentrations as a result of evaporation and the consequent increase in salinity, particularly where these lakes are closed and are fed by comparatively As-rich waters. For example, Mono Lake in California, USA, has As concentrations of the order of 10–20 mg L^{-1} (Maest *et al.*, 1992).

Azcue *et al.* (1994), in a survey of mining-affected Canadian lakes, emphasized the importance of sediment Fe oxyhydroxide sorption, which acts as an As sink, thereby reducing aqueous concentrations. This is not always the case, however. The mining-affected Moira Lake in Ontario, Canada, shows mean summer and winter concentrations of 62 and 22 µgL^{-1} As, respectively (Azcue and Nriagu, 1995). These elevated As concentrations were attributed to high fluvial inputs of As into the lake (c. 3.5 Mt per year), and limited sorption of this As onto lake sediment. Pit lakes, too, may hold considerable amounts of As. For example, measured concentrations in the alkaline Jamestown gold mine lake in California, USA, reach a maximum value of 1,200 µg L^{-1} (Savage *et al.*, 2009).

1.3.1.4   *Estuarine waters*

Baseline concentrations of As in ocean water range from 1 to 3.7 µg L^{-1} (e.g., Cullen and Reimer, 1989; Navarro *et al.*, 1993; Maher, 1985). These are similar to the baseline concentrations of As in estuarine water, which are typically 0.7–3.8 µg L^{-1} (Peterson and Carpenter, 1983; Abdullah *et al.*, 1995). These low estuarine concentrations are attributed to dilution by ocean and river water (Anninou and Cave, 2009). In estuaries where As concentrations are higher (e.g., the Loire, Seyler and Martin, 1990), the saltwater-freshwater interface has been found to be a zone of co-precipitation of Fe and As, as a result of the flocculation of poorly crystalline Fe oxide and oxyhydroxide precipitates (Cullen and Reimer, 1989). For this reason, in mining- and industry-affected estuarine environments As behavior in the water column is often non-conservative (i.e., it departs from simple mixing and consequent dilution), with As held in the estuarine environment rather than being moved out to the open ocean. For example, the Patos Lagoon (Brazil) estuarine sediments exhibit As concentrations of up to 50 µg kg^{-1}, while the freshwater portion of the lagoon exhibits concentrations of 7.7 µg kg^{-1}. The local fertilizer industry at the port of Rio Grande is the most probable source of the elevated As concentrations found in the estuarine sediments (Mirlean *et al.*, 2003).

1.3.1.5   *Groundwaters*

The baseline concentration range of As in groundwater is <0.5–10 µg L^{-1} (Table 1.4). Arsenic-contaminated groundwaters are common (Table 1.4), and the locations of many of these are shown Figure 1.2. These can be divided into three types: As-affected aquifers, geothermal waters and mining-affected waters. The As in both the first two types is derived mainly from natural sources (Welch *et al.*, 2000; Webster and Nordstrom, 2003; McArthur *et al.*, 2004), whereas that in the mining-affected areas is sourced from the anthropogenically induced breakdown of As-bearing minerals and from As-rich discharges. The characteristics of these groundwaters, and mechanisms by which they are enriched in As are discussed further in section 1.4.3.

1.3.2   *Arsenic in rocks*

Most rocks contain very low concentrations of As (Table 1.5). Igneous rocks, in particular, contain only up to c. 15 mg kg^{-1} As, except for gabbros and dolerites, which contain up to 113 mg kg^{-1} As. Metamorphic rocks, too, contain little As, except for fine-grained phyllites and slates, which have been reported to contain up to 143 mg kg^{-1} As (Boyle and Jonasson, 1973). Of the sedimentary rocks, phosphorites, Fe-formation and -rich sedimentary rocks and coals have been shown to contain very large amounts of As (e.g., up to 35,000 mg kg^{-1} in coal; Welch *et al.*, 1988) All of

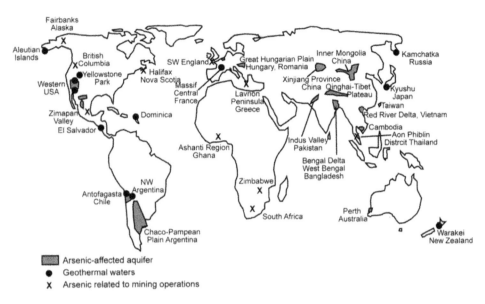

Figure 1.2.   Areas of known relatively high concentrations of As in groundwaters, associated with mining activities (Xs), geothermal activity (circles) or other natural causes (grey areas). See text for discussion. Adapted from Smedley and Kinniburgh (2002) and Ravenscroft *et al.* (2009).

these concentrations and the remainder shown in Table 1.5 reflect the minerals contained in the rocks, which have been discussed in section 1.2.4.

1.3.3   *Arsenic in soils and sediments*

Concentrations of As in soils and sediments vary widely (Table 1.5). Natural sediments, such as the alluvial sands in As-polluted aquifers in Bangladesh and elsewhere, have very low concentrations of As (generally <10 mg kg^{-1}; BGS and DPHE, 2001). Other natural soils, including peaty and bog soils and acid sulfate soils, contain more As owing to the weathering of primary As-bearing minerals. Even so, these soils still contain less than 100 mg kg^{-1} As on average. Sediments and soils contaminated by industrial activities contain the most As, up to several 1,000s of mg kg^{-1} (Table 1.5). This is due to the weathering of As-bearing minerals (e.g., alluvium draining the Río Tinto mines; Hudson-Edwards *et al.*, 1999) or the concentrating of As-bearing minerals and compounds into waste materials (e.g., mine tailings, Kossoff *et al.*, 2011; fly ash from coal combustion, Pandey *et al.*, 2011). Soil As contamination can also arise from the application of As-bearing products. For example, As has historically been used as a wood preservative in the form of chromated copper arsenate (CCA), and although the chemical is not now in use a significant number of loci of potential exposure remain (Nielsen *et al.*, 2011). Also, As concentrations were shown to exceed the industrial soil cleanup target level of 3.7 mg kg^{-1} in $>95\%$ of samples of soil from under 10 wooden bridges in Florida (Townsend *et al.*, 2003). The use of CCA is not illegal under all jurisdictions; it is reported to be an ongoing contaminant source in the Terengganu River basin (Malaysia) (Sultan and Shazili, 2009). Arsenic use as a pesticide has also been widespread. It has been estimated that 6,800 tonnes of As were applied to New Jersey (USA) soils between 1900 and 1980 (Murphy and Aucott, 1998). The use of inorganic As-based pesticides has since been banned in the USA, but the use of their organic equivalents, such as monosodium methanearsonate (MSMA) and disodium methanearsonate (DSMA), is still permitted under some circumstances (EPA, 2011). A particular area of concern is the application of As-bearing pesticides to golf courses. Concentrations of up to 302 mg kg^{-1} As have been reported for a golf course lakes in

Table 1.5.   Arsenic concentrations in rocks, soil, sediments and other solids. Adapted from Smedley and Kinniburgh (2002), with updating and additions.

Rocks, Soils and Sediments	Average or range of As (mg kg^{-1})	Reference
Rocks		
Igneous rocks		
Ultrabasic rocks	1.5 (0.03–15.8)	Onishi and Sandell (1955)
Basalt	1.5	Xie *et al.* (2011)
Gabbro, dolerite	1.5 (0.06–113)	Onishi and Sandell (1955)
Andesite, trachyte, latite	2.7 (0.5–5.8)	Baur and Onishi (1969)
Rhyolite	4.3 (3.2–5.4)	Ure and Berrow (1982)
Granite, aplite	1.3 (0.2–15)	Riedel and Eikmann (1986)
Metamorphic rocks		
Quartzitic	5.5 (2.2–7.6)	Riedel and Eikmann (1986)
Hornfels	5.5 (0.7–11)	Riedel and Eikmann (1986)
Phyllitic/slate	18 (0.5–143)	Boyle and Jonasson (1973)
Sedimentary rocks		
Mudstone/shale	0.55–3.2	Xie *et al.* (2011)
Limestone	0.89–3.39	Xie *et al.* (2011)
Phosphorite	21 (0.4–188)	Boyle and Jonasson (1973)
Iron formation and Fe-rich sedimentary rocks	1–2,900	Boyle and Jonasson (1973)
Evaporites (gypsum and anhydrite)	3.5 (0.1–10)	Riedel and Eikmann (1986)
Coals	0.3–35,000	Welch *et al.* (1988)
Sediments and soils		
Natural sediments		
Alluvial sand (Bangladesh)	2.9 (1.0–6.2)	BGS and DPHE (2001)
Alluvial mud/clay (Bangladesh)	6.5 (2.7–14.7)	BGS and DPHE (2001)
World average river sediment	5	Boyle and Jonasson (1973)
Lake sediment,	5.5 (0.9–44)	Cook *et al.* (1995)
British Columbia		
Volcanic ash, Argentina	6	Bundschuh *et al.* (2004)
Mining and industrially-affected		
Belgium, Germany and the Netherlands channel sediment	<5–69	De Vos *et al.* (1996)
River Tamar, UK, channel sediment	4.4–11,000	Rawlins *et al.* (2003)
Rio Tinto, Spain, alluvium	110–750	Hudson-Edwards *et al.* (1999)
Natural soils		
Various	7.2 (0.1–55)	Boyle and Jonasson (1973)
Peaty and bog soils	13 (2–36)	Boyle and Jonasson (1973)
Acid sulfate soils	6–41	Gustafsson and Tin (1994)
Other solids		
Mine tailings, Potosí, Bolivia	6,500–7,410	Kossoff *et al.* (2011)
Sewage sludge	9.8 (2.4–39)	Zhu and Tabatabai (1995)

central Florida (USA) compared to a concentration range of 0.1–3 mg kg^{-1} As in comparable sediments not associated with golf courses (Pichler *et al.*, 2008).

1.4   PROCESSES OF ARSENIC CYCLING IN THE ENVIRONMENT

1.4.1   *Sorption*

Although much As in the environment occurs in solid form (Table 1.1 and Table 1.5), it is often transported and cycled in the environment as aqueous species or as extremely fine particles that are generally nano-sized (i.e., $<10^{-9}$ m in diameter). Cycling of As between solids and liquids is largely controlled by the processes of sorption and desorption to mineral and organic matter surfaces.

The different types of sorption are illustrated in Figure 1.3. As in solution can form outer-sphere or inner-sphere complexes by the process of adsorption. In the former, the dissolved As is surrounded by water molecules, and does not directly bond with oxide surface atoms. By contrast, the As bonds directly with the surface in inner-sphere adsorption, and these bonds are generally regarded as stronger than outer-sphere bonds. Ion-exchange involves the exchange of the As atom for another atom in the mineral surface. Precipitation of As compounds at solid surfaces can occur either with material only from solution (surface precipitation), or additionally involving material within the solid (co-precipitation). Solid solution is a process involving incorporation of As within the full structure of minerals, such as those in Tables 1.1 and 1.2.

Adsorption processes involving As depend on factors such as temperature, solution pH and Eh, the presence of inorganic and organic oxidizing and reducing agents, competition with other species, and the surface area, morphology (i.e., the presence of defects and unsatisfied bonds) and charge of the solid surface. With regard to the latter, there is a unique 'zero point of charge' for each solid, known as pH_{ZPC}, which is the pH at which the surface has no charge and therefore does not adsorb As. Solid surfaces are positively charged below their pH_{ZPC} values, and negatively charged above them. The pH_{ZPC} for a number of common substances is shown in Table 1.6. This shows that at the near-neutral pH values of most As-polluted groundwaters (see section 1.4.3), minerals such as gibbsite, ferrihydrite and birnessite will be positively charged, and thus will sorb the negatively-charged As species that dominate (Fig. 1.1). The available data in Table 1.6 also show that ferrihydrite and organic matter have the highest surface areas and/or surface site densities, illustrating their high affinity for the sorption of As from water.

The degree of sorption of As also depends on the nature of the aqueous As species involved. As Figure 1.1 shows that the principal As(III) species As(OH)$_3$ is uncharged, while the principal As(V) species, $H_2AsO_4^-$ and $HAsO_4^{2-}$, are both charged. This suggests that in acid solutions As(V) is more strongly sorbed to positively charged mineral surfaces (e.g., the Fe oxyhydroxides) than As(III). Conversely, in alkaline solutions, where mineral surface sorption sites tend to be more negatively charged, the doubly charged As(V) species $HAsO_4^{2-}$ should be more mobile than the singularly charged As(III) equivalent, As $(OH)_4^-$. This hypothesis is supported by the experimental work of Dixit and Hering (2003), who found that sorption of As(V) onto Fe oxyhydroxides was more favored than that of As(III) below pH 5–6, whereas the converse applied above pH 7–8. Other more recent work, however, has indicated that As(III) may be sorbed more efficiently than As(V) on synthetic ferrihydrite (Zhu *et al.*, 2010).

In the natural environment competitive effects also constrain As sorption and hence mobility. In this regard, calcium, carbonate, chloride, magnesium, nitrate, phosphate and sulfate ions are known to compete with aqueous As species (Cheng *et al.*, 2009). It has been suggested, for example, that the high As concentrations in Bangladeshi groundwater may be at least in part due to carbonate displacement of sorbed As (Appelo *et al.*, 2002).

The adsorption capacities for As of the Fe oxyhydroxides ferrihydrite and goethite and the clay mineral montmorillonite, are illustrated in Figure 1.4. These data illustrate the differences between As(III) and As(V) sorption and their dependence on pH. Ferrihydrite shows very high capacity for sorption of both As(III) and As(V) compared to goethite and montmorillonite.

1.4.2 *Microbial controls on arsenic cycling*

Prokaryotic As metabolism has been detected in hydrothermal, temperate and cold environments and has been shown to be involved in the redox cycling of As (e.g., Rhine *et al.*, 2005; Inskeep *et al.*, 2007; Osborne *et al.*, 2010). In fact, bacteria are probably directly or indirectly responsible for most of the redox transformation of As in the environment. An illustration of this is given in Figure 1.5. As(V) sorbed to FeOOH can be released after reduction of Fe(III) to Fe(II) by Fe-reducing bacteria such as *Geobacter* (Fig. 1.5a). As(V)-respiring microbes can reduce aqueous

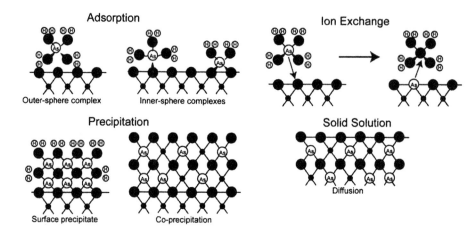

Figure 1.3.   Conceptual models for sorption of aqueous As with solid surfaces. Adapted from Brown (1990). Large filled circles are oxygen atoms, small circles filled with 'H' are hydrogen atoms, large circles with 'As' are arsenic atoms, small filled circles are other metal atoms, short lines between atoms are bonds, and long continuous lines are solid surfaces. See text for explanation.

Table 1.6.   pH at the zero point of charge (pH_{ZPC}) and surface area of common minerals and materials that are capable of sorbing As. Adapted from Langmuir (1997) and Ravenscroft et al. (2009).

Mineral/Material	pH_{ZPC}	Surface area (m^2 g^{-1})	Surface site density (nm^{-2})
Silica (amorphous)	3.5–3.9	53–292	4.2–12
Calcite	8.5–10.8	–	–
Gibbsite (α-Al(OH)$_3$)	9.84–10.0	120	2–12
Birnessite (α-MnO$_2$)	1.5–2.8	180	2–18
Pyrolusite (β-MnO$_2$)	4.6–7.3	–	–
Ferrihydrite (Fe(OH)$_3$)	8.5–8.8	250–600	20
Goethite (α-FeOOH)	5.9–6.7	45–169	2.6–16.8
Haematite (α-Fe$_2$O$_3$)	4.2–6.9	1.8	5–22
Kaolinite	$\leq$2–4.6	10–38	1.2–6.0
Montmorillonite (sodic)	$\leq$2–3	600–800	0.4–1.6
Organic matter	2*	260–1,300^	2.3^

*Algae and sewage effluent; ^soil humic material

As(V) to As(III), which can then be sorbed to FeOOH or Al(OH)$_3$ (Fig. 1.5b) in a process known as dissimilatory arsenate reduction (DARP). Fe-reducing DARP can also reduce both Fe(III) and As(V) to Fe(II) and As(III), respectively, at the mineral surface (Fig. 1.5c). It is possible that all three of these mechanisms operate at the same time, or that other, yet unrecognized, microbes are involved in the release of Fe and As. The involvement of microbes in the cycling of As is the subject of this book, and thus shall not be discussed further here.

1.4.3   Water-solid interactions

As pointed out by Ravenscroft et al. (2009), As can be mobilized from solids by four major mechanisms: reductive dissolution (RD), alkali desorption (AD), sulfide oxidation (SO) and geothermal arsenic (GT).

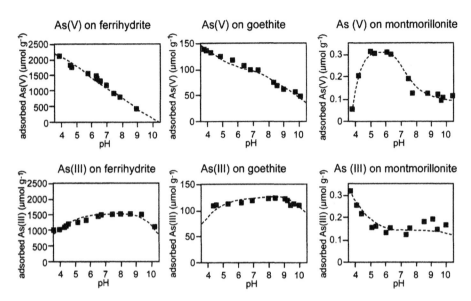

Figure 1.4. Adsorption of As(III) and As(V) per g of ferrihydrite, goethite and montmorillonite minerals. Data for ferrihydrite and goethite are adapted from Dixit and Hering (2003), and are for experiments conducted with 750 μg L^{-1} As, 0.01 M NaClO$_4$, 0.03 g L^{-1} ferrihydrite and 0.5 g L^{-1} goethite. Data for montmorillonite are adapted from Goldberg (2002), and are for experiments conducted with 1300 μg L^{-1} As and 40 g L^{-1} montmorillonite. Note that the y-axis scales are different for each mineral, indicating that ferrihydrite sorbs the most As, followed by goethite and then montmorillonite, for the experimental conditions reported. Note also that the maximum amount of As sorption occurs at different pHs for the different minerals and species of As.

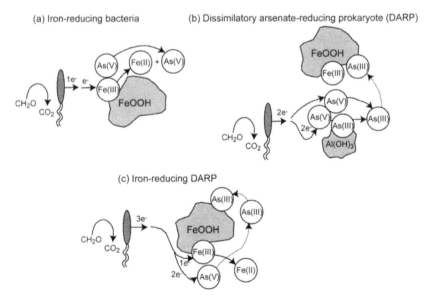

Figure 1.5. Possible mechanisms of As mobilization by microbes. Adapted from Oremland and Stolz (2005). See text for explanation.

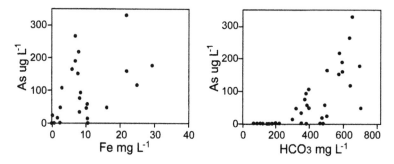

Figure 1.6.   Relationship of aqueous As to dissolved Fe and HCO_3 in Bangladesh groundwaters. Adapted from Nickson *et al.* (2000).

1.4.3.1   *Reductive dissolution*

Anoxic groundwaters are the major types of waters affected by RD leading to As pollution. As discussed above, Fe oxyhydroxides and oxides have high affinities for the sorption of As, but this As can be released to groundwaters by microbially-mediated redox reactions (Fig. 1.5; Gulens *et al.*, 1979; Nickson *et al.*, 1998, 2000). The As(V) that is released is reduced to As(III) by microbes (Fig. 1.5), or by other reductants such as sulfide and hydrogen (Inskeep *et al.*, 2002). This process involves the consumption of organic matter (Hem, 1977; Langmuir, 1997), and is represented as:

$$8FeOOH + CH_3COOH + 14H_2CO_3 \leftrightarrow 8Fe^{2+} + 16HCO_3^- + 12H_2O$$

The waters generated by RD are anoxic and of circum-neutral pH (6.3 to 7.8), and contain high concentrations of As(III), Fe(II), Mn, HCO_3^-, PO_4^{3-}, ammonium, dissolved organic carbon and, in places, H_2S, and low concentrations of NO_3^- and SO_4^{2-}. The high Mn is due to the reductive dissolution of Mn oxyhydroxides that are also present in the sediments, and the PO_4^{3-}, from the dissolution of the Fe oxyhydroxides to which it is sorbed. Bicarbonate, DOC and ammonium are generated by the degradation of organic matter.

The biogeochemical processes leading to RD are not obvious from data on aqueous solutions, owing to resorption reactions and the formation of secondary minerals in aquifer materials. This is illustrated in Figure 1.6, which demonstrates that aqueous As does not show a linear relationship with either aqueous Fe or aqueous HCO_3^-, as would be expected from the chemical reaction outlined above. The poor As-HCO_3^- correlation can be explained by the uptake of HCO_3^- in secondary carbonate minerals. The poor As-Fe relationship can be explained by the formation of secondary Fe sulfides (Fig. 1.7), magnetite, goethite, siderite and phosphates that sorb As to differing degrees, but also to resorption of aqueous As to unreduced Fe oxyhydroxides (Welch *et al.*, 2000; Horneman *et al.*, 2004; McArthur *et al.*, 2004). This is shown in more detail in Figure 1.8a, which conceptually depicts the release of 'kinetic As' due to the RD process. Low degrees of Fe oxyhydroxide (FeOOH) reduction result in release of As, but this is resorbed on the sediment, which stays brown. As reduction of FeOOH increases, the sediment turns grey due to the larger amount of Fe(II), and more As is release to groundwater because available sorption sites on the FeOOH surfaces are filled.

Intensive research has been carried out to determine the mechanism(s) of As release due to RD. A plausible model for As pollution in West Bengal and Bangladesh is outlined in Figure 1.8. Aquifers are developed in sands deposited on Pleistocene sediments produced by weathering of the Himalayas. Large amounts of organic matter are deposited in specific horizons both in the aquifer sands and in silts. A clay-rich palaeosol of low hydraulic conductivity separates the deeper sand aquifer from overlying organic-rich silts. Arsenic-polluted groundwaters are intersected in boreholes 3 and 4, which contain grey sands that have undergone RD. The deeper sands are

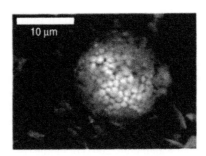

Figure 1.7.   Secondary As-bearing Fe sulfide from As-polluted aquifers, West Bengal, India.

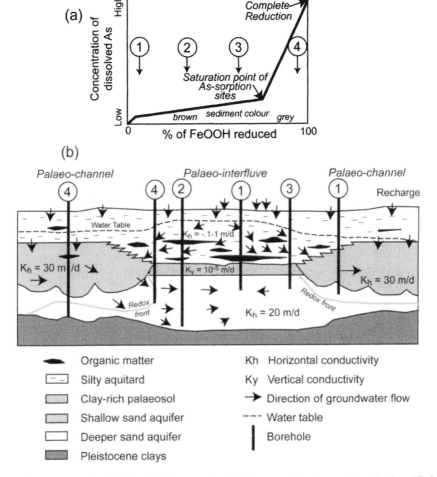

Figure 1.8.   Conceptual models for (a) Release of 'kinetic As' from partial to complete reduction of FeOOH. With low degrees of FeOOH reduction, the sediment remains brown, groundwater Fe concentrations increase, but As concentrations are low due to re-sorption. With higher degrees of FeOOH reduction, the sediments turn grey [reflecting high amounts of Fe(II)], and higher amounts of As are released, because all sorption sites on the unreduced FeOOH are saturated. The circled numbers 1 to 4 related to the boreholes shown in (b). Adapted from McArthur *et al.* (2004, 2008); (b) Groundwater flow in the JAM study area, West Bengal, India, showing relationship between organic matter, sedimentology, flow hydraulic conductivity and As concentration. See text for discussion. Adapted from McArthur *et al.* (2008).

As-poor because the organic matter required to drive RD is isolated from the sands by the clay-rich palaeosol. Owing to groundwater flow, however, the deeper sands in these boreholes (and especially borehole 2) are predicted to release As in the future because of the migration of the redox front and dissolved organic matter. This model, therefore, predicts that As pollution is related to the sedimentary architecture of the aquifers and the availability of organic matter.

Many uncertainties remain regarding the mechanism of As pollution to anoxic groundwaters. These include the source of the organic matter, which may be natural (e.g., McArthur et al., 2004) or human- (pond-)derived (Harvey et al., 2002, 2006), and the primary mineralogical host of the As, which is thought to be Fe oxyhydroxides (Nickson et al., 1998; McArthur et al., 2004), Mn oxyhydroxides (Smedley and Kinniburgh, 2002) or biotite (Breit et al., 2005).

1.4.3.2 Alkali desorption

The process of alkali desorption (AD) occurs in oxic, alkaline groundwaters in which desorption of As from minerals (largely Fe oxyhydroxides) is due to increases in pH and competition from other ions. As shown in Table 1.6, most mineral surfaces are negatively charged above pH 8.0, and so do not easily sorb negatively-charged As species from water, or may desorb As from their surfaces. Competition of As(V) with phosphate, sulfate and carbonate in these oxic waters can also explain the AD mechanism.

The AD mechanism has been invoked for As-polluted groundwaters in Argentina, the western USA, Spain, Russia and other locations in Asia (Smedley and Kinniburgh, 2002; Ravenscroft et al., 2009). The aquifer materials in these areas are largely different from those in areas affected by RD. In Argentina, for example, these are Tertiary to Mesozoic sedimentary rocks, older bedrock and volcanic loess, and it is the latter that is thought to be the source of most of the As (Smedley and Kinniburgh, 2002).

1.4.3.3 Sulfide oxidation

Sulfide minerals containing As (Table 1.1) are chemically unstable at the Earth's surface, and they weather in the presence of O_2 and $H_2O_{(l)}$, in reactions often mediated by bacteria and known as sulfide oxidation (SO). The most common As-containing sulfide mineral is As-bearing pyrite; this weathers to yield Fe(II), SO_4^{2-}, H^+ and As, in a series of linked reactions:

$$Fe(As, S)_{2(s)} + \frac{7}{2}O_{2(aq)} + H_2O_{(l)} \leftrightarrow Fe^{2+}_{(aq)} + 2(As, S)O_4^{2-}{}_{(aq)} + 2H^+_{(aq)}$$

$$Fe^{2+}_{(aq)} + \frac{1}{4}O_{2(aq)} + H^+_{(aq)} \leftrightarrow Fe^{3+}_{(aq)} + \frac{1}{2}H_2O_{(l)}$$

$$Fe(As, S)_{2(s)} + 14Fe^{3+}_{(aq)} + 8H_2O_{(l)} \leftrightarrow 15Fe^{2+}_{(aq)} + 2(As, S)O_4^{2-}{}_{(aq)} + 16H^+_{(aq)}$$

The other major form of As is arsenopyrite (Table 1.1), which has been shown to oxidise more rapidly than pyrite (Hudson-Edwards et al., 2005; Corkhill and Vaughan, 2009), releasing As to waters:

$$4FeAsS + 13O_2 + 6H_2O \leftrightarrow 4Fe^{2+}_{(aq)} + 4AsO_4^{3-}{}_{(aq)} + 4SO_4^{2-}{}_{(aq)} + 12H^+_{(aq)}$$

The As released to water by SO can be removed by the formation of secondary minerals (Table 1.2). For example, Figure 1.1d shows that introduction of Fe into the As system results in a large percentage of the oxic portion of the stability field being occupied by the Fe-hydroxyarsenate mineral scorodite. Arsenopyrite occupies a very restricted space at the extreme alkaline edge of the field, indicating both its unstable nature as a primary mineral, and its unlikely occurrence in any secondary assemblage that may form (e.g., Hudson-Edwards et al., 2005). According to the data in Figure 1.1d, scorodite can constitute the major solid sink acting to remove As from solution, given relatively oxic conditions and the presence of Fe. Other phases reported to sorb As from SO solutions include members of the jarosite family (Hudson-Edwards et al., 1999), Fe oxyhydroxides (Savage et al., 2000) and silicates (Mao et al., 2010).

1.4.3.4   *Geothermal arsenic*

Geothermal (GT) As is found in volcanic areas where water sources often have inputs from geothermal activity related to volcanic and/or hydrothermal processes (Hedenquist and Lowenstern, 1994). Such waters occur at colliding plate boundaries (subduction zones) at intraplate 'hot spots,' and along rift zones (Webster and Nordstrom, 2003). The high concentrations of As in these waters arise from dissolution of As gas or As-bearing minerals such as arsenopyrite or realgar in magmatic and hydrothermal waters, and subsequent mixing of these waters with meteoric waters (Ellis and Mahon, 1977; Welch *et al.*, 1988; Webster and Nordstrom, 2003; Hudson-Edwards and Archer, 2012). Both As(III) and As(V) are found in these geothermal waters, depending on pH, redox potential and the availability of As(III)-oxidizing bacteria (Langner *et al.*, 2001; Webster and Nordstrom, 2003). The GT waters have chemical signatures arising from mixing of meteoric waters, deep-aquifer Na-Cl brines and interactions with hydrothermal vapors (Ellis and Mahon, 1977; Aiuppa *et al.*, 2006).

ACKNOWLEDGEMENTS

The authors thank their many colleagues for many useful discussions on As geochemistry and mineralogy over the years, and in particular Maria Alfredsson, Jane Archer, Bill Dubbin, Heather Jamieson, John McArthur, David Polya, Joanne Santini, David Vaughn and Kate Wright.

REFERENCES

Abdullah, M.I., Shiyu, Z. & Mosgren, K.: Arsenic and selenium species in the oxic and anoxic waters of the Oslofjord, Norway. *Mar. Pollut. Bull.* 31 (1995), pp. 116–126.
Aiuppa, A., Avino, R., Brusca, L., Calior, S., Chiodini, G., D'Alessandro, W., Favara, R., Federico, C., Ginevra, W., Inguaggiato, S., Longo, M., Pecoraino, G. & Valenz, M.: Mineral control of arsenic content in thermal waters from volcano-hosted hydrothermal systems: insights from island of Ischia and Phlegrean Fields, Campanian Volcanic Province, Italy. *Chem. Geol.* 229 (2006), pp. 313–330.
Andreae, M.O.: Arsenic in rain and the atmospheric mass balance of arsenic. *J. Geophys. Res.* 85 (1980), pp. 4512–4518.
Andreae, M.O., Byrd, T.J. & Froelich, O.N.: Arsenic, antimony, germanium and tin in the Tejo estuary, Portugal: modelling of a polluted estuary. *Environ. Sci. Technol.* 17 (1983), pp. 731–737.
Anninou, P. & Cave, R.R.: How conservative is arsenic in coastal marine environments? A study in Irish coastal waters. *Estuar. Coast. Shelf Sci.* 82 (2009), pp. 515–524.
Appelo, C.A.J., Van der Weiden, M.J.J., Tournassat, C. & Charlet, L.: Surface complexation of ferrous iron and carbonate on ferrihydrite and the mobilization of arsenic. *Environ. Sci. Technol.* 36 (2002), pp. 3096–3103.
Azcue, J.M. & Nriagu, J.O.: Impact of abandoned mine tailings on the arsenic concentrations in Moria Lake, Ontario. *J. Geochem. Explor.* 52 (1995), pp. 81–89.
Azcue, J.M., Mudroch, A., Rosa, F. & Hall, G.E.M.: Effects of abandoned gold mine tailings on the arsenic concentrations in water and sediments of Jack of Clubs Lake, BC. *Environ. Technol.* 15 (1994), pp. 669–678.
Azcue, J.M., Mudroch, A., Rosa, F., Hall, G.E.M., Jackson, T.A. & Reynoldson, T.: Trace elements in water, sediments, porewater, and biota polluted by tailings from an abandoned gold mine in British Columbia, Canada. *J. Geochem. Explor.* 52 (1995), pp. 25–34.
Barbaris, B. & Betterton, E.A.: Initial snow chemistry survey of the Mogollon Rim in Arizona. *Atmos. Environ.* 30 (1996), pp. 3093–3103.
Bauer, W.H. & Onishi, B.-M.-H.: Arsenic. In: K.H. Wedepohl (ed.) *Handbook of geochemistry.* Springer-Verlag, Berlin (1969), pp. 33-A-1-33-0-5.
Benson, L.V. & Spencer, R.J.: A Hydrochemical Reconnaissance Study of the Walker River Basin, California and Nevada. USGS Open File Rep. 83–740, United States Geological Survey, Denver, 1984.
BGS & DPHE: Arsenic contamination of groundwater in Bangladesh (four volumes). In: D.G. Kinniburgh & P.L. Smedley (eds): *BGS technical report WC/00/19.* British Geological Survey, Keyworth, 2001.

Bowell, R.J. & Parshley, J.: Arsenic cycling in the mining environment. Characterization of waste, chemistry, and treatment and disposal, proceedings and summary report on the U.S. EPA workshop on managing arsenic risks to the environment, May 1–3. Denver, Colorado, USA, 2001.

Boyle, R.W. & Jonasson, I.R.: The geochemistry of As and its use as an indicator element in geochemical prospecting. *J. Geochem. Explor.* 2 (1973), pp. 251–296.

Breit, G.N., Lowers, H.A., Foster, A.L., Perkins, R.B., Yount, J.C., Whitney, J.W., Uddin, M.N.I. & Muneem, A.: Redistribution of arsenic and iron in shallow sediment of Bangladesh. In: *Symposium on the behaviour of arsenic in aquifers, soils and plants: implications for management*, Dhaka, 16–18 January. Centro Internacional de Mejoramiento de Maiz y Trigoand the U.S. Geological Survey, 2005.

Brown, G.E. Jr.: Spectroscopic studies of chemisorptions reaction mechanism at oxide-water interfaces. In: M.F. Hochella Jr. and A.F. White (eds): Mineral-water interface geochemistry. *Reviews in Mineralogy,* 23 (1990), pp. 81–95.

Bruckard, W.J., Davey, K.J., Jorgensen, F.R.A., Wright, S., Brew, D.R.M., Haque, N. & Vance, E.R.: Development and evaluation of an early removal process for the beneficiation of arsenic-bearing copper ores. *Min. Engineer.* 23 (2010), pp. 1167–1173.

Bundschuh, J., Farias, B., Martin, R., Storniolo, A. Bhattacharya, P., Cortes, J., Bonorino, G. & Albonuy, R.: Groundwater arsenic in the Chaco-Pampean Plain, Argentina: case study from Robles county, Santiago del Estero Province. *Appl. Geochem.* 19 (2004), pp. 231–243.

Charnock, J.M., Polya, D.A., Gault, A.G. & Wogelius, R.A.: Direct EXAFS evidence for incorporation of As^{5+} in the tetrahedral site of natural andraditic garnet. *Am. Mineral.* 92 (2007), pp. 1856–1861.

Chatterjee, A., Das, D., Mandal, B.K., Chowdhury, T.R., Wamanta, G. & Chakraborti, D.: Arsenic in groundwater in 6 districts of West Bengal, India – the biggest arsenic calamity in the world 1. Arsenic species in drinking water and urine of the affected people. *Analyst* 120 (1995), pp. 643–650.

Chauhan, S., D'Cruz, R., Fauqi, S., Singh, K.K., Varma, S., Singh, M. & Karthik, V.: Chemical warfare agents. *Environ. Toxicol. Pharmacol.* 26 (2008), pp. 113–122.

Cheng, H., Hu, T., Luo, J., Xu, B. & Zhao, J.: Geochemical processes controlling fate and transport of arsenic in acid mine drainage (AMD) and natural systems. *J. Hazard. Mat.* 165 (2009), pp. 13–26.

Clark, L.A.: The Fe-As-S system – phase relations and applications. *Econ. Geol.* 55 (1960), pp. 1345–1381.

Cook, S.J., Levson, V.M., Giles, T.R. & Jackaman, W.: A comparison of regional lake sediment and till geochemistry surveys – a case-study from the Fawnie Creek area, central British Columbia. *Explor. Min. Geol.* 4 (1995), pp. 93–110.

Corkhill, C.L. & Vaughan, D.J.: Arsenopyrite oxidation: a review. *Appl. Geochem.* 24 (2009), pp. 2342–2361.

Craw, D., Chappell, D. & Reay, A.: Environmental mercury and arsenic sources in fossil hydrothermal systems, Northland, New Zealand. *Environ. Geol.* 39 (2000), pp. 875–887.

Crecelius, E.A.: The geochemical cycle of arsenic in Lake Washington and its relation to other elements. *Limnol. Oceanog.* 20 (1975), pp. 441–451.

Cullen, W.R.: *Is arsenic an aphrodisiac?* RSC Publishing, Cambridge, 2008.

Cullen, W.R. & Reimer, K.J.: Arsenic speciation in the environment. *Chem. Rev.* 89 (1989), pp. 713–764.

Das, D., Chatterjee, A., Mandal, B.K., Samanta, G., Chakraborti, D.E. & Chanda, B.: Arsenic in ground-water in 6 districts of West Bengal, India – the biggest arsenic calamite in the world. 2. Arsenic concentration in drinking-water, hair, nails, urin, skin-scale and liver-tissue, biopsy of the affected people. *Analyst* 120 (1995), pp. 917–924.

Del Razo, L.M., Arellano, M.A. & Cebrián, M.E.: The oxidation states of arsenic in well-water from a chronic arsenicism area of northern Mexico. *Environ. Pollut.* 64 (1990), pp. 143–153.

De Vos, W., Ebbing, J., Hindel, R., Schalic, J., Swennen, R. & Van Keer, I.: Geochemical mapping based on overbank sediments in the heavily industrialised border area of Belgium, Germany and the Netherlands. *J. Geochem. Explor.* 56 (1996), pp. 91–104.

Dixit, S. & Hering, J.G.: Comparison of arsenic (V) and arsenic (III) sorption onto iron oxide minerals: implications for arsenic mobility. *Environ. Sci. Technol.* 37 (2003), pp. 4182–4189.

Dousová, B., Erbanová, L. & Novák, M.: Arsenic in atmospheric deposition at the Czech-Polish border: two sampling campaigns 20 years apart. *Sci. Tot. Environ.* 387 (2007), pp. 185–193.

DPHE/BGS/MML: *Groundwater studies for arsenic contamination in Bangladesh. Phase I: Rapid Investigation Phase.* BGS/MML Technical Report to Department for International Development, UK,1999, 6 volumes.

Drahota, P. & Filippi, M.: Secondary arsenic minerals in the environment: a review. *Environ. Int.* 35 (2009), pp. 1243–1255.

Duker, A.A., Carranza, E.J.M. & Hale, M.: Arsenic geochemistry and health. *Environ. Int.* 31 (2005), pp. 631–641.

Edmunds, W.M., Cook, J.M., Kinniburgh, D.G., Miles, D.L. & Trafford, J.M.: *Trace-element occurrence in British groundwaters.* Res. Report SD/89/3, British Geological Survey, Keyworth, 1989.

Ellis, A.J. & Mahon, W.A.J.: *Chemistry and geothermal systems.* Academic Press, New York, 1977.

Environmental Protection Agency (EPA): *Organical arsenicals*, 2001.

Fawcett, W.E. & Jamieson, H.E.: The distinction between ore processing and post-depositional transformation on the speciation of arsenic and antimony in mine waste and sediment. *Chem. Geol.* 283 (2011), pp. 109–118.

Feng, J., Clement, R. & Raynor, M.: Characterization of high-purity arsine and gallium arsenide epilayers grown by MOCVD. *J. Crystal Growth* 310 (2008), pp. 4780–4785.

Foster, A.L., Brown Jr., G.E., Tingle, T.N. & Parks, G.A.: Quantitative arsenic speciation in mine tailings using X-ray absorption spectroscopy. *Am. Mineral.* 83 (1998), pp. 553–568.

Froelich, P.N., Kaul, L.W., Byrd, J.T., Andreae, M.O. & Roe, K.K.: Arsenic, barium, germanium, tin, dimethyl-sulfide and nutrient biogeochemistry in Charlotte Harbour, Florida, a phosphorus-enriched estuary. *Estuar. Coast. Shelf Sci.* 20 (1985), pp. 239–264.

Goldberg, S.: Competitive adsorption of arsenate and arsenite on oxides and clay minerals. *Soil Sci. Soc. Am. J.* 66 (2002), pp. 413–421.

Graeme, K.A. & Pollack, C.V. Jr.: Heavy metal toxicity, part I: arsenic and mercury. *J. Emerg. Medicine* 16 (1998), pp. 45–56.

Gulens, J., Champ, D.R. & Jackson, R.E.: Influence of redox environments on the mobility of arsenic in ground water. In E.A. Jenne (ed.): *ACS Symposium series 93. chemical modeling in aqueous systems.* American Chemical Society, Washington, D.C., 1979, pp. 81–95.

Gustafsson, J.P. & Tin, N.T.: Arsenic and selenium in some Vietnamese acid sulfate soils. *Sci. Tot. Environ.* 151 (1994), pp. 153–158.

Harvey, C.F., Swartz, C.H., Badruzzaman, A.B.M., Keon-Blute, N., Yu, W., Ali, M.A., Jay, J., Beckie, R., Niedan, V., Brabander, D., Oates, P.M., Ashfaque, K.H., Islam, S., Hemond, H.F. & Ahmed, M.F.: Arsenic mobility and groundwater extraction in Bangladesh. *Science* 298 (2002), pp. 1602–1606.

Harvey, C.F., Ashfaque, K.N., Yu, W., Badruzzaman, A.B.M., Ali, M.A., Oates, P.M., Michael, H.A., Neumann, R.B., Beckie, R., Islam, S. & Ahmed, M.F.: Groundwater dynamics and arsenic contamination in Bangladesh. *Chem. Geol.* 228 (2006), pp. 112–136.

Hedenquist, J.W. & Lowenstein, J.B.: The role of magmas in the formation of hydrothermal ore deposits. *Nature* 370 (1994), pp. 519–527.

Hem, J.D.: Reactions of metal ions at surface of hydrous iron oxides. *Geochim. Cosmochim. Acta* 41 (1977), pp. 527–538.

Hochella Jr., M.F., Moore, J.N., Golla, U. & Putnis, A.: A TEM study of samples from acid mine drainage systems: metal-mineral association with implications for transport. *Geochim. Cosmochim. Acta* 63 (1999), pp. 3395–3406.

Hopenhayn, C.: Arsenic in drinking water: impacts on human health. *Elements* 2 (2006), pp. 103–107.

Horneman, A., van Geen, A., Kent, D.V., Mathe, P.E., Zheng, Y., Dhar, R.K., O'Connell, S., Hoque, M.A., Aziz, Z., Shamsudduha, M., Seddique, A.A. & Ahmed, K.M.: Decoupling of As and Fe release to Bangladesh groundwater under reducing conditions. Part I: Evidence from sediment profiles. *Geochim. Cosmochim. Acta* 68 (1994), pp. 3459–3473.

Howard, A.G., Apte, S.C., Comber, S.D.W. & Morris, R.J.: Biogeochemical control of the summer distribution and speciation of arsenic in the Tamar estuary. *Estuar. Coast. Shelf Sci.* 27 (1988), pp. 427–443.

Hsu, K.-H., Froines, J.R. & Chen, C.-J.: Studies of arsenic ingestion from drinking water in northeastern Taiwan: chemical speciation and urinary metabolites. In: C.O. Abernathy, R.L. Calderon & W.R. Chappell (eds): *Arsenic exposure and health effects.* Chapman Hall, London, 1997, pp. 190–209.

Hudson-Edwards, K.A. & Archer, J.: Geochemistry of As-, F- and B-bearing waters in and around San Antonio de los Cobres, Argentina, and implications for drinking and irrigation water quality. *J. Geochem. Explor.* 112 (2012), pp. 276–284.

Hudson-Edwards, K.A., Schell, C. & Macklin, M.G.: Mineralogy and geochemistry of alluvium contaminated by metal mining in the Rio Tinto area, southwest Spain. *Appl. Geochem.* 14 (1999), pp. 1015–1030.

Hudson-Edwards, K.A., Jamieson, H.E., Charnock, J.M. & Macklin, M.G.: Arsenic speciation in waters and sediments of ephemeral floodplain pools, Río Guadiamar, Aznalcóllar, Spain. *Chem. Geol.* 219 (2005), pp. 175–192.

Inskeep, W.P., McDermott, T.R. & Fendorf, S.: Arsenic (V)/(III) cycling in soils and natural waters: chemical and microbiological processes. In: W.T. Frankenberger (ed): *Environmental chemistry of arsenic*. Marcel Dekker, New York, 2002, pp. 185–215.

Inskeep, W.P., Macur, R.E., Hamamura, N., Warelow, T.P., Ward, S.A. & Santini, J.M.: Detection, diversity and expression of aerobic bacterial arsenite oxidase genes. *Environ. Microbiol.* 9 (2007), pp. 934–943.

Kintz, P., Ginet, M., Marques, N. & Cirimele, V.: Arsenic speciation of two specimens of Napoleon's hair. *Forensic Sci. Int.* 170 (2007), pp. 204–206.

Kocourková, E., Sracek, O., Houzar, S., Cempírek, J., Losos, Z., Filip, J. & Hršelová, P.: Geochemical and mineralogical control on the mobility of arsenic in a waste rock pile at DiouháVes, Czech Republic. *J. Geochem. Explor.* 110 (2011) 61–73.

Kossoff, D., Hudson-Edwards, K.A., Dubbin, W.E. & Alfredsson, M.A.: Incongruent weathering of Cd and Zn from mine tailings: a column leaching study. *Chem. Geol.* 281 (2011), pp. 52–71.

Kuhlmeier, P.D.: Partitioning of arsenic species in fine-grained soils. *J. Air Waste Manag. Assoc.* 47 (1997a), pp. 481–490.

Kuhlmeier, P.D.: Sorption and desorption of arsenic from sandy soils: column studies. *J. Soil Contam.* 6 (1997b), pp. 21–36.

Kwong, Y.T.J., Beauchemin, S., Hossain, M.F. & Gould, W.D.: Transformation and mobilization of arsenic in the historic Cobalt mining camp, Ontario, Canada. *J. Geochem. Explor.* 92 (2007), pp. 133–150.

Langmuir, D.: *Aqueous environmental geochemistry*. Prentice-Hall, New York, 1997.

Langner, H.W., Jackson, C.R., McDermott, T.R. & Inskeep, W.P.: Rapid oxidation of arsenite in a hot spring ecosystem, Yellowstone National Park. *Environ. Sci. Technol.* 35 (2001), pp. 3302–3309.

Lattanzi, P., Da Pelo, S., Musu, E., Atzei, D., Elsener, B., Fantauzzi, M. & Rossi, A.: Enargite oxidation: a review. *Earth-Sci. Rev.* 86 (2008), pp. 62–88.

Lièvremont, D., Bertin, P.H. & Lett, M.-C.: Arsenic in contaminated waters: biogeochemical cycle, microbial metabolism and biotreatment processes. *Biochimie* 91 (2009), pp. 1229–1237.

Lowers, H.A., Breit, G.N., Foster, A.L., Whitney, J., Yount, J. & Uddin, M.: Arsenic incorporation into authigenic pyrite, Bengal Basin sediment, Bangladesh. *Geochim. Cosmochim. Acta* 71 (2007), pp. 2699–2717.

Luo, Z.D., Zhang, Y.M., Ma, L., Zhang, G.Y., He, X., Wilson, R., Byrd, D.M., Griffiths, J.G., Lai, S., He, L., Grumski, K. & Lamm, S.H.: Chronic arsenicism and cancer in Inner Mongolia – consequences of well-water arsenic levels greater than $50 \, \mu g \, l^{-1}$. In: C.O. Abernathy, R.L. Calderon & W.R. Chappell (eds): *Arsenic exposure and health effects*. Chapman Hall, London, 1997, pp. 55–68.

Maest, A., Pasilis, S., Miller, L. & Nordstrom, D.: Redox geochemistry of arsenic and iron in Mono Lake, California, USA. In: Y.K. Charaka & A.S. Maest (eds): *Proc. 7th Int. Symp. Water-Rock Interaction*. A.A. Balkema, Rotterdam, 1992, pp. 507–511.

Maher, W.A.: Arsenic in coastal waters of South Australia. *Water Res.* 19 (1985), pp. 933–934.

Mao, M., Lin, J. & Pan, Y.: Hemimorphite as a natural sink for arsenic in zinc deposits and related mine tailings: evidence from single-crystal EPR spectroscopy and hydrothermal synthesis. *Geochim. Cosmochim. Acta* 74 (2010), pp. 2943–2956.

Matschullat, J.: Arsenic in the geosphere – a review. *Sci. Tot. Environ.* 249 (2000), 297–312.

McArthur, J.M., Banerjee, D.M., Hudson-Edwards, K.A., Mishra, R., Purohit, R., Ravenscroft, P., Cronin, A., Howarth, R.J., Chatterjee, A., Talukder, A., Lowry, D., Houghton, S. & Chadha, D.K.: Natural organic matter in sedimentary basins and its relation to arsenic in anoxic ground water: the example of West Bengal and its worldwide implications. *Appl. Geochem.* 19 (2004), 1255–1293.

McArthur, J.M., Ravenscroft, P., Banerjee, D.M., Milsom, J., Hudson-Edwards, K.A., Sengupta, Bristow, C., Sarkar, A., Tonkin, S. & Purohit, R.: How palaeosols influence groundwater flow and arsenic pollution: a model from the Bengal Basin and its worldwide implication. *Water Res. Res.* 44 (2008), p. W11411.

McCleskey, R.B., Nordstrom, D.K., Susong, D.D., Ball, J.W. & Taylor, H.E.: Source and fate of inorganic solutes in the Gibbon River, Yellowstone National Park, Wyoming, USA. II. Trace element chemistry. *J. Volcanol. Geotherm. Res.* 196 (2010), pp. 139–155.

Meng, X., Bang, S. & Korfiatis, G.P.: Effects of silicate, sulfate, and carbonate on arsenic removal by ferric chloride. *Water Res.* 34 (2000), pp. 1255–1261.

Mirlean, N. & Roisenberg, A.: The effect of emissions of fertilizer production on the environment contamination by cadmium and arsenic in southern Brazil. *Environ. Pollut.* 143 (2006), 335–340.

Mirlean, N., Ahdrus, V.E., Baisch, P., Griep, G. & Casartelli, M.R.: Arsenic pollution in Patos Lagoon estuarine sediments, Brazil. *Mar. Pollut. Bull.* 46 (2003), pp. 1480–1484.

Mohan, D. & Pittman, C.U.: Arsenic removal from water/wastewater using adsorbents; a critical review. *J. Hazard. Mat.* 142 (2007), pp. 1–53.

Murphy, E.A. & Aucott, M.: An assessment of the amounts of arsenical pesticides used historically in a geographical area. *Sci. Tot. Environ.* 218 (1998), pp. 89–101.

Navarro, M., Sanchez, M., Lopez, H. & Lopez, M.C.: Arsenic contamination levels in waters, soils and sludges in southeast Spain. *Bull. Environ. Contam. Toxicol.* 50 (1993), pp. 356–362.

Nickson, R.T., McArthur, J.M., Burgess, W., Ahmed, K.M., Ravenscroft, P. & Rahman, M.: Arsenic poisoning of Bangladesh groundwater. *Nature* 395 (1998), p. 338.

Nickson, R.T., McArthur, J.M., Ravenscroft, P., Burgess, W.G. & Ahmed, K.M.: Mechanism of arsenic poisoning of groundwater in Bangladesh and West Bengal. *Appl. Geochem.* 15 (2000), pp. 403–413.

Nicolli, H.B., Suriano, J.M., Peral, M.A.G., Ferpozzi, L.H. & Baleani, O.A.: Groundwater contamination with arsenic and other trace-elements in an area of the Pampa, province of Córdoba, Argentina. *Environ. Geol. Water Sci.* 14 (1989), pp. 3–16.

Nicolli, H.B., Bundschuh, J., Garcia, J.W., Falcón, C. M. & Jean, J.-S.: Sources and controls for the mobility of arsenic in oxidizing groundwaters from loess-type sediments in arid/semi-arid dry climates – Evidence from the Chaco-Pampean plain (Argentina). *Water Res.* 44 (2010), pp. 5589–5604.

Nielsen, S.S., Petersen, L.R., Kjeldsen, P. & Jakobsen, R.: Amendment of arsenic and chromium polluted soil from wood preservation by iron residues from water treatment. *Chemosphere* 84 (2011), pp. 383–389.

Nordstrom, D.K. & Archer, D.G.: Arsenic thermodynamic data and environmental geochemistry. In: K.G. Stollenwerk (ed): *Arsenic in ground water: geochemistry and occurrence.* Kluwer, Boston 2003, pp. 1–25.

Onishi, H. & Sandell, E.G.: Geochemistry of arsenic. *Geochim. Cosmochim. Acta* 7 (1955), pp. 1–33.

Oremland, R.S. & Stolz, J.F.: Arsenic, microbes and contaminated aquifers. *Trends Microbiol.* 13 (2005) 45–49.

Osborne, T.H., Jamieson, H.E., Hudson-Edwards, K.A., Nordstrom, D.K., Walker, S.R., Ward, S.A. & Santini, J.M.: Microbial oxidation of arsenite in a subarctic environment: diversity of arsenite oxidase genes and identificaiton of a psychrotolerant arsenite oxidiser. *BMC Microbiol.* 10 (2010), p. 205.

Pandey, V.C., Singh, J.S., Singh, R.P., Wingh, N. & Yunus, M.: Arsenic hazards in coal fly ash and its fate in Indian scenario. *Res. Conservation Recycl.* 55 (2011), pp. 819–835.

Pascua, C., Charnock, J., Polya, D.A., Sato, T., Yokoyama, S. & Minato, M.: Arsenic-bearing smectite from the geothermal environment. *Min. Mag.* 69 (2005), pp. 897–906.

Peters, S.C., Blum, J.D., Karagas, M.R., Chamberlain, C.P. & Sjostrom, D.J.: Sources and exposure of the New Hampshire population to arsenic in public and private drinking water supplies. *Chem. Geol.* 228 (2006), pp. 72–84.

Peterson, M.L. & Carpenter, R.: Biogeochemical processes affecting total arsenic and arsenic species distributions in an intermittently anoxic Fjord. *Mar. Chem.* 12 (1983), pp. 295–321.

Pichler, T., Veizer, J. & Hall, G.E.M.: Natural input of arsenic into a coral reef ecosystem by hydrothermal fluids and its removal by Fe(III) oxyhydroxides. *Environ. Sci. Technol.* 33 (1999), pp. 1373–1378.

Pichler, T., Brinkmann, R. & Scarzella, G.I.: Arsenic abundance and variation in golf course lakes. *Sci. Tot. Environ.* 394 (2008), pp. 313–320.

Plant, J.A., Kinniburgh, D.G., Smedley, P.L., Fordyce, F.M. & Klinck, B.A.: Arsenic and Selenium. In: H.D. Holland and K.K. Turekian (eds): *Treatise on geochemistry.* Elsevier, 2007, Chapter 9.02, pp. 17–66.

Raven, K.P., Jain, A. & Loeppert, R.H.: Arsenite and arsenate adsorption on ferrihydrite: kinetics, equilibrium, and adsorption envelopes. *Environ. Sci. Technol.* 32 (1998), pp. 344–349.

Ravenscroft, P., Brammer, H. & Richards, K.: *Arsenic pollution: a global synthesis.* Wiley-Blackwell, Chichester, UK, 2009.

Rawlins, B.G., O'Donnell, K. & Ingham, M.: *Geochemical survey of the tamar catchment/(South West England).* British Geological Survey report CR/03/027, 2003.

Rhine, E.D., Garcia-Dominguez, E., Phelps, C.D. & Young, L.Y.: Environmental microbes can speciate and cycle arsenic. *Environ. Sci. Technol.* 39 (2005), pp. 9569–9573.

Riedel, F.W. & Eikmann, T.: Natural occurrence of arsenic and its compounds in soils and rocks. *Wissensch, Umwelt* 3–4 (1986), pp. 108–117.

Romero, L., Alonso, H., Campano, P., Fanfani, L., Cidu, R., Dadea, C., Keegan, T., Thornton, I. & Farago, M.: Arsenic enrichment in waters and sediments of the Rio Loa (Second Region, Chile). *Appl. Geochem.* 18 (2003), pp. 1399–1416.

Root, R.A., Vlassopoulos, D., Rivera, N.A., Rafferty, M.T., Andrews, C. & O'Day, P.A.: Speciation and natural attenuation of arsenic and iron in a tidally influenced shallow aquifer. *Geochim. Cosmochim. Acta* 73 (2009), pp. 5528–5553.

Rosa, C.J.P., McPhie, J. & Relvas, J.M.R.S.: Type of volcanoes hosting the massive sulfide deposits of the Iberian Pyrite Belt. *J. Volcanol. Geotherm. Res.* 194 (2010), pp. 107–126.

Roussel, C., Néel, C. & Bril, H.: Minerals controlling arsenic and lead solubility in an abandoned gold mine tailings. *Sci. Tot. Environ.* 263 (2000), pp. 209–219.

Ruiz, C. & Arribas, A.: Mineralogy and geochemistry of the Masa Valverde blind massive sulfide deposit, Iberian Pyrite Belt (Spain). *Ore Geol. Rev.* 9 (2002), pp. 1–22.

Sarmiento, A.M., Nieto, J.M., Casiot, C., Elbaz-Poulichet, F. & Egal, M.: Inorganic arsenic speciation at river basin scales: The Tinto and Odiel Rivers in the Iberian Pyrite Belt, SW Spain. *Environ. Pollut.* 157 (2009), pp. 1202–1209.

Savage, K.S., Tingle, T.N., O'Day, P.A., Waychunas, G.A. & Bird, D.K.: Arsenic speciation in pyrite and secondary weathering phases, Mother Lode Gold District, Tuolumne County, California. *Appl. Geochem.* 15 (2000), pp. 1219–1244.

Savage, K.S., Ashley, R.P. & Bird, D.K.: Geochemical evolution of a high arsenic, alkaline pit-lake in the Mother Lode Gold District, California. *Econ. Geol.* 104 (2009), pp. 1171–1211.

Schwedt, G. & Rieckhoff, M.: Analysis of oxothio arsenic species in soil and water. *J. PraktischeChemi-Chemiker-Zeitung* 338 (1996), pp. 55–59.

Scudlark, J.R. & Church, T.M.: The atmospheric deposition of arsenic and association with acid precipitation. *Atmos. Environ.* 22 (1988), pp. 937–943.

Senior, G.D., Smith, L.K., Silvester, E. & Bruckard, W.J.: The flotation of gersdorffite in sulfide nickel systems – a single mineral study. *Int. J. Min. Process.* 93 (2009), pp. 165–171.

Seyler, P. & Martin, J.-M.: Biogeochemical processes affecting arsenic species distribution in a permanently stratified lake. *Environ. Sci. Technol.* 23 (1989), pp. 1258–1263.

Seyler, P. & Martin, J.-M.: Distribution of arsenite and total dissolved arsenic in major French estuaries: dependence on biogeochemical processes and anthropogenic inputs. *Mar. Chem.* 29 (1990), pp. 277–294.

Seyler, P. & Martin, J.-M.: Arsenic and selenium in a pristine river-estuarine system: the Krka, Yugoslavia. *Mar. Chem.* 34 (1991), pp. 137–151.

Sharma, V.K. & Sohn, M.: Aquatic arsenic: toxicity, speciation, transformations, and remediation. *Environ. Int.* 35 (2009), pp. 743–759.

Simon, G., Huang, H., Penner-Hahn, J.E., Kesler, S.E. & Kao, L.S.: Oxidation state of gold and arsenic in gold-bearing arsenian pyrite. *Am. Mineral.* 84 (1999), pp. 1071–1079.

Smedley, P.L. & Kinniburgh, D.G.: A review of the source, behaviour and distribution of arsenic in natural waters. *Appl. Geochem.* 17 (2002), pp. 517–568.

Smedley, P.L. & Kinniburgh, D.G.: Arsenic in groundwater and the environment. In O. Selinus (ed.): *Essentials of medical geology. Impacts of the natural environment on public health.* Elsevier, Burlington, MA, 2004, Chapter 11, pp. 263–299,

Smedley, P.L., Nicolli, H.B., Macdonald, D.M.J., Barros, A.J. & Tullio, J.O.: Hydrogeochemistry of arsenic and other inorganic constituents in groundwaters from La Pampa, Argentina. *Appl. Geochem.* 17 (2002), pp. 259–284.

Stewart, F.H.: Data of Geochemistry, 6th ed. Chap. Y. Marine Evaporites. *US Geol. Surv. Prof. Pap.* 440-Y, 1963.

Sultan, K. & Shazili, N.A.: Distribution and geochemical baselines of major, minor and trace elements in tropical topsoils of the Terengganu River basin, Malaysia. *J. Geochem. Explor.* 103 (2009), pp. 57–68.

Sun, G., Liu, J., Luong, T.V., Sun, D. & Wang, L.: Endemic arsenicosis; a clinical diagnostic with photo illustrations. UNICEF East Asia and Pacific Regional Office, Bangkok, Thailand, 2004.

Townsend, T., Solo-Gabriele, H., Tolaymat, T., Stook, K. & Hosein, N.: Chromium, copper, and arsenic concentrations in soil underneath CCA-treated wood structures. *Soil Sed. Contam.* 12 (2003), pp. 779–798.

Turpeinen, R., Pantsar-Kallio, M. & Kairesalo, T.: Role of microbes in controlling the speciation of arsenic and production of arsines in contaminated soils. *Sci. Tot. Environ.* 285 (2002), pp. 133–145.

Ure, A. & Berrow, M.: The elemental constituents of soils. In: H.J.M. Bowen (ed.): *Environmental chemistry.* Royal Society of chemistry, London, 1982, Chapter 3, pp. 94–203.

Varsányi, I., Fodré, Z. & Bartha, A.: Arsenic in drinking water and mortality in the southern Great Plain, Hungary. *Environ. Geochem. Health* 13 (1991), pp. 14–22.

Vaughan, D.J.: Arsenic. *Elements* 2 (2006), pp. 71–75.

Walker, S.R., Parsons, M.B., Jamieson, H.E. & Lanzirotti, A.: Arsenic minerlogy of near-surface tailings and soils: influences on arsenic mobility and bioaccessibility in the Nova Scotia gold mining disctricts. *Can. Mineral.* 47 (2009), pp. 533–556.

Webster, J.G. & Nordstrom, D.K.: Geothermal arsenic. In: A.H. Welch & K.G. Stollenwerk, (eds): *Arsenic in groundwater: geochemistry and occurrence.* New York: Springer-Verlag (2003), pp. 101–126.

Welch, A.H., Lico, M.G. & Hughes, J.L.: Arsenic in groundwater of the western United States. *Ground Water* 26 (1988), pp. 333–347.

Welch, A.H., Westjohn, D.B., Helsel, D.R. & Wanty, R.B.: Arsenic in groundwater of the United States. *Ground Water* 38 (2000), pp. 589–604.

White, D.E., Hem, J.D. & Waring, G.A.: Data of Geochemistry, 6th ed. M. Flesicher (ed.). Chapter F. Chemical Composition of Sub-Surface Waters. *US Geol. Surv. Prof. Pap.* 440-E, 1963.

Wilson, F.H. & Hawkins, D.B.: Arsenic in streams, stream sediments and ground water, Fairbanks area, Alaksa. *Environ. Geol.* 2 (1978), pp. 195–202.

Xie, X., Want, Y., Ellis, A., Su, C., Li, J. & Mengdi, L.: The sources of geogenic arsenic in aquifers at Datong basin, northern China: constraints from isotopic and geochemical data. *J. Geochem. Explor.* 110 (2011), pp. 155–166.

Ye, L., Cook, N. J., Ciobanu, C. L., Yuping, L., Qian, Z., Tiegeng, L., Wei, G., Yulong, Y. & Danyushevskiy, L.: Trace and minor elements in sphalerite from base metal deposits in South China: A LA-ICPMS study. *Ore Geol. Rev.* 39 (2011), pp. 188–217.

Yu, W.H., Harvey, C.M. & Harvey, C.F.: Arsenic in groundwater in Bangladesh: a geostatistical and epidemiological framework for evaluating health effects and potential remedies. *Wat. Res. Res.* 39 (2003), p. 1146.

Yudovich, Ya.E. & Ketris, M.P.: Arsenic in coal: a review. *Int. J. Coal. Geol.* 61 (2005), pp. 141–196.

Zhu, B.J. & Tabatabai, M.A.: An alkaline oxidation method for determination of total arsenic and selenium in sewage sludges. *J. Environ. Qual.* 24 (1995), pp. 622–626.

Zhu, J., Pigna, M., Cozzolino, V., Caporale, A.G. & Violante, A.: Sorption of arsenite and arsenate on ferrihydrite: Effect of organic and inorganic ligands. *J. Hazard. Mat.* 189 (2010), pp. 564–571.

CHAPTER 2

Giant Mine, Yellowknife, Canada: Arsenite waste as the legacy of gold mining and processing

Mackenzie Bromstad & Heather E. Jamieson

2.1 INTRODUCTION

In this chapter we report research on the characterization of the arsenic (As) waste problem left after the closure of Giant Mine – a waste legacy that includes one of the largest concentrations of arsenite at the surface of the Earth. We discuss the transformation and mobility of the various species of As, and examine factors influencing the long-term stability of various As-bearing compounds.

The most common As-bearing mineral is arsenopyrite (FeAsS), which is relatively common in many rocks, including coals and metal ores (Reimann et al., 2009). Quartz-carbonate vein gold deposits hosted in greenstone belts, such as the deposit at Giant Mine in northern Canada, often contain 5–10 wt% sulfide minerals, including arsenopyrite and As-bearing pyrite. Most of the gold at Giant Mine occurs as a refractory phase, meaning that it is incorporated submicroscopically within sulfide minerals, mainly arsenopyrite (Canam, 2006).

Gold is usually extracted from ore using cyanide leaching, but refractory gold deposits require an additional processing step to liberate the submicroscopic gold and make it available for leaching. At Giant Mine, the ore was roasted, an oxidation process which converts FeAsS to porous roaster iron oxides (ROs), or calcine (Walker et al., 2005, 2011). The two major gases produced from roasting were As vapor and sulfur dioxide (SO_2) (SRK, 2002a). The roasting process oxidized As present in FeAsS from As(–I) to arsenite [As(III)] as shown by Equation 2.1:

$$2FeAsS + 5O_2 \rightarrow Fe_2O_3 + As_2O_3 + 2SO_2 \tag{2.1}$$

As a result, As precipitated from the vapors as As trioxide (As_2O_3) (INAC, 2007), which is highly soluble and one of the most toxic forms of As to humans (Ruby et al., 1999). Originally As vapors were allowed to vent freely into the atmosphere, but in 1951 the first of many generations of gas capture technology was implemented, in the form of an Cottrell electrostatic precipitator (ESP), to reduce As emissions by capturing As_2O_3 dust (SRK, 2002a). The As-rich dust produced by roasting contained approximately 60% As, 80% of which was As_2O_3 (SRK, 2002c). A smaller amount of As in the dust also occurred as a mixture of As(III) and arsenate [As(V)] oxidation states within ROs (Walker et al., 2005).

As_2O_3 dusts and fine RO particles were released aerially from the roaster stack throughout the lifetime of the mine (1949–1999) and deposited in the surrounding area; this continued even after the ESP was installed. As_2O_3 dusts also accumulated from the ESP baghouse (fabric-based exhaust filter) and from cleaning the roaster. In 1951 it was decided that the best way of storing As_2O_3 dusts was in underground storage chambers. These were reliant on permafrost to keep As_2O_3 from dissolving and entering the groundwater. By 1962 five chambers had been excavated and filled near the ESP baghouse, and dust storage switched to mined-out stopes (shafts or tunnels) in permafrost ground. There was increasing concern during this storage process that permafrost would recede owing to mine workings and that As_2O_3 chambers could not be kept dry; in the late 1970s strong observational evidence indicated that the permafrost had indeed been receding and groundwater movement increasing.

Prior to the 1980s, the mine operators attempted to maintain permafrost in As_2O_3 storage chambers by pumping cold air into them. Monitoring involved checking visually for ice on the walls. However, beginning with the chambers constructed in the 1980s these precautions were discontinued and requirement for permafrost abandoned. The last four chambers constructed were partially above the permafrost zone and no ice was observed in them (SRK, 2002a).

In total, approximately 237,000 tonnes of As_2O_3-containing dust were stored underground (INAC, 2007) in chambers with a total volume of 220,640 m^3 (SRK, 2005c). Approximately 2.3 million tonnes of tailings have also been used as backfill underground; half of this consists of calcine. There are now concerns regarding As_2O_3 storage, including the immediate stability of stopes used to store As_2O_3, as well as the presence of water in some storage chambers, and temperature fluctuation from $-4°C$ and $+5°C$. The key issue to be addressed during the remediation for the underground As_2O_3 is the potential for planned re-flooding of the mine that would lead to uncontrolled release of As (INAC, 2007). A human health and ecological risk assessment has estimated that without remediation, flooding the mine could result in As release as high as 16,000 kg/year ($kg\,a^{-1}$), well above the proposed target of no more than 2,000 $kg\,a^{-1}$ to limit human health and ecological effects (SRK, 2002a).

In addition to concerns with underground As_2O_3, final remediation of the site needs to address old mine buildings and infrastructure, contaminated sediments in Baker Creek (which flows through the mine site), tailings spills, stored tailings, the calcine pond, waste rock (INAC, 2007) and an estimated 20,000 tonnes of aerially deposited As_2O_3-bearing roaster emissions (Wrye, 2008).

Researchers at Queen's University have been studying various aspects of the Giant Mine site for more than 12 years. Some of the questions considered include: Does ore roasting affect the long-term stability of mine waste? What is the importance of disposal environment on arsenite-bearing mine waste? Does the presence or absence of acid mine drainage exert an important control on the release of As to drainage waters?

2.2   BACKGROUND

2.2.1   *Physiographic and geological setting*

Giant Mine is located approximately 5 km north of Yellowknife, Northwest Territories (Fig. 2.1). The deposit lies within the Yellowknife Supergroup, part of the Canadian Shield Slave Structural Province. The Yellowknife supergroup consists of Archaean metavolcanic and metasedimentary rocks, intruded by younger granitoids. Several early Proterozoic gabbro and diabase dikes crosscut the area, and several fault lines divide the volcanic and granitoid rock units (Kerr and Wilson, 2000). The eastern half of the Giant Mine property lies principally over the Kam Group of the Yellowknife Bay Formation, which consists of variolitic pillowed and massive flows, dominated by basalts and metamorphosed to greenschist facies (Bleeker, 2007).

The deposit at Giant Mine occurs in schist and shear zones within this greenstone belt, specifically as quartz-sulfide bands alternating with carbonate-sericite schists resulting from four deformational events. Gold is primarily hosted as a refractory phase within arsenopyrite, and to a lesser extent within pyrite. Some free gold was present but did not make up the bulk of the deposit (Canam, 2006).

Yellowknife and Giant Mine lie within a zone of discontinuous permafrost. The most prevalent sediment present in the Yellowknife area is glacial till, which is generally discontinuous and less than 2 m thick. Glacial Lake McConnell covered the entire area of the Giant Mine property during the last glacial retreat 10,000 years ago. The lake included the basins of Great Slave, Great Bear, and Athabaska Lakes, and was approximately 280 m deep. During the time of Lake McConnell, Yellowknife would have been under approximately 80 m of water. The glaciolacustrine sediments overlying till on much of the Giant Mine property attest to this. Organic deposits often overlie the glaciolacustrine sediments of formerly submerged regions in low-lying areas, and can be 1 m thick. Wetlands occur in some low-lying areas. The Yellowknife area includes abundant outcrop, up to 75% in some areas (Kerr and Wilson, 2000). Approximately 30% of the Giant Mine surface property consists of outcrop (Wrye, 2008), much of it denuded of vegetation by historical SO_2

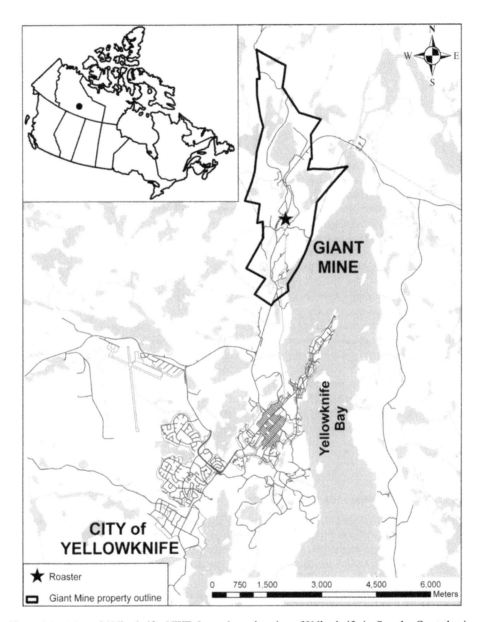

Figure 2.1.   Map of Yellowknife, NWT. Inset shows location of Yellowknife in Canada. Created using data from Government of Canada (2009), INAC (2007) and Mapsof.net (2011).

emissions from the roaster (Bleeker, 2007). Most soil pHs are near-neutral. What vegetation there is on outcrops is confined to outcrop crevices (~10% of outcrop); there is scrub forest in areas with more soil, and wetlands in areas with a lack of drainage (INAC, 2007). Outcrop vegetation varies with wind direction and distance from the roaster.

The prevailing wind in the Yellowknife area is from the east and south, and secondarily from the northwest (SENES, 2005). The climate is cool and dry. Air temperature records beginning in 1943 indicate an air temperature range from −50°C to 32°C, with an average annual temperature of −4.5°C. Annual precipitation averaged 281 mm between 1971 and 2000, while annual evaporation averaged 415 mm between 1994 and 2002 (INAC, 2007).

Yellowknife Bay is part of Great Slave Lake, the fifth-largest lake in Canada in terms of surface area, and the deepest. Surface water in the area is covered with ice at least seven months of the year.

2.2.2   *History of mining, processing and waste management at Giant Mine*

From 1949 to 1999 ore was processed by roasting a sulfide concentrate produced by flotation, and leaching the resulting calcine with cyanide (Walker *et al.*, 2005, 2011; Fawcett and Jamieson, 2011; INAC, 2007).

The roasting was carried out at relatively low temperature (500°C) and conducted with air-deficient conditions typical of the methods used to remove arsenopyrite (Walker *et al.*, 2011). The first ESP was installed in 1951 to capture As_2O_3 dust, probably motivated by the loss of gold in stack emissions. Until the roaster ceased operations in 1999, As_2O_3 was captured using an ESP and baghouse and stored underground, but a significant quantity of dust still escaped in stack emissions (Wrye, 2008). Sulphur dioxide was never scrubbed from stack emissions.

In the first two years, tailings, which were a mixture of flotation waste and calcine, were discharged directly onto the shore of Yellowknife Bay. Beginning in 1951, tailings were deposited into several former lakes. Later, tailings dams and several impoundments were constructed (Fig. 2.2). From 1956 to 1978 tailings were also used as backfill into the mine workings (SRK, 2005d).

Prior to sedimentation controls, a significant amount of fine tailings, including ESP dust and calcine, flowed over the ice-covered tailings dams each spring and was deposited at the upstream end of Baker Pond, a natural water body located along Baker Creek (Fawcett and Jamieson, 2011).

Water treatment designed to remove As began in 1957 with plant updates in 1967 and again in 1981. Discharged sludge was stored in a former tailings pond (INAC, 2007).

A purpose-built storage pond for calcine waste was built between 1950 and 1954 on a natural clay deposit near Baker Creek with the intention of reprocessing calcine for additional gold. The reprocessing was not deemed economically viable and was shut down after 3 months in operation. The pond was used for calcine tailings storage until 1971, after which it was covered in a layer of composted manure and clay, later augmented with soil removed from one of the surface pits (INAC, 2004).

2.2.3   *Mine management, ownership and responsibility for remediation*

The original claims for Giant Mine were staked in 1935 by Burwash Yellowknife Mines Limited. Several owners operated the mine, which was a major employer for the community, until 1999, when it went into receivership and a court-appointed receiver transferred property control from Royal Oak Mines Inc. to the Canadian government, specifically the Department of Indian Affairs and Northern Development (DIAND). Miramar Giant Mine Ltd., who operated Con mine, a second large gold operation just south of the city, immediately purchased the mine from DIAND under an agreement indemnifying Miramar for existing environmental liabilities at Giant Mine. Miramar continued to mine ore from Giant at a reduced scale, and trucked it to Con site for processing until 2004 (INAC, 2007). A Northwest Territories court assigned Miramar Giant Mine Ltd. into bankruptcy in 2006 (Terriplan Consultants, 2007), at which time Giant Mine became "orphaned and abandoned." DIAND has developed a comprehensive remediation plan which, at the time of writing, is undergoing environmental assessment.

2.3   ARSENIC AND ARSENITE IN MINE WASTES AND SURROUNDING AREA

During most of the years that Giant Mine operated, there was little legislation and few emission controls, resulting in a complex legacy of arsenite- and arsenate-bearing solid wastes as well as contaminated surface and underground waters.

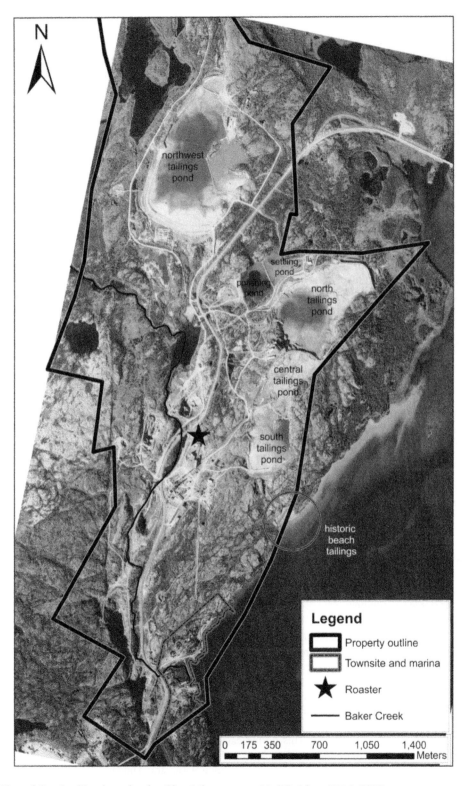

Figure 2.2.   Satellite photo showing Giant Mine property. Modified from INAC (2007).

2.3.1 *Water*

2.3.1.1 *Surface water: streams and lakes*
Input of As to surface streams and lakes originates from runoff and freshet that comes into contact with mine waste and contaminated soil. Mine water from underground workings is pumped to surface tailings impoundments and influences surface water downstream. Arsenic-bearing sediments can act as both sources and sinks for As in overlying water. There is a marked difference in As concentrations upstream of Giant Mine, through the mine site, and downstream of Giant Mine. The highest concentrations in streams are located in side streams west, north, and southeast of the roaster stack, in the direction of the prevailing wind. Historical roaster deposition in these areas could thus be providing a source of As to Baker creek through these side stream sediments and soils (SRK, 2005b).

2.3.1.2 *Mine water*
Water in the underground mine workings originates either from relatively clean sources, such as direct infiltration from precipitation and streams, or from tailings pond waters. For representative As concentrations in mine waters see Table 2.1. Interaction with underground As_2O_3 storage chambers produces a distinct change in the composition of infiltrating water, increasing the total As concentration and the As(III). In other areas of the mine, the water is dominated by arsenate. Flow discharge from the mine to the northwest tailings pond varies between seasons. In the winter it averages $1300 \, L \, min^{-1}$, while during freshet it averages $2600 \, L \, min^{-1}$ (SRK, 2005a).

2.3.2 *Solid arsenic-bearing material*

Table 2.2 contains a list of solid materials on the mine property that contain significant amounts of As.

2.3.2.1 *Mine tailings*
The tailings deposited at Giant Mine are a mixture of flotation tails (crushed rock remaining after sulfides were removed for roasting), calcine (iron oxide-rich roaster waste) and ESP residue (As trioxide-rich particles). Walker *et al.* (2011) have shown that although the calcine comprised a minor component of the discharge, it provides the highest As loading. The As in both calcine and ESP components includes As(III) and As(V) compounds. Arsenic in the flotation component is mostly arsenopyrite with some grains exhibiting oxidation rims of As(V)-bearing Fe oxyhydroxide. Roaster-generated iron oxides found in the calcine component include maghemite and hematite and contain both As(III) and As(V) (Walker *et al.*, 2005, 2011).

The dominant minerals in the tailings are quartz, dolomite, chlorite, muscovite and calcite (Walker *et al.*, 2011). Acid mine drainage is unlikely to be a problem since the sulfides were mostly destroyed during roaster oxidation and there is abundant carbonate. The potential for the release of As from tailings to aqueous solution depends on the stability of the As-hosting phases under storage conditions.

2.3.2.2 *Roaster waste (calcine, ESP dust, underground storage, stack emissions)*
Calcine, the oxidized product of roasting flotation tailings, consists mainly of gold-bearing ROs (maghemite) remaining after most S and As has been removed. In calcine samples, As(V) and As(III) are the dominant oxidation states, with some As(−I) (Walker *et al.*, 2005, 2011; Fawcett and Jamieson, 2011). ROs in calcine have been found to contain <0.5% to 7% As (Walker *et al.*, 2005). Most of the calcine in Giant Mine was deposited in combination with flotation tailings, but approximately $34,000 \, m^3$ was separated, as part of an experiment in secondary gold recovery, and is now stored and covered. This buried calcine is near-neutral in pH and contains high concentrations of As, on average $6189 \, mg \, kg^{-1}$ with $1.02 \, mg \, L^{-1}$ As water soluble (INAC, 2004).

Arsenic-rich ESP dust collected in the roaster baghouse was stored underground and on average contains 60% wt. As. Of that As, approximately 80% is As_2O_3 (INAC, 2007). The next most

Table 2.1. Aqueous As concentrations in waters[a] at Giant Mine.

	Mean As $(mg\,L^{-1})$	Median As $(mg\,L^{-1})$	Range As $(mg\,L^{-1})$	Notes	n[b]
Surface streams and lakes					
Baker creek background levels[4]	0.012	–[c]	0.0079–0.017	Upstream of mine, Baker Creek catchment	3
Baker creek, upstream of mine[4]	0.065	0.026	0.02–0.22	Influenced by historic roasting	7
Baker creek, adjacent to mine[4]	0.055	0.041	0.02–0.099		9
Side streams W,N,SE of roaster[4]	0.23	0.121	0.056–1.44	Reflect historic roaster deposition; feed into Baker creek	16
Northwest tailings pond[5,2]	11.2	9.7	4.4–26.8	Most recent tailings deposition. Highest sample[5] is 13 mg L⁻1 greater than next highest	8
North tailings pond[5]	3.69	–	2.64–4.73		2
Settling pond[5]	0.002	–	0.001–0.003		2
Yellowknife Bay porewaters[3]	–	–	<0.001–0.95		–
Surface seeps					
Southeast of mill[4]	–	–	0.65–2	Undisturbed areas, atmospheric As only	3
C-shaft and C-dry area[4]	–	–	0.13–0.89	As from oxidation of arsenopyrite	6
Mill area[4]	–	3.4	0.26–24.9 0.13–1.1[6]	Highest As near baghouse	8
Open pits[4]	–	–	0.27–1.11 0.2–4.1[7]	As from waste rock	2
Mine water					
Annual average mine waste water, short term[1]	33	–	–	Includes tailings runoff. Before remediation methods begin.	12
Annual average mine waste water, long term[1]	2.5	–	–	Includes tailings runoff. After remediation implementation	12
Seepage from NW tailings pond[2]	5	–	–		–
Seepage from contact with mine walls[2]	0.4	0.04	0.01–0.7	In contact with soils, bedrock, and mine walls. Above water table.	19
Seepage of tailings backfill[2]	5.5	4	0.09–20	Above water table	8
Seepage of tailings backfill, As_2O_3 chamber influence[2]	41.8	–	41.5–42.1	Above water table	2
Seepage of waste rock backfill[2]	0.94	0.95	0.23–1.63	Above water table	4
Seepage of As_2O_3 chambers[2]	3,350	3,460	2,280–4,210	Above water table	15

[a]CCME drinking water and protection of aquatic life guideline $= 0.005\ mg\,L^{-1}$; [b]"n" indicates number of samples; [c]"–" indicates either that the source omitted the relevant data or that the calculation was unnecessary due to small sample size; [1]SENES 2005 (SD L1); [2]SRK 2005a; [3]Andrade *et al.* 2010; [4]SRK 2005b; [5]Fawcett *et al.* 2011 manuscript; [6]Royal Oak 1998 quoted in SRK 2005b; [7]Golder 2000 quoted in SRK 2005b.

Table 2.2.   Solid arsenic concentrations in various media at Giant Mine.

	Average As (mg kg^{-1})	Median As (mg kg^{-1})	Range As (mg kg^{-1})	Notes	n[a]
Tailings					
South tailings pond[9]	2,384	2,440	1,760–2,990		8
Central tailings pond[9]	2,628	2,650	1,325–3,850		7
North tailings pond[9,13]	3,110	2,945	1,500–5,080	Lowest values for subaerial tailings	16
Northwest tailings pond[9,13]	2,930	3,280	338–4,875	Most recent tailings have highest values (4,480–4,875)	8
Water treatment sludge[9]	26,400	–[b]	10,500–42,300		2
Underground tailings (1967–1976)[10]	1,495	1,270	875–3,315		11
Underground tailings (1959–1970)[10]	2,918	2,890	1,190–5,935		9
Subaerial historic shoreline tailings[12,13]	1,448	1,371	810–2,240	Originally deposited 1,949–1,951. Mace (1998)[8] gives 65–1,016 mg kg^{-1} As for entire beach	4
Submerged historic shoreline tailings[12,13]	1,135	–	1,070–1,200	Originally deposited 1949–1951. Mace (1998)[8] gives 65–1016 mg kg^{-1} As for entire beach	2
Roaster waste					
Calcine[3]	6,189	6,175	280–>10,000	two samples >10,000 mg kg^{-1} As	11
ROs in calcine residue[12]	2.9%	–	<0.5%–7.6%	As hosted within ROs, (mostly) maghemite	32
ROs in shoreline tailings[12]	1.3%	–	<0.5%–2.9%	As hosted within ROs, (mostly) maghemite. Produced by earlier roasting method than ROs in calcine residue tested	15
Average ESP baghouse dust[4]	60%	–	–		–
Pre-1964 ESP baghouse dust[4]	46%	48%	28%–67%		33
Post-1964 ESP baghouse dust[4]	65%	68%	41%–74%		18
Soils[c]					
Mill area[1,2]	1,824	1,170	21–11,800	Near roaster and calcine buildings	156
Calcine pond[3]	324	115	20–1,350	overburden and soils below calcine layer in calcine pond	9
West of tailings settling pond[1]	3,376	2,550	7–25,500	experienced tailings spills multiple times; 25,500 is outlier, next highest value is ~10,000	38

(*Continued*)

Table 2.2. Continued.

	Average As (mg kg^{-1})	Median As (mg kg^{-1})	Range As (mg kg^{-1})	Notes	n[a]
Soils[c]					
Historic foreshore tailings area[1,2]	473	46	7–3,660	Soil around tailings beach, does not include tailings	38
Townsite[1,2]	1,224	322	10–16,600		89
Away from mine activity[1,2]	299	49	6–3,280	Approximate; samples in areas with no known anthropogenic activity or As input other than natural sources and roaster deposition	99
Entire site[1,2]	1,197	270	6–25,500	25,500 is an outlier, ~10,000 is next closest measurement	586
Sediments					
Upstream of site[5]	177	–	–		–
Creek channel through mine site[5]	1,643	–	–		–
Baker Pond[6]	1,161 1,736[7]	1,080	332–2,000	All values from single core	22
Baker Pond mounded tailings and sediments[13]	2,700	–	2,520–2,880		2
Inside breakwater[6]	249	245	26–580	All values from single core	24
Vegetated inside breakwater[6,11]	1,430	1,410	33–3,700	Mixture of core samples, bulk samples, and bulk samples of sediment effected by plant roots	32
Yellowknife Bay immediately outside the breakwater[5]	1,275	–	1,110–1,440[8]		2
Yellowknife Bay deep sediments[14]	–	–	25–1,310 630–2,800[15]	For top 20 cm of sediments; Mudroch et al. (1989, quoted by Andrade *et al.*, 2010) values for "top few cm" ~20 years earlier	5 core

[a]"n" indicates number of samples; [b]"–" indicates either data omitted by the source, or a calculation made irrelevant due to small sample size; [c]GNWT guideline for industrial soil for Giant Mine is 340 mg kg^{-1} As; [1]Average and median calculated from data in Golder 2005, including a compilation of EBA 1994 and 1998, Golder 2001, ESG 2000, ESG and Queen's University 2000, ESG 2001, INAC 2004; [2]average and median calculated from data in Wrye 2008; [3]INAC 2004; [4]SRK 2002c, compilation of Lakefield 2002, CANMET 2000, Royal Oak Mines 1998, Giant Mine Laboratory 1982; [5]INAC 2007; [6]average and median calculated from Fawcett 2008; [7]Mace 1998 quoted by Golder 2001; [8]Golder 2001; [9]average and median calculated from Miramar Giant Mine, quoted by Golder 2001; [10]SRK 2002b; [11]average and median calculated from Stephen 2011; [12]Walker 2006; [13]Walker *et al.* 2011; [14]Andrade *et al.* 2010; [15]Mudroch *et al.* 1989 quoted by Andrade *et al.* 2010.

prevalent As hosts in ESP dust are Fe arsenates and ROs. Owing to the preponderance of As_2O_3 as an As host, As in ESP dust occurs overwhelmingly as As(III), with some As(V) and negligible As(–I) (Fawcett and Jamieson, 2011).

The concentration of As in ESP dust varied with the age of the As gas capturing technology in use, as did the Au content. In general, the older dust (pre-1964) had less As and two orders of magnitude more Au than the newer (post-1964) dust. Solubility also differs between old and new dust; solubility experiments found older dust solubility is approximately $15\,g\,L^{-1}$ at 25°C, while newer dust measures $8.3–10.8\,g\,L^{-1}$ As at 25°C. The older dust is only slightly less soluble than that of reagent-grade As_2O_3, however, newer dust has a solubility about half that of reagent-grade As_2O_3. The difference in solubility between reagent-grade As_2O_3 and old and new ESP dust is thought to be related to the antimony (Sb) that occurs in solid solution with some of the As_2O_3 created in the Giant Mine roaster as a consequence of the chemistry of the host minerals being roasted. The more Sb in the As_2O_3 structure, the less soluble it is. Newer ESP dust contains more Sb in solid solution than older ESP dust (SRK, 2002c). See Table 2.2 for different As concentrations in dust.

ESP dust is stored in a series of 10 purpose built chambers and five mined-out stopes. The storage chambers are located near the central workings of the mine, close to the source of the dust, and are mostly between 20 and 75 m below the surface. They are sealed with concrete bulkheads to isolate the dust from other mine workings. Approximately 237,000 tonnes of dust are stored underground at Giant Mine. While the mine was in operation the dust was assayed regularly for As and Au content; over time, with changes in technology and practice, As content in newly produced dust increased and Au content declined (INAC, 2007).

Stack emissions consisted of SO_2 and As-bearing gases, both produced by ore roasting (see Equation 2.1) (INAC, 2007). The As-gases condensed into a dust with As-hosting phases consisting mostly of As_2O_3 in the mineral structure of arsenolite, with some As-bearing ROs in the form of maghemite (discussed above in connection to calcine) (Walker *et al.*, 2005). Of the approximately 20,000 tonnes of As emissions released by the roaster stack during its lifetime (Wrye, 2008), approximately 7,900 tonnes of As were emitted from 1949 to 1952, when the first As gas capturing technology was implemented (Walker *et al.*, 2005).

2.3.2.3   *Soil*

Soils both contaminated and uncontaminated by anthropogenic As can contain naturally occurring As from the mineral arsenopyrite. Arsenopyrite and/or its weathering products are present in elevated concentrations in some soils around Giant Mine because of the naturally high arsenopyrite content in some of the bedrock hosting the Giant Mine gold deposit (Kerr, 2001). The Government of the Northwest Territories (GNWT) guideline for As content in industrial soils of $340\,mg\,kg^{-1}$, above which remediation is recommended, is based on a background level of As in Yellowknife of $150\,mg\,kg^{-1}$ (Risklogic, 2002; GNWT, 2003). The worldwide average As content in all soils is $5\,mg\,kg^{-1}$; the average for Canada is $6.6\,mg\,kg^{-1}$ (Reimann *et al.*, 2009). The industrial soil guideline for all of Canada determined by the Canadian Council for Ministers of the Environment (CCME) is $12\,mg\,kg^{-1}$, based on a background level of $10\,mg\,kg^{-1}$ As (CCME, 2007).

Soils at Giant Mine have been contaminated by As from a multitude of sources. Some representative As concentrations in different areas of the mine site can be seen in Table 2.2. Soils located closer to mining operations can have As contamination resultant of inputs such as waste rock, tailings, calcine residue, ESP dust, etc., as well as aerial input from roaster emissions (Golder, 2005). Soils located farther away from day-to-day mining can be impacted by anthropogenic As from aerial roaster emissions. As discussed in section 2.3.2.2, roaster emissions contained As in the form of As-bearing iron oxides and As_2O_3. These distinctively identifiable roaster-produced species have been found persisting in surface soils, even over 50 years after the start of operation of the roaster (Walker *et al.*, 2005; Wrye, 2008). Of nine soils samples taken of the overburden and soil around the calcine pond in 2004, four had As concentrations greater than $200\,mg\,kg^{-1}$ (INAC, 2004).

2.3.2.4 *Sediment*

Sediment in Baker Creek, which flows through the property, has been contaminated with As from several sources, including both intentional and accidental spring decanting of tailings impoundments, accidental spills and aerial emissions (Fig. 2.2). A breakwater constructed where Baker Creek reaches Yellowknife Bay has concentrated the sediments in this area, which is largely vegetated by cattails (*Typha latifolia*) and horsetails (*Equisetum fluviatile*) (Fawcett, 2009). Sediment in Yellowknife Bay itself has been affected by mining activities (Murdoch *et al.*, 1989 quoted by Andrade *et al.*, 2010; Andrade *et al.*, 2010). Farther north, historic tailings that were deposited in the first two years of mine operation have been largely eroded into Yellowknife Bay (Andrade *et al.*, 2010). Most of the As in sediment would include a significant arsenite component from roaster waste.

2.4 TRANSFORMATION AND REMOBILIZATION OF ARSENIC SPECIES

The stability of As species in aqueous solution depends on environmental factors such as pH and Eh, and one species can be transformed into another either abiotically or through microbial action. As-bearing compounds are also subject to dissolution, oxidation, or reduction. In the mine waste environment, these transformations can result in potential re-mobilization of As. We now describe two examples of post-depositional transformation of As species.

2.4.1 *Microbial oxidation of arsenite to arsenate*

In most of the mine waters, dissolved As is in the form of arsenate. The exception is where infiltrating waters have interacted with the As_2O_3 chambers. Seeps from rock fractures and drillholes near and below the chambers produce water that contains, typically, 3,000 mg L^{-1} total dissolved As, 50 to 70% of which is As(III) (SRK, 2005a). The pH of these waters ranges from slightly acidic to near-neutral, and the temperature is 2 to 10°C. Where the As(III)-rich water drips slowly down stope walls, thick gelatinous biofilms are formed (Fig. 2.3). Osborne *et al.*

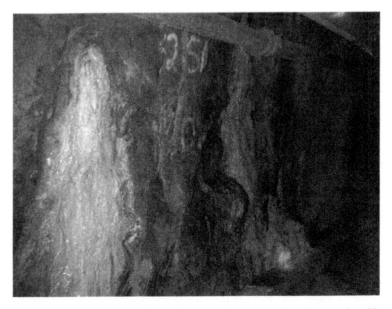

Figure 2.3. Biofilms formed from arsenite-bearing groundwater seeps in underground workings at Giant Mine (Photo courtesy of S. Simpson, Lorax Consulting).

(2010) isolated and described a psychrotolerant As(III)-oxidizing bacterium from one of these biofilms and showed that it grows at a broad range of temperatures (4 to 25°C). Samples of water, biological material and mineral precipitates were taken from a well-developed biofilm formed under a seeping drillhole located 150 m below one of the chambers storing arsenic trioxide. The liquid samples from the top and bottom of the biofilm were pH-neutral (6.9 and 6.5) and similar in total As concentration (717 and 686 mg L^{-1} for samples filtered on site to less than 0.22 microns). Liquid from the top of the biofilm contained mostly As(III) (96% of total As) whereas that from the bottom contained less (As(III) made up only 66% of total As), although the microbial diversity was similar (Osborne *et al.*, 2010).

The mineral precipitates intimately associated with the biofilm were shown, using X-ray absorption methods, to be a mixture of As(V) and As(III) compounds. Synchrotron-based micro-X-ray diffraction (XRD) was used to identify the As(V) as yukonite [$Ca_7Fe_{12}(AsO_4)_{10}(OH)_{20}\cdot15H_2O$] (Walker *et al.*, 2009) and the As(III) as either claudetite (As_2O_3) or manganarsite ($Mn_3As_2O_4(OH)_4$)]. These microcrystalline precipitates formed a coating on the rock wall under the biofilm. It is likely that the Ca- and Fe-bearing minerals in the rock, notably ankerite, partially dissolved as a result of As(III) oxidation, which lowers the pH:

$$2H_3AsO_3 + O_2 \rightarrow HAsO_4^{2-} + H_2AsO_4^- + 3H^+ \qquad (2.2)$$

The solution is rapidly neutralized by dissolution of the Ca-Fe carbonate minerals. The minerals that precipitate contain As(III) and As(V) from seepage waters and Ca, Fe and Mn dissolved from wallrock. Microcrystalline gypsum is also present, evidence of sulfide oxidation under slightly acidic conditions. The neutralizing role of wallrock carbonate was demonstrated by the low pH (4) of an unfiltered water sample that was stored for 18 days in contact with the biofilm but separated from wallrock.

Both field and experimental evidence indicate that effective microbially-mediated As(III) oxidation at low temperatures is occurring at Giant Mine (Osborne *et al.*, 2010). This is most obvious in the very As(III)-rich underground workings but may also occur in the stream and lake sediments at the surface.

2.4.2 *Arsenic cycling in lake sediments*

Much of the As-rich mine waste that has been introduced to sediments in Baker Creek and Yellowknife Bay – through tailings discharges, spills and roaster stack emissions – is in the form of As(III)-bearing compounds. The fate of this As(III) has been shown to depend on the environment of deposition (Andrade *et al.*, 2010; Jamieson *et al.*, 2011; Fawcett *et al.*, 2011).

In shallow (<2 m), oxygenated sediments that contain significant amounts of material eroded from the shoreline tailings, As is hosted by ROs which have similar As(III)/As(V) to those on shore (Walker *et al.*, 2005; Jamieson *et al.*, 2011). The lack of microbial populations, and organic matter, and the circum-neutral pore water pH is thought to favor the stability of roaster oxides in these shallow sediments (Andrade, 2006; Andrade *et al.*, 2010); but the fact that As(III) has persisted both onshore and in shallow sediments for more than 50 years of exposure requires explanation. Walker *et al.*, (2011) have proposed that maghemite, the dominant RO, stabilizes As(III) in its crystal framework in association with structural defects. The lake water in the area has low concentrations of As, but the sediment pore water contains up to 3 mg L^{-1} total As, much of it As(III) (Andrade, 2006). The origin of the As(III) in the pore water is not fully understood; it may be a result of interaction of the pore water with the submerged tailings, or from groundwater migrating from the large tailings impoundments.

In deeper (12 m water depth) sediments in Yellowknife Bay, very small RO grains are found with As(III)/As(V) higher than those in the shallow submerged tailings. This is thought to be due either to preferential leaching of As(V) which may be more loosely bound in the maghemite structure, or to reduction of As(V) to As(III) in the ROs.

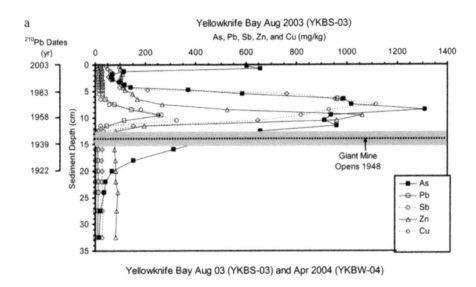

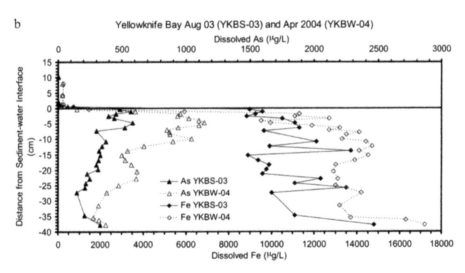

Figure 2.4. Sediment and pore water data from Yellowknife Bay. The upper figure shows metal(loid) concentration in sediment core and the associated timeline. The lower figure shows dissolved Fe and As concentrations from the same area. Details may be found in Andrade *et al.* (2010).

In profile, the pattern of pore water and coexisting sediment from the Yellowknife Bay site shows a distinctive pattern of post-depositional mobility of As (Fig. 2.4). High-resolution sampling of sediment core (which was also dated isotopically) revealed an As-rich sediment horizon formed during the early years of mine operations, when emissions were uncontrolled, overlain by cleaner sediments and topped by a very thin veneer of As-Fe-Mn oxyhydroxide. Dialysis arrays were used to sample pore water from a location very close to the core sample site; the analysis suggests that dissolved As is migrating upwards from the mine waste horizon but is captured in the oxyhydroxide layer at the surface. In other words, the surface water is protected from the As leached from buried mine waste by sorption on Fe-Mn precipitates at the sediment-water interface. This oxyhydroxide veneer is vulnerable to reductive dissolution should organic matter deposition increase in Yellowknife Bay.

Fawcett and Jamieson (2011) documented post-deposition reduction of As(V) to As(III) in sediments from Baker Pond. Some As is attenuated as As(III)-S phases, either a discrete As sulfide mineral or As(III) sorbed to the surface of Fe sulfide.

The As(III)/As(V) speciation of sediment pore waters from the deep Yellowknife Bay site is complex and thought to be controlled by pH-dependent sorption mechanisms and microbially-driven seasonal transformations of authigenic Fe-sulfides, as described in Andrade *et al.* (2010).

2.5   SITE REMEDIATION

At the time of writing, a comprehensive remediation plan is undergoing environmental assessment. The most challenging aspect of the site is considered to be the As_2O_3 stored in underground chambers.

Approximately 60 different plans for remediation of the underground As_2O_3 at Giant Mine were originally considered. Several alternatives were thoroughly investigated and vetted by a team of independent experts. They considered various options that involved leaving the As_2O_3 dust in place underground, and others in which the As_2O_3 dust was removed to another more secure location, possibly reprocessing it for gold, were considered. Through a combination of short- and long-term environmental risk assessments, worker health risks, monetary cost analyses, and public discussion, the final recommendation was that the As_2O_3 dust be left in place, frozen, and monitored for perpetuity (INAC, 2007).

The chambers will be kept frozen using a combination of active and passive freezing methods. Heat will be passively removed in winter and actively removed during the summer. It is expected to take 5 to 10 years to establish this frozen block; in the meantime, unstable bulkheads and crown pillars in danger of short-term failure are being stabilized (INAC, 2007).

The underground workings will be flooded to a level below the frozen chambers. Openings of the underground workings to the surface will be plugged (INAC, 2007). This is necessary because open pits can cause several problems: pit walls pose a hazard to future site use; open pits lie over back-filled stopes of questionable long-term stability; water from the flooded mine could intrude into the pits, forming pools accessible from the surface; and Baker Creek could flow into the pits and underground. The pits will be backfilled with any available clean backfill material and the interaction of contaminated mine water with the surface environment will be prevented (INAC, 2007).

Waste rock on the surface is not expected to cause significant problems. Drainage is expected to remain near neutral and in compliance with water regulations. Most surface waste rock has been used in road construction and these areas, unless needed for future access, will be demolished and revegetated. Tailings will be covered and revegetated to limit public access to tailings and dust, as well as As-rich runoff. The tailings covers will consist of two layers, one of crushed rock from a local quarry, and one of borrowed local silt and silty clay to accommodate revegetation efforts. The goal for tailings is for revegetation of tailings areas to create a suitable area for future traditional or recreational use. Settling and polishing ponds will be treated in a similar way to tailings ponds as regards covering. However, the sludge and water chemistry will be monitored for increased As, and the sludge will be stabilized under the covers with a filter cloth and possibly backfill of contaminated soils. The calcine pond will be left as is, as the current soil over the calcine layer has been deemed sufficient to isolate it. The historical foreshore tailings will be left alone when submerged, and for beached tailing the cover currently in place will be extended to prevent, or at lease limit, wave erosion (INAC, 2007).

Water treatment will continue to address mine water expressed in open pits, as well as tailings runoff and dam seepage. Existing diversions in Baker Creek will be made permanent, or re-routed if appropriate. Treated water will no longer be discharged into the creek after construction of a new water treatment plant, and a partially breached weir will be removed so that the creek can return to its natural level. The pond immediately below the mill, which contains contaminated sediments, will be cut off from Baker Creek and backfilled. Options for Baker Pond sediments are still being investigated and include rerouting channels, removal of sediments, or backfilling to create a wetland (INAC, 2007).

All surface material with As concentrations above the industrial land use criteria will be either excavated and disposed of or covered with clean fill material. Waste rock will be backfilled into the frozen zone of one of the pits. Contaminated soil will be backfilled into the frozen zone of one of the pits and into the tailings and sludge ponds. Spilled tailings will be excavated and put in one of the existing tailings ponds (INAC, 2007).

All buildings will be demolished. All non-hazardous waste (i.e., surface infrastructure) will be disposed of in a former quarry near the northwest tailings pond and made into a landfill (INAC, 2007).

2.6 SUMMARY

The planned remediation at Giant will reduce but not eliminate As release from the site. Specifically, the total As (mostly in the form of dissolved As) released will decrease from the current level of approximately 500 kilograms per year to less than 200 kilograms per year. Without this remediation, assuming that Giant Mine were allowed to flood as is the case for most underground mines after closure, As release could increase to many thousands of kilograms per year.

Following remediation, Baker creek will still be contaminated with As, and there will be a potential for adverse effects on bottom-feeding fish and terrestrial animals living in the area. There will need to be some restriction on access and activities within the remediated site to limit the risk to human health.

Giant Mine represents one of the largest concentrations of arsenite at the Earth's surface. The original form of As associated with mineralization is As(–I) in arsenopyrite, but mining and processing of this large gold deposit has oxidized much of the sulfide-hosted As to As(III) and As(V). In addition, crushing and roasting has transformed As that was physically locked in bedrock into As in fine grains, dust and water (Walker *et al.*, 2011), increasing mobility and bioavailability.

It is important to note that throughout its lifetime, Giant Mine was operated according to the best practices of the day. At its inception, roaster stack emissions and release of tailings to streams and lakes were not regulated. Roasting, which was used to precondition gold-bearing sulfide for cyanide leaching, was an effective method of extracting gold but resulted in large quantities of fine-grained arsenic trioxide. Modern gold mines usually employ pressure oxidation, which results in a more manageable arsenate waste product, but roasters are still used in some cases, albeit with stricter controls and better waste containment than those at the time of Giant Mine operation.

The lessons learned from research on the nature of As-bearing mine waste at Giant include: (1) understanding processing history is critical to characterizing mine waste, (2) As can be mobile even in the absence of acid mine drainage, and (3) the depositional environment affects the long-term stability of As-bearing compounds. Overall, a large, long-lived mine, especially one operated before the 1970s, leaves a complex environmental legacy. Finally, it is important to note that even the very high cost of the proposed remediation plan is outweighed by the monetary value of the gold extracted over the lifetime of Giant Mine (Bullen and Robb, 2006).

ACKNOWLEDGEMENTS

Research conducted by graduate students at Queen's University has been an important source of information for this chapter, and we acknowledge the work of Stephen Walker, Claudio Andrade, Skya Fawcett and Lori Wrye. Access and support from the Giant mine Remediation Team, Indian and Northern Affairs Canada was appreciated, as was advice from staff at SRK Consulting Ltd. Hendrik Falck of the Northwest Territories Geoscience Office provided a helpful perspective. We appreciated all these individuals and organizations but any errors or omissions are the responsibility of the authors.

REFERENCES

Andrade, C.F., Jamieson, H.E., Kyser, T.K., Praharaj, T. & Fortin, D.: Biogeochemical redox cycling of arsenic in mine-impacted lake sediments and co-existing pore waters near Giant Mine, Yellowknife Bay, Canada. *Appl. Geochem.* 25 (2010), pp. 199–211.

Bleeker. W.: Guide to the "Giant Section" of the Yellowknife greenstone belt: a transverse through middle and upper members of the Yellowknife Bay Formation and syn-orogenic conglomerates of the Jackson Lake Formation, field trip guidebook, *Geological Association of Canada (GAC) Mineralogical Association of Canada (MAC) Joint Annual Meeting*, Yellowknife, Canada, 2007, 29 pp.

Bullen, W. & Robb, M.: Economic contribution of gold mining in the Yellowknife mining district. In: C.D. Anglin, H. Falck, D.F. Wright & E.J. Ambrose (eds): *Gold in the Yellowknife Greenstone Belt, Northwest Territories: Results of the EXTECH III Multidisciplinary Research Project.* Geological Association of Canada, Mineral Deposits Division, 2006, pp. 38–49.

Canadian Council of Ministers of the Environment (CCME) (2007): Canadian soil quality guidelines for the protection of environmental and human health: Summary tables. Updated September, 2007. In: Canadian Council of Ministers of the Environment (1999): Canadian environmental quality guidelines, Winnipeg, Canada.

Canam, T.W.: Discovery, mine production, and geology of the Giant Mine. In: C.D. Anglin, H. Falck, D.F. Wright & E.J. Ambrose (eds): *Gold in the Yellowknife Greenstone Belt, Northwest Territories: Results of the EXTECH III Multidisciplinary Research Project.* Geological Association of Canada, Mineral Deposits Division Special Publication No. 3, St. John's, Canada, 2006, pp. 188–196.

Fawcett, S.E.: *Speciation and mobility of antimony and arsenic in mine waste and the aqueous environment in the region of the Giant Mine, Yellowknife, Canada.* PhD Thesis, Queen's University, Kingston, Ontario, Canada, 2009.

Fawcett, S.E. & Jamieson, H.E.: The distinction between ore processing and post-depositional transformation on the speciation of arsenic and antimony in mine waste and sediment. *Chem. Geology* 283:3–4 (2011), pp. 109–118.

Fawcett, S.E., H.E. Jamieson, D.K. Nordstrom & McCleskey, R.B.: Arsenic and antimony geochemistry of mine wastes and associated waters and sediments at the Giant Mine, Yellowknife, Northwest Territories, Canada. Revised Manuscript, 2011.

Golder Associates (Golder) (2005): Distribution of arsenic in surficial materials: Giant mine, In: INAC (2007): Giant Mine remediation plan, supporting document I1. 184 pp.

Golder Associates (Golder) (2001): Geochemistry of mine wastes, Giant Mine Site, Yellowknife, NT, In: INAC (2007): Giant Mine remediation plan, supporting document B2. 440 pp.

Government of Canada: CanVec, Canada, geographic vector database, 2009. Available at: http://geogratis.ca/geogratis/en/product/search.do?id=5460AA9D-54CD-8349-C95E-1A4D03172FDF 1.1. Accessed 12 August 2011.

Government of Northwest Territories (GNWT) (2003): Remediation criteria for arsenic in the Yellowknife area soils and sediment In: GNWT (2003): Environmental guidelines for contaminated site remediation. Available at: www.enr.gov.nt.ca/_live/documents/content/siteremediation.pdf. Accessed 11 May 2011.

Indian and Northern Affairs Canada (INAC) and SRK Consulting Engineers and Scientists (SRK) (2004): Characterization of soil and groundwater in calcine and mill areas, Giant Mine. In: INAC (2007): Giant Mine remediation plan, supporting document K2. 56 pp.

Indian and Northern Affairs Canada (INAC): Giant Mine remediation plan. Report of the Giant Mine remediation team-Department of Indian Affairs and Northern Development as submitted to the Mackenzie Valley Land and Water Board, Yellowknife, Canada, 2007, 260 pp. Available at: http://www.mvlwb.ca/mv/Registry/Forms/FolderView.aspx?RootFolder=%2fmv%2fRegistry%2f2007%2fMV2007L8-0031%2fremediationplan&FolderCTID=&View={9F47DAB1-E4D7-4B6D-A050-461E63EB7AFE}. Accessed 12 May 2011.

Jamieson, H.E., Walker, S.R., Andrade, C.F., Wrye, L.A., Rasmussen, P.E., Lanzirotti, A. & Parsons, M.B.: Identification and characterization of arsenic and metal compounds in contaminated soil, mine tailings, and house dust using synchrotron-based microanalysis. *Human Ecological Risk Assessment: An International Journal* 17:6 (2011), pp. 1292–1309.

Kerr, D.E. & Wilson, P.: Preliminary surficial geology studies and mineral exploration considerations in the Yellowknife area, Northwest Territories. Geological Survey of Canada, *Current Research* 2000-C3, 2000, 8 pp.

Kerr, D.E.: Till geochemistry, Yellowknife area, NWT. Geological Survey of Canada, Open File D4019, Natural Resources Canada, Ottawa, 2001.

Mapsof.net: Canada Provinces Blank, 2011. Available at: http://mapsof.net/canada/static-maps/png/canada-provinces-blank. Accessed 15 August 2011.

Osborne, T.H., Jamieson, H.E., Hudson-Edwards, K.A., Nordstrom, D.K., Walker, S.R., Ward, S.A. & Santini, J.M.: Microbial oxidation of arsenite in a subarctic environment: diversity of arsenite oxidase genes and identification of a psychotolerant arsenite oxidizer. *BMC Microbiology* 10 (2010), 205.

Reimann, C., Matschullat, J., Birke, M. & Salminen, R.: Arsenic distribution in the environment: The effects of scale: *Appl. Geochem.* 24:7 (2009), pp. 1147–1167.

Risklogic Scientific Services, Inc. (Risklogic): Determining natural (background) arsenic soil concentrations in Yellowknife, NWT, and deriving site-specific human-health based remediation objectives for arsenic in the Yellowknife area: Final Report, 2002, 34 pp.

Ruby, M.V., Schoof, R., Brattin, W., Goldade, M., Post, G., Harnois, M., Mosby, D.E., Casteel, S.W., Berti, W., Carpenter, M., Edwards, D., Craigin, D. & Chappell, W.: Advances in evaluating the oral bioavailability of inorganics in soil for use in human health risk assessment. *Amer. Chem. Soc.* 33:21 (1999), pp. 3697–3705.

SENES Consultants Limited (SENES) (2005): Water Treatment Update. In: INAC (2007): Giant Mine Remediation Plan, supporting document L1. 23 pp.

SRK Consulting Engineers and Scientists (SRK): Final Report: Giant Mine arsenic trioxide management alternatives. Report submitted to INAC, Yellowknife, Canada, 2002a,140pp. Available at: http://www.mvlwb.ca/mv/Registry/Forms/FolderView.aspx?RootFolder=%2fmv%2fRegistry%2f2007%2fMV2007L8-0031%2ftrioxide&FolderCTID=&View={9F47DAB1-E4D7-4B6D-A050-461E63EB7AFE}. Accessed 12 September 2011.

SRK Consulting Engineers and Scientists (SRK) (2002b): Giant Mine geochemical characterization of other sources. In: SRK (2002a): Giant Mine arsenic trioxide management alternatives – Final Report, supporting document 4. 108 pp.

SRK Consulting Engineers and Scientists (SRK) (2002c): Giant Mine arsenic trioxide dust properties. In: SRK (2002a): Giant Mine arsenic trioxide management alternatives – Final Report, supporting document 5. 63 pp.

SRK Consulting Engineers and Scientists (SRK) (2005a): Giant Mine underground mine water chemistry. In: INAC (2007): Giant Mine Remediation Plan, supporting document B4. 124 pp.

SRK Consulting Engineers and Scientists (SRK) (2005b): Giant Mine surface water chemistry. In: INAC (2007): Giant Mine Remediation Plan, supporting document B6. 40 pp.

SRK Consulting Engineers and Scientists (SRK) (2005c): Arsenic trioxide chamber drilling and testing program (2004). In: INAC (2007): Giant Mine Remediation Plan, supporting document D2. 221 pp.

SRK Consulting Engineers and Scientists (SRK) (2005d): Giant Mine Remediation Plan: Tailings and Sludge Containment Areas. In: INAC (2007): Giant Mine Remediation Plan, supporting document K1. 233 pp.

Stephen, C.C.: *The role of typha latifolia in controlling the mobility of metals and metalloids in northern mine-contaminated aquatic environments.* B.Sc. H. Thesis, Queen's University, Kingston, Ontario, Canada, 2011.

Terriplan Consultants (2007): Giant Mine abandonment and restoration: preliminary identification of the issues and potential impacts on the city of Yellowknife, Final Report. 58 pp.

Walker, S.R., Jamieson, H.E., Lanzirotti, A., Andrade, C.F. & Hall, G.E.M. The speciation of arsenic in iron oxides in mine wastes from the Giant Gold Mine, N.W.T.: Application of synchrotron micro-XRD and micro-XANES at the grain scale. *The Canadian Mineralogist* 43:4 (2005), pp. 1205–1224.

Walker, S.R.: *The solid-phase speciation of arsenic in roasted and weathered sulfides at the Giant Gold Mine, Yellowknife, NWT: Application of synchrotron microXANES and microXRD at the grain scale.* PhD Thesis, Queen's University, Kingston, Ontario, Canada, 2006.

Walker, S.R. Jamieson, H.E., Lanzirotti, A., Hall, G.E. & Peterson, R.C. The effect of ore roasting or arsenic oxidation state and solid phase speciation in gold mine tailings. *Appl. Geochem.* (2011) in press.

Wrye, L.: *Distinguishing between natural and anthropogenic sources of arsenic in soils from the Giant Mine, Northwest Territories, and the North Brookfield Mine, Nova Scotia.* M.Sc. Thesis, Queen's University, Kingston, Ontario, Canada, 2008.

CHAPTER 3

Genotoxic and carcinogenic risk of arsenic exposure.
Influence of interindividual genetic variability

Ricard Marcos & Alba Hernández

3.1 INTRODUCTION

Arsenic (As) is a naturally occurring element classified as a human carcinogen on the basis of strong epidemiological evidence (IARC, 2004; ATSDR, 2008; IARC, 2009).

It is released into the environment by several processes and from several sources: e.g. natural weathering of As-rich geological forms, pesticide use, mining, the extraction and burning of fossil fuels, and solid wastes incineration. Owing to the passage of water through contaminated rock and soils, As can be found in the drinking water of millions of people worldwide, sometimes at concentrations much higher than those considered safe. This causes major health problems in a number of regions. Chronic exposure to As, mainly via drinking As-contaminated water, has been associated in humans with an increased incidence of cancer of the lung, skin, urinary bladder, liver and other organs (Cantor *et al.*, 2006), as well as with other pathologies such as keratosis, vascular diseases, peripheral neuropathy and diabetes (Abernathy *et al.*, 1999).

Environmental As can be found in both inorganic and organic forms. Once in the body, inorganic As (i-As) is rapidly absorbed and distributed in the blood to tissues where it is bio-transformed, mainly in the liver, to its organic methylated [MMA(V) and MMA(III)] and dimethylated [DMA(V) and DMA(III)] forms prior to excretion in the urine. Some of the ingested i-As and its metabolites are retained in the body, mainly in keratin-rich biological derivatives of the ectoderm, such as hair and nails (Bencko *et al.*, 1971). Organic forms of As, such as arsenobetaine and arsenocholine (arsenosugars), are present in some foods, especially in seafood and some types of grain, but they are considered non-toxic owing to their inability to react with cellular compounds and because they are easily excreted (Buchet *et al.*, 1994).

Although methylation was initially considered to be a means of detoxifying the harmful i-As, its role was redefined when the toxicity of the organic As metabolites, such as monomethylarsonous acid [MMA(III)], was found to be much higher than that of their inorganic parental species both *in vitro* and *in vivo* (Styblo *et al.*, 2002; Kligerman and Tennant, 2007). This finding demonstrated a straight correlation between As metabolism and the toxic, genotoxic and carcinogenic potential of As-exposure.

Susceptibility to As-induced health effects differs greatly between individuals. Part of the variation has been attributed to genetic variation in the ability to metabolize As. It is reasonable to expect that mutations affecting enzyme activity might influence the retention and distribution of the more toxic As metabolites and, consequently, lead to differences in As susceptibility, including the risk of cancer. It is for this reason that most of the studies related to As individual susceptibility have been focused on genes somehow related to As metabolism, and have relied on urinary As excretion patterns as an indicator of As methylation (reviewed in Hernández and Marcos, 2008).

In this chapter we wish to focus on the genotoxic and carcinogenic potential of i-As and on genetic variation in susceptibility to its genotoxic and carcinogenic effects.

3.2 CARCINOGENIC RISK

The serious health effects of chronic exposure to As appear about 5–20 years after the beginning of the exposure. These effects go from general discomfort, depigmentation, hyperkeratosis, neuropathies and hepatopathies to gangrene, cancer and death (Ferrecchio *et al.*, 2000).

Epidemiological studies of humans have demonstrated an association between chronic As exposure and cancer of the skin, lung and urinary bladder, and possibly liver, kidney, and prostate (IARC, 2004, 2009). The data for skin, lung and urinary bladder are widely agreed to demonstrate an etiological role for As exposure, although the data on the other organs are considered more controversial. In contrast, experimental studies of animals (adult mice, rats, hamsters, rabbits, dogs, and monkeys) have consistently yielded negative results for i-As as a carcinogen when administered alone (IARC, 2004; NRC, 1999). On the basis of these observations, many authors have suggested that As might act as a co-carcinogen or tumor promoter rather than an initiating direct carcinogen with a no-threshold effect (reviewed in Wang and Rossman, 1996).

Carcinogenesis is a complex procedure involving both genotoxic and non genotoxic mechanisms and, although the exact mechanisms remain unclear, several pathways have been proposed to explain As carcinogenicity.

First, As may act as a co-carcinogen by inhibiting DNA repair enzymes (Li and Rossman, 1989). It has been shown to impair both nucleotide excision repair (NER) and base excision repair (BER), probably by interacting with zinc finger motifs of DNA repair enzymes (Hartwig *et al.*, 2003; Rossman, 2003). Well-studied examples are poly(ADP-ribose) polymerase-1 (*PARP-1*), the formamidopyrimidine-DNA glycosylase (*FPG*), and xeroderma pigmentosum complementation group A (*XPA*) (Hartwig *et al.*, 2003). Alternatively, arsenic may interfere with DNA repair by less direct mechanisms (Rossman, 2003). As exposure has been associated with decreased expression of the DNA excision repair cross-complement 1 (*ERCC1*) in human populations in the USA and Mexico (Andrew *et al.*, 2006) and, in Mexico, a reduced ability to repair DNA was observed in children exposed to As, relative to that in healthy children (Méndez-Gómez *et al.*, 2008).

Second, As-mediated carcinogenesis may include epigenetic mechanisms. Thus, As exposure has been shown to alter the methylation pattern, of both global DNA and gene promoters, as well as the chromatin structure (histone acetylation, methylation and phosphorylation), and the miRNA expression (Ren *et al.*, 2011).

Third, i-As and its organic derivatives are cytotoxic and are capable of triggering apoptosis. Trivalent forms of As have been found to induce apoptosis in several cellular systems with involvement of membrane-bound cell death receptors, activation of caspases, release of calcium stores and changes of the intracellular glutathione level. As and its organic derivatives are also known to cause calcium ion deregulation. Calcium homeostasis plays a critical role in apoptosis, and the As-mediated calcium disturbances lead to genomic damage and apoptotic cell death, which can result in cancer (Florea *et al.*, 2005).

Finally, As carcinogenic effects may involve oxidative stress. Several *in vivo* and *in vitro* studies have shown that As compounds induced the formation of reactive oxygen species and increased the generation of 8-oxo-7,8-dihydro-2'-deoxyguanosine (8-oxo-dG) (Hei and Filipic, 2004), a commonly used biomarker for oxidative DNA damage. Hepatic lipid peroxidation and glutathione depletion are observed in chronically i-As-treated animals (Mazumder, 2005). Expression of a number of oxidative stress-related genes, such as heme oxgenase-1 (*HO1*) and metallothionein (*MT*), often rises after acute, high-dose i-As *in vivo* exposure (Liu *et al.*, 2001). Increases in hepatic DNA 8-oxo-dG levels, a biomarker for oxidative DNA damage, have been associated with hepatocarcinogenesis induced by methylated arsenicals (Kinoshita *et al.*, 2007). A recent study has shown that As biomethylation appears to be obligatory for As-induced oxidative DNA damage at subtoxic doses and linked in some cells with the accelerated transition to an *in vitro* cancer phenotype (Kojima, 2009).

In addition, we would like to mention that the hypothesis of As-induced cancer being a stem cell (SC)-based disease is certainly gaining strength. Findings show that As is able to impact human SC population dynamics *in vitro* by blocking differentiation and creating more key targets for transformation. In fact, some observations indicate that during *in vitro* malignant transformation, As causes a remarkable survival selection of SCs, generating a noticeable overabundance of cancer SCs (CSCs). Thus, As impacts key, long-lived SC populations as critical targets to cause or facilitate later oncogenic events in adulthood (Tokar *et al.*, 2011).

3.3 GENOTOXIC RISK

The mechanisms of action of As are complex, including both non-genotoxic and genotoxic effects. Both types of mechanisms have been proposed to be implicated in the carcinogenic outcomes. Many different genotoxic endpoints have been reported from the human populations exposed to As through drinking water in various countries, confirming the results obtained in both *in vivo* and *in vitro* studies that have shown that As can cause DNA damage induction (Basu *et al.*, 2001).

As has been considered a non-mutagenic carcinogen, because of the negative results obtained in assays testing for point mutations in bacteria and mammalian cells. However, the mouse lymphoma assay has yielded evidence of mutagenesis (Soriano *et al.*, 2007).

While opinion is still divided on whether As can induce gene mutation, there is general agreement that it can cause chromosomal mutation: numerous studies using the chromosome aberrations (CA) test in metaphase cells, and the micronuclei (MN) test have shown As to be a clastogen (Basu *et al.*, 2001; Hughes, 2002; Hei and Filipic, 2004). Recently, As-induced DNA breakage has been detected using the comet assay (Guillamet *et al.*, 2004; Graham-Evans *et al.*, 2004; Mouron *et al.*, 2006). This assay detects not only DNA breaks but also alkali-labile sites and transient breaks induced during DNA repair (Collins, 2004). It can also reveal As-induced oxidative damage (Schwerdtle *et al.*, 2003) and the synergistic effects of UV radiation on chromosome mutation (Andrew *et al.*, 2006).

There is thus some evidence that As can cause gene mutation, and very strong evidence that it can induce chromosomal mutation – that it can act as a clastogen. There is now also direct evidence for genotoxic effects on humans exposed to As, both environmentally (Martínez *et al.*, 2004) and occupationally (Palus *et al.*, 2005). All this would indicate that As compounds are genotoxic, acting by several mechanisms and inducing a wide range of genotoxic effects.

As indicated in the previous section the genotoxic effects of As can also be indirect, by affecting repair activity. Thus, it has been demonstrated that As can potentiate the genotoxicity of other organic mutagen-carcinogens, particularly polycyclic aromatic hydrocarbons (PAHs), including benzo[*a*]pyrene (BAP) and ultraviolet radiation (Basu *et al.*, 2001; Danaee *et al.*, 2004).

3.4 GENETIC POLYMORPHISMS AFFECTING CARCINOGENIC RISK

As indicated in section 3.2, cancer induction related to As exposure is a complex process and several mechanisms may be involved. Thus, genes involved in DNA repair, apoptosis and oxidative damage seem to be clear candidates to be included in studies of As-induced cancer. In practice, however, the studies describing the modulating effects of different genes on As effects have focused on a varied number of genes, with varied functions, and little information is available for direct comparison of published studies.

Although most of the studies on the effects induced by As exposure indicate the importance of methylation in the toxicity and carcinogenicity of As, few of the genes involved have been studied. Glutathione S-transferase omega genes (*GSTO1* and *GSTO2*) have been considered, but no studies have included the arsenic (III) methyltransferase (*AS3MT*) gene. The product of *AS3MT* catalyzes the methylation of As, and the polymorphism Thr287 variant has been associated with an increased percentage of the most toxic and genotoxic As metabolite MMA(III) and a reduced percentage of DMA(III) in urine (Hernández *et al.*, 2008; Valenzuela *et al.*, 2009).

To clarify the current position of work on genetic factors modulating the cancer risk associated with As exposure, in this section we review the available studies of genetic variation in the risks of As-induced cancers of the bladder, the skin and the lungs. No data are available for other cancers.

3.4.1   Bladder cancer

Urothelial carcinoma (UC) has been associated with exposure to As in drinking water (Negri and La, 2001), although only a few studies have been conducted to determine whether there is generic variation in UC incidence.

Among the different candidate genes that have been selected, there are the glutathione S-transferase omega genes (GSTO1 and GSTO2). They have been considered to be involved in As reduction and, consequently, they may have an important role in As metabolism and in its potential carcinogenic risk. Variation in GSTO1 and GSTO2 has been studied in a recent hospital-based case-control study with 149 UC cases and 251 healthy controls (Chung et al., 2011). In this study, the homozygous variant genotype of the GSTO2 polymorphism Asn142Asp (having Asp replacing Asn at position 142) was inversely associated with UC risk. In addition, those individuals carrying at least one copy of the wild-type allele (Asn) present higher levels of MMA(III), that is considered the most reactive specie of As.

A previous study evaluated also the possible role of GSTO1 and GSTO2 polymorphisms, together with that of the CYP2E1 gene, in patients with urothelial cancer. CYP genes in the P450 cytochrome enzyme family are involved in the metabolism of many foreign (xenobiotic) compounds. In this case the Ala419Asp polymorphism for GSTO1 was used, while for GSTO2 the selected polymorphisms were the Asn424Asp and a transition polymorphism at position 183 (A183G). For the CYP2E1 gene, the selected polymorphisms were the RsaI polymorphism (C1053T) and a 96-bp insertion in the 5' flanking region. This hospital-based case-control study consisted of 520 histological confirmed UC cases, and 520 age- and gender-matched cancer-free controls. Significantly increased UC risks were found for study subjects with high As exposure. The analysis of individual polymorphisms only revealed an association for the GSTO2 A183G polymorphism, with increased risk of cancer for the homozygous variant genotypes; in addition, significantly increased UC risk was also observed in study subjects with one specific diplotype of GSTO1-GSTO2. It must be indicated that the highest UC risk was found for those individuals presenting different environmental risk factors linked to the development of urothelial carcinoma (cigarette smoking, alcohol consumption, As and occupational exposures) and carrying two or more risk genotypes/diplotypes of CYP2E1, GSTO1 and GSTO2. This would suggest a significant joint additive effect of cigarette smoking, alcohol consumption, As and occupational exposures and risk genotypes/diplotypes of CYP2E1, GSTO1 and GSTO2 on UC risk (Wang et al., 2009).

The modulating role of genes involved in DNA repair has also been studied for UC patients. The underlying hypothesis is that such genes can modify the As–cancer relationship, possibly because As impairs DNA repair capacity. The importance of DNA damage and repair pathways for As carcinogenicity is highlighted by its co-carcinogenic potential in the presence of UV irradiation and other mutagenic-carcinogenic agents (Rossman et al., 2001). Only one study has been conducted looking for association between genes involved in NER, As exposure and UC cancer (Andrew et al., 2009). This was a population-based study with 549 controls and 342 bladder cancer cases; the role of polymorphisms of X-ray cross-complementation gene XRCC3 (Thr241Met, IVS7-14A>G, Ex2+2A>G) and ERCC2 (Lys751Gln, Asp312Asn, IVS19-70C>T) was studied. Gene–environment interaction with As exposure was observed in relation to bladder cancer risk for individuals carrying the variant allele (Met) of the double-strand break-repair gene XRCC3 Thr241Met, in comparison to homozygous wild type (Thr/Thr) genotypes. Haplotype analysis confirmed the association of the XRCC3 241Met allele. Thus, the conclusion of this study is that double-strand break repair genotype may enhance As associated bladder cancer susceptibility (Andrew et al., 2009).

UC incidence related to As exposure has also been associated with polymorphisms in genes regulating the cell cycle, such as P53, the cyclin kinase inhibitor P21 and the cyclin CCND1. Such genes have a role in the maintenance of genomic stability and it is well known that As may cause genome instability (Hartwig et al., 2003). Thus, a hospital-based case-controlled study was conducted to explore the relationships between the P53 codon 72 (Arg72Pro), P21 codon 31 (Ser31Arg) and CCND1 (G870A) polymorphisms and UC risk. Results indicated that subjects carrying the *P21* Arg/Arg genotype had a slight but significant increase in UC risk; however, there was no association of *P53* or *CCND1* polymorphisms with UC risk. Nevertheless, significant effects were observed in terms of a combination of the three gene polymorphisms on the UC risk (Chung *et al.*, 2008). It must be indicated that subjects with the *P21* Arg/Arg present increasing levels of total As, i-As and MMA concentration in urine, and decreasing proportion of DMA that, as has already been pointed out, constitute the more important risk factors for As exposure.

To explore the distribution of the methylation capability in patients with different stages and grades of UCs, two studies have been recently conducted. In the first, 100 UC cases were included and the role of polymorphisms in the heme oxygenase-1 (*HO-1*) and in the NADPH quinone oxidoreductase-1 (*NQO1*) genes was studied (Huang *et al.*, 2008). In the second, polymorphisms in the myeloperoxidase (*MPO*) and sulfotransferase 1A1 (*SULT1A1*) genes were evaluated in 112 patients at different stages and grades of UC. Differential effects of the As methylation capability were found among patients with different stages of UC in both studies; however, urinary As concentrations were borderline significantly increased with the progress of UC, regardless of whether or not they had been exposed to As from drinking water. A significantly different distribution of the *HO-1* poly (GT)$_n$ repeat alleles was found in subjects with different UC stage, whereas this tumor type was not related to the *NQO1* (Pro609Ser) genotypes (Huang *et al.*, 2008). On the other hand, both the *MPO* (G463A, in the promoter region) and *SULT* (Arg213His) polymorphisms were found to modify the As methylation profile and UC progression (Huang *et al.*, 2009).

3.4.2 Skin cancer

Hyperpigmentation and hyperkeratosis are characteristic dermatological lesions induced by long-term exposure to As (Chen and Lin, 1994), presenting significant association with skin cancer. Different studies have tried to link skin cancer incidence, related to As exposure, with genetic variations at different loci. In West Bengal (India) more than 6 million people are exposed to high levels of As through drinking water and, since only 15–20% of the exposed individuals show As-induced skin lesions, it is assumed that genetic variation might play an important role in As toxicity and carcinogenicity.

Although an important number of studies relating the effects on dermatological lesions have been published, only a few studies have been conducted to find associations between skin cancer, As exposure, and genetic polymorphisms. In a nested case-control study on 67 As-induced skin cancer patients and 241 matched controls in southwest Taiwan, three glutathion S-transferase genes related with the metabolism of xenobiotics were analyzed (*GSTM1*, *GSTT1* and *GSTP1*). Although no single polymorphism was associated with risk, this was significantly associated with the combination of the three *GST* genotypes. Thus, those individuals who had at least one null variant genotype for the *GSTM1* and *GSTT1*, and the variant allele of the Ile105Val *GSTP1* polymorphism, presented a relative risk of developing skin cancers around 5-fold; compared with those carrying the wild-type genotypes of all three *GSTs* (Chen *et al.*, 2005).

Polymorphisms in different genes involved in DNA repair have also been studied by the same group. The rationale for this selection is based on the close relationship between As exposure, DNA repair and cancer. The studied genes were *XPD* (Arg156Arg) and 8-oxoguanine DNA glycosylase *OGG1* (Ser326Cys). In a nested case-control study on 66 As-induced skin cancer patients and 239 matched controls, a significant association with As-induced skin cancer was found for genetic polymorphisms of the DNA repair gene *XPD*. The relative risk of developing skin cancer was around 2-fold for individuals homozygous for the variant allele (Arg156Arg), after adjustment for age, gender, and cumulative As exposure. No associations were observed for the other gene (Chen *et al.*, 2005).

The roles of polymorphisms in NER genes such as *XPA* (A23G) and *XPD* (Asp312Asn and Lys751Gln) have also been evaluated to test their modulating role on non-melanoma skin cancer (NMSC) related to As exposure. A large population-based controls analysis included 880 cases of basal cell carcinoma (BCC), 666 cases of squamous cell carcinoma (SCC), and 780 controls.

Results indicated that there was an increased BCC risk associated with high As exposure among those carrying the homozygous variant G/G for *XPA*. For *XPD*, individuals having the variant allele at both loci (312Asn and 751Gln) occurred less frequently among BCC and SCC cases compared with controls, for both case groups. In the stratum of subjects who had variants for both *XPD* polymorphisms, there was a 2-fold increased risk of SCC associated with elevated As (Applebaum *et al.*, 2007).

It is interesting to point out that many of the genetic factors described for skin cancer have also been involved in the development of premalignant skin lesions due to i-As exposure. Thus, different studies have identified genetic variations in *GST*, *MPO*, catalase *(CAT)*, *P53*, *XPD* and *XRCC1*, that are associated with As-induced skin lesions (De Chaudhuri *et al.*, 2006; Banerjee *et al.*, 2007; Breton *et al.*, 2007). These identified genes are related to cellular glutathione levels, oxidative stress, the NER pathway, control of cell growth or maintenance of genomic stability, which would indicate a certain overlapping in the mechanisms involved in premalignant skin lesions and skin cancer induction.

3.4.3 *Lung cancer*

Lung cancer is considered one of the main causes of death caused by As exposure, with this carcinogen acting as etiological agent in cancer that occurs in subjects who have never smoked (Mead, 2005). Nevertheless, this type of cancer has not received the attention of investigators with respect to the study of individual genetic factors modulating its risk.

In fact, only one study using the classical single-nucleotide polymorphisms (SNP) approach has been conducted (Adonis *et al.*, 2005). This study evaluated the role of the null genotype of *GSTM1* and the T3801C polymorphism of the *CYP1A1* genes as modulating factors in lung cancer patients from Northern Chile. In males, the T3801C change, characteristic of the *CYP1A1*2A* allele, was associated with a highly significant increase in estimated relative lung cancer risk. Relative lung cancer risk for the combined *CYP1A1*2A/GSTM1* null genotypes was also significant, which increased with the smoking habit. Authors conclude that the cancer mortality rate for As-associated cancers at least partly might be related to differences in As biotransformation.

A second study aimed to link lung cancer risk with urinary As and polymorphisms in folate-metabolizing genes, by using two intronic polymorphisms of the cystathionine beta-synthase *(CBS)* gene. Methylation of i-As relies on folate-dependent one-carbon metabolism and this was the rational for selection of this gene. Cbs is an important enzyme involved in the conversion of homocysteine to cystathionine, a precursor of cysteine and glutathione. The study was conducted in 45 lung cancer cases and 75 controls from As-exposed areas in Cordoba, Argentina. The analysis was limited to subjects with metabolite concentrations above detection limits; the mean %MMA was higher in cases than in controls and the lung cancer odds ratio for subjects with %MMA in the upper tertile, compared to those in the lowest tertile, was significant. Although the study was too small for a definitive conclusion, this provides an indication that lung cancer risks might be highest in those individuals with a high %MMA who also carried the *CBS* rs234709 (C/T at intron 5) and rs4920037 (G/A at intron 12) variant alleles (Steinmaus *et al.*, 2010). These results add to the increasing body of evidence that variation in As metabolism plays an important role in As-disease susceptibility, which may be extensible to other genes involved in As metabolism.

From the general overview of the studies reported in the previous sections, it is difficult to establish a clear role for any particular gene or polymorphism in any one particular cancer type. The main difficulty in reaching a general conclusion is that different genes have been used in the association studies on the different cancers. Another problem with the reported studies is the small sample sizes in most of them and the low statistical significance of the results obtained. It is remarkable that so few genes related to the main cancer mechanisms (as reported in section 3.2) have been included in these studies. Variation in repair genes seems to be linked to cancer risk, although surprisingly NER genes seem to have a more important role than BER, despite the assumed association of As exposure with oxidative damage – the repair of which should normally involve BER.

Table 3.1. Genetic variants modulating As-associated carcinogenic risk.

Health outcome	Author(s)	Cases/Controls	Genetic variant	Gene category
Bladder cancer	Chung *et al.*, 2011	251/149	*GSTO2* Asp142Asn	Metabolism
	Wang *et al.*, 2009	520/520	*GSTO2* A183G	
	Huang *et al.*, 2009	112/0	*HO-1* poli(GT)$_n$	
			MPO G463A	
	Huang *et al.*, 2008	100/0	*SULT* Arg213His	
	Andrew *et al.*, 2009	342/549	*XRCC3* Thr241Met	DNA repair
	Chung *et al.*, 2008	170/402	*P21* Ser31Arg	Cell cycle
Skin cancer	Chen *et al.*, 2005	67/241	*GSTM* null or *GSTT1* null + *GSTP1* Ile104Val	Metabolism
	Chen *et al.*, 2005	66/239	*XPD* Arg156Arg	DNA repair
	Applebaum *et al.*, 2007	1546/780	*XPA* A23G	
			XPD Asp312Asn	
			XPD Lys751Gln	
Lung cancer	Adonis *et al.*, 2005	108/108	*CYP450-1A1* T3801C	Metabolism
	Steinmaus *et al.*, 2010	45/75	*CBS* rs234709	
			CBS rs4920037	

In addition, other sources of exposure to other genotoxic agents generally found in As-contaminated areas are never taken into account. This is an important weakness of these studies because of the co-mutagenic action of As.

A summary of the results reported in the previous sections is presented in Table 3.1.

3.5   GENETIC POLYMORPHISMS AFFECTING GENOTOXIC RISK

In spite of the large number of studies conducted to determine genotoxic risk in exposed populations, few studies have considered the role of variations in specific genes as biomarkers of individual sensitivity. It is remarkable that no association studies have been carried out with genes involved in As metabolism, mainly taking into account that the As metabolite MMA(III) is considered highly genotoxic and, consequently, its accumulation into the body can increase the risk of DNA damage.

Only five studies have been found linking levels of genetic damage in people exposed to As and its modulation by genetic polymorphisms. The first evaluated the role of *GST* alleles on the levels of cytogenetic damage (CA and MN) in people exposed to As, and with symptoms of toxicity, in West Bengal, India. The products of such genes catalyze the conjugation of xenobiotics with glutathione and are, therefore, important for maintaining cellular genomic integrity. In this study a total of 422 unrelated As-exposed subjects (244 skin-symptomatic and 178 asymptomatic) were evaluated. Results show that symptomatic individuals had higher level of cytogenetic damage compared to asymptomatic individuals, and asymptomatic individuals had significantly higher genotoxicity than unexposed individuals. *GSTM1* individuals carrying at least a functional allele had significantly higher risk of As-induced skin lesions, but no modulating effects of *GST* alleles on the cytogenetic damage were reported (Ghosh *et al.*, 2006).

Many studies support the role of As in altering one or more DNA repair processes. In this context, a study was conducted with 37 individuals from New Hampshire, USA, and 16 from Sonora, Mexico. As exposure was associated with decreased expression of the *ERCC1* gene in isolated lymphocytes, at the mRNA and protein levels. In addition, lymphocytes from As-exposed individuals showed higher levels of DNA damage, as measured by the comet assay, both at baseline and after a 2-acetoxyacetylaminofluorene (2-AAAF) challenge. 2-AAAF is a well-known genotoxic and carcinogenic agent that interacts with DNA, producing DNA adducts that are repaired by the NER mechanism. These data provide further evidence to support the

Table 3.2.   Genetic variants modulating As-associated genotoxic risk.

Biomarker	Author(s)	Subjects	Genetic variant	Association	Gene category
CA	Banerjee *et al.*, 2007	318	*XRCC2* Lys751Glu	Increased risk for 751Lys carriers	DNA repair
	De Chaudhuri *et al.*, 2008	349	*XRCC3* Thr241Met	Increased risk for 241Thr carriers	
	Kundu *et al.*, 2011	421	*P53* Pro72Arg	Increased risk for 72Arg carriers	Cell cycle

ability of As to inhibit the DNA repair machinery, which is likely to enhance the genotoxicity and mutagenicity of other directly genotoxic compounds, as part of a co-carcinogenic mechanism of action (Andrew *et al.*, 2006).

Further effects of the association between As exposure and DNA repair genotype on genetic damage have been investigated in a group of 318 unrelated As-exposed subjects (165 with hyperkeratosis and 153 without any As-induced skin lesions), drinking water contaminated with As to a similar extent in West Bengal, India. In this case the frequency of CA was evaluated according to the genetic characteristic with respect to the polymorphisms Lys751Glu at the *ERCC2* gene, involved in NER repair. A statistically significant increase in both CA per cell and percentage of aberrant cells was observed in the wild type homozygous (Lys/Lys) individuals compared to those with at least one variant (Glu) allele. This was observed in each of the two study groups, and also, in the total study population. This would indicate that *ERCC2* Lys/Lys genotype at codon 751 is significantly associated with As-induced premalignant hyperkeratosis, which is possibly due to sub-optimal DNA repair capacity of this genotype (Banerjee *et al.*, 2007).

In another study carried out in West Bengal India, the distribution of CA in individuals with keratosis was studied, as well as its association with one polymorphism in the *P53* gene (Pro72Arg). This study comprises 349 unrelated exposed individuals (162 individuals with keratosis and 187 individuals without As-specific skin lesions). The results showed that health effects (i.e. peripheral neuropathy, conjunctivitis and respiratory illness) and CA were significantly higher in the keratotic group compared to individuals with no skin lesions. Moreover, individuals with the arginine homozygous genotype (Arg/Arg) showed increased levels of CA compared to individuals with other genotypes. These results would suggest that individuals with keratosis are more susceptible to As-induced health effects and to genetic damage, and that the arginine variant of *P53* can further influence the repair capacity of As-exposed individuals, leading to increased accumulation of CA (De Chaudhuri *et al.*, 2008).

The modulating effect of polymorphisms on repair genes is also the subject of a recent study (Kundu *et al.*, 2011). In this case the role of a homologous recombination repair gene, the X-ray repair cross-complementing group 3 (*XRCC3*) was studied determining the effect of the Thr241Met polymorphism on the CA level. The rationale for the selection of this gene is that impairment of the homologous recombination pathway may lead to erroneous rejoining of broken DNA strands, resulting in genomic instability. A case-control study was conducted in West Bengal, India, involving 206 cases with As-induced skin lesions and 215 controls without As induced skin lesions, but having similar As exposure. Results indicated that the presence of at least one Met allele was protective against the development of skin lesions. A significant correlation was observed between protective genotype and decreased frequency of CA. Thus, the results indicate the protective role of the Met allele against chromosome instability (Kundu *et al.*, 2011).

A summary of the reported studies is indicated in Table 3.2.

3.6   CONCLUSIONS

The main conclusion of this work is that several genetic polymorphisms have been found to be associated with different types of genetic damage and As-induced cancers (Tables 3.1 and 3.2).

Nevertheless, the numbers of individuals selected in the different studies are in general quite low, which reduces the significance of the associations found. The low sample size, the few genes and polymorphisms used, and the lack of repeated studies indicate that more studies are required to draw firm conclusions attributing defined roles and risks to different particular genetic polymorphisms.

ACKNOWLEDGEMENTS

This work has been carried out in the frame of the projects funded by grants from the Generalitat de Catalunya (2009SGR-725) and the Spanish Ministry of Education and Science (SAF2011-23146).

REFERENCES

Abernathy, C.O., Liu, Y.P., Longfellow, D., Aposhian, H.V., Beck, B., Fowler, B., Goyer, R., Menzer, R., Rossman, T., Thompson, C. & Waalkes, M.: Arsenic: health effects, mechanisms of actions, and research issues. *Environ. Health Perspect.* 107:7 (1999), pp. 593–597.

Adonis, M., Martínez, V., Marín, P. & Gil, L.: CYP1A1 and GSTM1 genetic polymorphisms in lung cancer populations exposed to arsenic in drinking water. *Xenobiotica* 35:5 (2005), pp. 519–530.

Andrew, A.S., Burgess, J.L., Meza, M.M., Demidenko, E., Waugh, M.G., Hamilton, J.W. & Karagas, M.R.: Arsenic exposure is associated with decreased DNA repair *in vitro* and in individuals exposed to drinking water arsenic. *Environ. Health Perspect.* 114:8 (2006), pp. 1193–1198.

Andrew, A.S., Mason, R.A., Kelsey, K.T., Schned, A.R., Marsit, C.J., Nelson, H.H. & Karagas, M.R.: DNA repair genotype interacts with arsenic exposure to increase bladder cancer risk. *Toxicol. Lett.* 187:1 (2009), pp. 10–14.

Applebaum, K.M., Karagas, M.R., Hunter, D.J., Catalano, P.J., Byler, S.H., Morris, S. & Nelson, H.H.: Polymorphisms in nucleotide excision repair genes, arsenic exposure, and non-melanoma skin cancer in New Hampshire. *Environ. Health Perspect.* 115:8 (2007), pp. 1231–1236.

ATSDR. Agency for Toxic Substances and Diseases Registry. Toxicological profile for arsenic. Atlanta, GA, USA, 2008.

Banerjee, M., Sarkar, J., Das, J.K., Mukherjee, A., Sarkar, A.K., Mondal, L. & Giri, A.K.: Polymorphism in the *ERCC2* codon 751 is associated with arsenic-induced premalignant hyperkeratosis and significant chromosome aberrations. *Carcinogenesis* 28:3 (2007), pp. 672–676.

Basu, A., Mahata, J., Gupta, S. & Giri, A.K.: Genetic toxicology of a paradoxical human carcinogen, arsenic: a review. *Mutat. Res.* 488:2 (2001), pp. 171–194.

Bencko, V., Dobisova, A. & Macaj, M.: Arsenic in the hair of a non-occupationally exposed population. *Atmos. Environ.* 5:4 (1971), pp. 275–279.

Breton, C.V., Zhou, W., Kile, M.L., Houseman, E.A., Quamruzzaman, Q., Rahman, M., Mahiuddin, G. & Christiani, D.C. Susceptibility to arsenic-induced skin lesions from polymorphisms in base excision repair genes. *Carcinogenesis* 28:7 (2007), pp. 1520–1525.

Buchet, J.P., Pauwels, J. & Lauwerys, R.: Assessment of exposure to inorganic arsenic following ingestion of marine organisms by volunteers. *Environ. Res.* 66:1 (1994), pp. 44–51.

Cantor, K.P., Ward, M.H., Moore, L. & Lubin, J.: Water contaminants. In: D. Schottenfeld & J.F. Fraumeni Jr. (eds): *Cancer epidemiology and prevention.* Oxford University Press, New York, USA, 2006, pp. 382–384.

Centeno, J.A., Mullick, F.G., Martinez, L., Page, N.P., Gibb, H., Longfellow, D., Thompson, C., Ladich, E.R.: Pathology related to chronic arsenic exposure. *Environ. Health Perspect.* 110:Suppl 5 (2002), pp. 883–886.

Chen, C.J. & Lin, L.J.: Human carcinogenicity and atherogenicity induced by chronic exposure to inorganic arsenic. In: J.O. Nriagu (ed.): *Arsenic in the environment, Part II: Human health and ecosystem effects.* Advances in Environmental Science and Technology, vol. 27. John Wiley & Sons, Inc., New York, USA, 1994, pp. 109–131.

Chen, C.J., Hsu, L.I., Wang, C.H., Shih, W.L., Hsu, Y.H., Tseng, M.P., Lin, Y.C., Chou, W.L., Chen, C.Y., Lee, C.Y., Wang, L.H., Cheng, Y.C., Chen, C.L., Chen, S.Y., Wang, Y.H., Hsueh, Y.M., Chiou, H.Y. & Wu, M.M.: Biomarkers of exposure, effect, and susceptibility of arsenic-induced health hazards in Taiwan. *Toxicol. Appl. Pharmacol.* 206:2 (2005), pp. 198–206.

Chen, J.G., Chen, Y.G., Zhou, Y.S., Lin, G.F., Li, X.J., Jia, C.G., Guo, W.C., Du, H., Lu, H.C., Meng, H., Zhang, X.J., Golka, K. & Shen. J.H.: A follow-up study of mortality among the arseniasis patients exposed to indoor combustion of high arsenic coal in Southwest Guizhou Autonomous Prefecture, China. *Int. Arch. Occup. Environ. Health.* 81:1 (2007), pp. 9–17.

Chiu, H.F., Ho, S.C., Wang, L.Y., Wu, T.N., Yang, C.Y.: Does arsenic exposure increase the risk for liver cancer? *J. Toxicol. Environ. Health A.* 67:19 (2004), pp. 1491–1500.

Chung, C.J., Pu, Y.S., Su, C.T., Huang, C.Y. & Hsueh, Y.M.: Gene polymorphism of glutathione S-transferase omega 1 and 2, urinary arsenic methylation profile and urothelial carcinoma. *Sci. Total Environ.* 409:3 (2011), pp. 465–470.

Chung, C.J., Huang, C.J., Pu, Y.S., Su, C.T., Huang, Y.K., Chen, Y.T. & Hsueh, Y.M.: Polymorphisms in cell cycle regulatory genes, urinary arsenic profile and urothelial carcinoma. *Toxicol. Appl. Pharmacol.* 232:2 (2008), pp. 203–209.

Collins, A.R.: The comet assays for DNA damage and repair: principles, applications, and limitations. *Mol. Biotechnol.* 26:3 (2004), pp. 249–261.

Danaee, H., Nelson, H.H., Liber, H., Little, J.B. & Kelsey, K.T.: Low dose exposure to sodium arsenite synergistically interacts with UV radiation to induce mutations and alter DNA repair in human cells. *Mutagenesis* 19:2 (2004), pp. 143–148.

De Chaudhuri, S., Mahata, J., Das, J.K., Mukherjee, A., Ghosh, P., Sau, T.J., Mondal, L., Basu, S., Giri, A.K. & Roychoudhury, S.: Association of specific p53 polymorphisms with keratosis in individuals exposed to arsenic through drinking water in West Bengal, India. *Mutat. Res.* 601:1/2 (2006), pp. 102–112.

De Chaudhuri, S., Kundu, M., Banerjee, M., Das, J.K., Majumdar, P., Basu, S., Roychoudhury, S., Singh, K.K. & Giri, A.K.: Arsenic-induced health effects and genetic damage in keratotic individuals: involvement of p53 arginine variant and chromosomal aberrations in arsenic susceptibility. *Mutat. Res.* 659:1/2 (2008), pp. 118–125.

Ferreccio, C., Gonzalez, C., Milosavjlevic, V., Marshall, G., Sancha, A.M. & Smith, A.H.: Lung cancer and arsenic concentrations in drinking water in Chile. *Epidemiology* 11:6 (2000), pp. 673–679.

Florea, A.M., Yamoah, E.N. & Dopp, E.: Intracellular calcium disturbances induced by arsenic and its methylated derivatives in relation to genomic damage and apoptosis induction. *Environ. Health Perspect.* 113:6 (2005), pp. 659–664.

Ghosh, P., Basu, A., Mahata, J., Basu, S., Sengupta, M., Das, J.K., Mukherjee, A., Sarkar, A.K., Mondal, L., Ray, K. & Giri, A.K. Cytogenetic damage and genetic variants in the individuals susceptible to arsenic-induced cancer through drinking water. *Int. J. Cancer* 118:10 (2006), pp. 2470–2478.

Graham-Evans, B., Cohly, H.H., Yu, H. & Tchounwou, P.B.: Arsenic-induced genotoxic and cytotoxic effects in huma keratinocytes, melanocytes and dendritic cells. *Int. J. Environ. Res. Public Health* 1:2 (2004), pp. 83–89.

Guillamet, E., Creus, A., Ponti, J., Sabbioni, E., Fortaner, S. & Marcos, R.: *In vitro* DNA damage by arsenic compounds in a human lymphoblastoid cell line (TK6) assessed by the alkaline Comet assay. *Mutagenesis* 19:2 (2004), pp. 129–135.

Hartwig, A., Blessing, H., Schwerdtle, T. & Walter, I.: Modulation of DNA repair processes by arsenic and selenium compounds. *Toxicology* 193:1/2 (2003), pp. 161–169

Hei, T.K. & Filipic, M.: Role of oxidative damage in the genotoxicity of arsenic. *Free Radic. Biol. Med.* 37:5 (2004), pp. 574–581.

Hernández, A. & Marcos, R.: Genetic variations associated with interindividual sensitivity in the response to arsenic exposure. *Pharmacogenomics* 9:8 (2008), pp. 1113–1132.

Hernández, A., Xamena, N., Sekaran, C., Tokunaga, H., Sampayo-Reyes, A., Quinteros, D., Creus, A. & Marcos, R.: High arsenic metabolic efficiency in *AS3MT* $_{287}$*Thr* allele carriers. *Pharmacogene. Genom.* 18:4 (2008), pp. 349–355.

Huang, S.K., Chiu, A.W., Pu, Y.S., Huang, Y.K., Chung, C.J., Tsai, H.J., Yang, M.H., Chen, C.J. & Hsueh, Y.M.: Arsenic methylation capability, heme oxygenase-1 and NADPH quinone oxidoreductase-1 genetic polymorphisms and the stage and grade of urothelial carcinomas. *Urol. Int.* 80:4 (2008), pp. 405–412.

Huang, S.K., Chiu, A.W., Pu, Y.S., Huang, Y.K., Chung, C.J., Tsai, H.J., Yang, M.H., Chen, C.J. & Hsueh, Y.M.: Arsenic methylation capability, myeloperoxidase and sulfotransferase genetic polymorphisms and the stage and grade of urothelial carcinoma. *Urol. Int.* 82:2 (2009), pp. 227–234.

Hughes, M.F.: Arsenic toxicity and potential mechanisms of action. *Toxicol. Lett.* 133:1 (2002), pp. 1–16.

International Agency for Research on Cancer (IARC): Some drinking-water disinfectants and contaminants, including arsenic. *IARC Monographs on the Evaluation of Carcinogenic Risks to Humans,* Vol. 84, Lyon, France, 2004.

International Agency for Research on Cancer (IARC). International Agency for Research on Cancer special report: policy, a review of human carcinogens—part C: metals, arsenic, dusts, and fibres. *Lancet Oncol.* 10:5 (2009), pp. 453–455.

Kligerman, A.D. & Tennant, A.H.: Insights into the carcinogenic mode of action of arsenic. *Toxicol. Appl. Pharmacol.* 222:3 (2007), pp. 281–288.

Kinoshita, A., Wanibuchi, H., Wei, M., Yunoki, T. & Fukushima, S.: Elevation of 8-hydroxydeoxyguanosine and cell proliferation via generation of oxidative stress by organic arsenicals contributes to their carcinogenicity in the rat liver and bladder. *Toxicol. Appl. Pharmacol.* 221:3 (2007), pp. 295–305.

Kojima, C., Sakurai, T., Waalkes, M.P. & Himeno, S.: Cytolethality of glutathione conjugates with monomethylarsenic or dimethylarsenic compounds. *Biol. Pharm. Bull.* 28:10 (2005), pp. 1827–1832.

Kundu, M., Ghosh, P., Mitra, S., Das, J.K., Sau, T.J., Banerjee, S., States, J.C. & Giri, A.K.: Precancerous and non-cancer disease endpoints of chronic arsenic exposure: the level of chromosomal damage and *XRCC3* T241M polymorphism. *Mutat. Res.* 706:1/2 (2011), pp. 7–12.

Li, J.H. & Rossman, T.G.: Inhibition of DNA ligase activity by arsenite: a possible mechanism of its comutagenesis. *Mol. Toxicol.* 2:1 (1989), pp. 1–9.

Liu, J. & Waalkes, M.P.: Liver is a target of arsenic carcinogenesis. *Toxicol. Sci.* 105:1 (2008), pp. 24–32.

Liu, J., Kadiiska, M.B., Liu, Y., Lu, T., Qu, W. & Waalkes, M.P.: Stress-related gene expression in mice treated with inorganic arsenicals. *Toxicol. Sci.* 61:2 (2001), pp. 314–320.

Martínez, V., Creus, A., Venegas, W., Arroyo, A., Beck, J.P., Gebel, T.W., Surrallés, J. & Marcos, R.: Evaluation of micronucleus induction in a Chilean population environmentally exposed to arsenic. *Mutat. Res.* 564:1 (2004), pp. 65–74.

Mazumder, D.N.: Effect of chronic intake of arsenic-contaminated water on liver. *Toxicol. Appl. Pharmacol.* 206:2 (2005), pp. 169–175.

Mead, M.N.: Arsenic: in search of an antidote to a global poison. *Environ. Health Perspect.* 113:6 (2005), pp. A378–A386.

Méndez-Gómez, J., García-Vargas, G.G., López-Carrillo, L., Calderón-Aranda, E.S., Gómez, A., Vera, E., Valverde, M., Cebrián, M.E. & Rojas, E.: Genotoxic effects of environmental exposure to arsenic and lead on children in region Lagunera, Mexico. *Ann. NY Acad. Sci.* 1140 (2008), pp. 358–367.

Moore, M.M., Honma, M., Clements, J., Bolcsfoldi, G., Cifone, M., Delongchamp, R., Fellows, M., Gollapudi, B., Jenkinson, P., Kirby, P., Kirchner, S., Muster, W., Myhr, B., O'Donovan, M., Oliver, J., Omori, T., Ouldelhkim, M.C., Pant, K., Preston, R., Riach, C., San, R., Stankowski Jr., L.F., Thakur, A., Wakuri, S. & Yoshimura, I.: Mouse lymphoma thymidine kinase gene mutation assay: International workshop on genotoxicity tests workgroup report – Plymuth, UK 2002. *Mutat. Res.* 540:2 (2003), pp. 127–140.

Mouron, S.A., Grillo, C.A., Dulout, F.N. & Golijow, C.D.: Induction of DNA strand breaks, DNA-protein cross-links and sister chromatid exchanges by arsenite in a human lung cell line. *Toxicol. In Vitro* 20:3 (2006), pp. 279–285.

Negri, E. & La, V.C.: Epidemiology and prevention of bladder cancer. *Eur. J. Cancer Prev.* 10:1 (2001), pp. 7–14.

NRC (National Research Concil): Arsenic in the drinking water. Washington, DC, USA. National Academy, 1999.

Palus, J., Lewinska, D., Dziubaltowska, E., Stepnik, M., Beck, J., Rydzynski, K. & Nilsson, R.: DNA damage in leukocytes of workers occupationally exposed to arsenic in copper smelters. *Environ. Mol. Mutagen.* 46:2 (2005), pp. 81–87.

Ren, X., McHalle, C.M., Skibola, C.F., Smith, A.H., Smith, M.T. & Zhang, L. An emerging role for epigenetic disregulation in arsenic toxicity and carcinogenesis. *Environ. Health Perpect.* 119:1 (2011), pp. 11–19.

Rossman, T.G.: Mechanism of arsenic carcinogenesis: an integrated approach. *Mutat. Res.* 2003 533:1/2 (2003), pp. 37–65.

Rossman, T.G., Uddin, A.N., Burns, F.J. & Bosland, M.C.: Arsenite is a cocarcinogen with solar ultraviolet radiation for mouse skin: an animal model for arsenic carcinogenesis. *Toxicol. Appl. Pharmacol.* 176:1 (2001), pp. 64–71.

Schwerdtle, T., Walter, I., Mackiw, I. & Hartwig, A.: Induction of oxidative DNA damage by arsenite and its trivalent and pentavalent methylated metabolites in cultured human cells and isolated DNA. *Carcinogenesis* 24:5 (2003), pp. 967–974.

Soriano, C., Creus, A. & Marcos, R.: Gene-mutation induction by arsenic compounds in the mouse lymphoma assay. *Mutat. Res.* 634:1/2 (2007), pp. 40–50.

Steinmaus, C., Yuan, Y., Kalman, D., Rey, O.A., Skibola, C.F., Dauphine, D., Basu, A., Porter, K.E., Hubbard, A., Bates, M.N., Smith, M.T. & Smith, A.H.: Individual differences in arsenic metabolism and lung cancer in a case-control study in Cordoba, Argentina. *Toxicol Appl Pharmacol.* 247:2 (2010), pp. 138–145.

Styblo, M., Drobna, Z., Jaspers, I., Lin, S. & Thomas, D.J.: The role of biomethylation in toxicity and carcinogenicity of arsenic: a research update. *Environ. Health. Perspect.* 110:suppl 5 (2002), pp. 767–771.

Tokar, E.J., Qu,W. & Waalkes, M.P.: Arsenic, stem cells, and the developmental basis of adult cancer. *Toxicol. Sci.* 120:S1 (2011), pp. S192–S203.

Valenzuela, O.L., Drobná, Z., Hernández-Castellanos, E., Sánchez-Peña, L.C., García-Vargas, G.G., Borja-Aburto, V.H., Stýblo, M. & Del Razo, L.M.: Association of *AS3MT* polymorphisms and the risk of premalignant arsenic skin lesions. *Toxicol. Appl. Pharmacol.* 239:2 (2009), pp. 200–207.

Wang, Z. & Rossman, T.G.: The carcinogenicity of arsenic. In: L.W. Chang, L. Magos & T. Suzuki (eds): *Toxicology of metals.* Lewis Publishers, Boca Raton, FL, USA, 1996, pp. 221–229.

Wang,Y.H., Yeh, S.D., Shen, K.H., Shen, C.H., Juang, G.D., Hsu, L.I., Chiou, H.Y. & Chen, C.J.: A significantly joint effect between arsenic and occupational exposures and risk genotypes/diplotypes of *CYP2E1*, *GSTO1* and *GSTO2* on risk of urothelial carcinoma. *Toxicol. Appl. Pharmacol.* 241:1 (2009), pp. 111–118.

CHAPTER 4

Overview of microbial arsenic metabolism and resistance

John F. Stolz

4.1 INTRODUCTION

The study of the role of arsenic (As) in microbial metabolism continues to be an evolving story with an interesting history. The volatilization of As was first explored by Bartolomeo Gosio in the 1890s and further investigated in the 1950s by Challenger (as reviewed in Bentley and Chasteen, 2002). Gosio experimented with a variety of fungi and concluded that the volatile gas they exuded, later identified as trimethylarsine (TMAs) by Challenger, was toxic to mice, rats, and rabbits and, by extension, humans (This conclusion, however, has been challenged recently by Cullen and Bentley (2005) as an "urban myth", as they cite more recent but less well publicized studies on the toxicity, or lack thereof, of TMAs). Arsenite [As(III)] oxidation also has been known for almost a century: Green reported on a strain of *Bacillus* exhibiting this activity in 1918. Investigations in the late 1970s uncovered widespread As resistance (Nakahara *et al.*, 1977), but it was only in the 1980s that the mechanisms of resistance were deciphered and found, paradoxically, to involve arsenate [As(V)] reduction (as reviewed in Mukhopadhyay *et al.*, 2002; Rosen, 2002; Silver and Phung, 2005). It was the discovery that some organisms were not only resistant but actually able to gain energy from the transformation of As oxyanions, however, that has led to the full appreciation of microbial As metabolism and its role in the biogeochemical cycle (Oremland and Stolz, 2003; Rhine *et al.*, 2005; Oremland *et al.*, 2005). We now know of dozens of prokaryotes that can use As(III) as an electron donor in either aerobic or anaerobic respiration, or As(V) as an electron acceptor. Furthermore, the advent of highly sensitive mass spectrometry (MS) (e.g., inductively coupled-MS) and nano-secondary ion MS (SIMS) has resulted in the discovery of a wide range of biological arsenicals including arsenolipids, arsenosugars, and arsenothiols. Thus, As can play a more central role in microbial metabolism then hitherto appreciated, and microbial activity can significantly impact the forms of As in the environment, influencing its speciation, mobility, and toxicity.

4.2 ARSENIC RESISTANCE

Microorganisms have evolved a variety of methods to tolerate As (Tsai *et al.*, 2009). The mechanisms involved are based on three basic characteristics of As: As(III) is more toxic than As(V) as it can bind to thiol groups in proteins rendering them inactive; As(III) can enter the cell through aquaglyceroporins (Rosen, 2002) but may also be generated in the cell through reductive processes; As(V) is an analogue of phosphate, entering cells through the phosphate transport system and interfering with oxidative phosphorylation and other reactions involving phosphate. In that context, we note that there are genes annotated as phosphate/phosphite transport proteins in operons encoding arsenic oxidoreductases (e.g., dissimilatory arsenate reductase Arr and aerobic arsenite oxidase Aio), suggesting that there may be As-specific transporters in addition to the well characterized As(III)-specific transporter ACR3 (see below). Specific mechanisms known to provide As resistance include the ArsC system, methylation, and As(III) oxidation as described in brief below.

4.2.1   *The* ars *operon*

The ArsC system confers resistance through a somewhat paradoxical process. At its core is an As(III)-specific exporter, ArsB (or ACR3), which removes the As from the cell. This transport can be passive or active, which in the latter case involves an associated ATPase (ArsA). An As(V) reductase, ArsC, a cytoplasmic 13–15 kDa protein related to tyrosine phosphate phosphatases, facilitates the reduction of As(V) when a suitable electron donor, such as reduced thioredoxin or glutaredoxin, is provided. The genes involved in resistance were originally discovered on plasmids, but have since been identified on the chromosomes of a diverse group of organisms, including Archaea, Bacteria, yeasts, and protoctists. A search of the prokaryotes in the Joint Genome Institute's database in 2011 (http://img.jgi.doe.gov/cgi-bin/w/main.cgi) yielded 540 genomes containing a gene annotated as *arsC*, 1,000 genes identified as "arsenate reductase," and some homologs annotated as "tyrosine phosphatase". Given the diversity of species that have been sequenced by JGI – the database includes many environmental isolates as well as clinical isolates and pathogenic species – the significant number of ArsC homologs suggests that arsenic resistance is indeed widespread however most environmental studies have focused on arsenic-contaminated environments (Achour *et al.*, 2007; Bachate *et al.*, 2009; Cai *et al.*, 2009; Kaur *et al.*, 2009; Saltikov and Olsen, 2002). The genes detected are normally clustered in an *ars* operon, and contain the regulatory element *arsR*, but variations have been seen where the operon lacks *arsC*, and others are known with multiple copies of *arsC* either in tandem or located at different places on the chromosome. These variations may be the result of selection due to exposure to the different As forms (organic or inorganic) and oxidation states [As(III) or As(V)]. The *ars* operon may be constitutively expressed at low levels, but up-regulated upon exposure, under the regulatory control of ArsR and ArsD (Rosen, 2002).

4.2.2   *Methylation*

Methylation is another mechanism that can confer resistance to As. A range of methylated As species exist, with the arsenic in the +5, +3 or −3 oxidation state. They include monomethyl arsonate [MMA(V)], methylarsonite [MMA(III)], dimethylarsinate [DMA(V)], dimethylarsinite [DMA(III)], and trimethylarsine oxide (TMAO) as well as several volatile arsines (−3), including monomethylarsine, dimethylarsine, and trimethylarsine (TMAs). As previously mentioned, Gosio established that fungi could generate methylated As. Challenger proposed a scheme in which As(V) was eventually transformed to TMAs (Bentley and Chasteen, 2002). In this scheme, As(V) is first reduced to As(III) then methylated, and each methylation step results in the reoxidation of the As, thus requiring a reductive step to As(III) prior to further methylation. Several different enzymes have been identified with methylase activity, such as in the pathway requiring methyl cobalamin (as reviewed in Stolz *et al.*, 2006). However, many of the enzymes involved are yet to be characterized. ArsM, a S-adenosine methyltransferase, was identified from a strain of *Rhodobacter sphaeroides*; it conferred resistance to As and generated trimethylarsine (Qin *et al.*, 2006). Although first identified as part of an *ars* operon, it has been found associated with the operon encoding the respiratory arsenate reductase (Arr) in *Alkaliphilus oremlandii*. Recently, homologs of this enzyme have been found in a thermoacidophilic eukaryotic alga from Yellowstone (*CmarsM7, CmarsM8*) and shown to methylate As(III) to MMA(III), DMA(V), TMAO, and TMAs (Qin *et al.*, 2009). In addition, *arsM* homologs were found in three species of cyanobacteria (*Microcystis* sp. PCC7806, *Nostoc* sp. PCC7120, *Synechocystis* sp. PCC6803) of which two (SSArsM, NsArsM) were heterologously expressed and shown to methylate As(III) to TMAs (Yin *et al.*, 2011).

4.3   ARSENIC IN ENERGY GENERATION

As alluded to in the previous section, As oxyanions may participate in electron transfer reactions ultimately leading to the generation of ATP. As(III) can serve as an electron donor in respiration (either aerobic or anaerobic) and photoautotrophy (see Chapters 5 and 6) and As(V) can serve as an electron acceptor in anaerobic respiration. The transformation of certain organoarsenicals may also generate energy.

4.3.1   *Arsenite oxidation*

The microbial oxidation of As(III) was first demonstrated in a bacillus in 1918 (Green, 1918). Long recognized as a mechanism of detoxification, it has only recently been linked to energy generation. Phylogenetically widespread, aerobic As(III) oxidation, or organisms that contain homologues of the aerobic As(III) oxidase (*aioBA*) genes, have been detected in more than 35 strains representing the two prokaryotic domains of Life and at least 28 genera, including members of the Crenarcheaota, Aquificales, and Thermus, as well as Alpha-, Beta-, and Gamma-proteobacteria. Most of these organisms are aerobic heterotrophs or chemolithautotrophs (see Chapter 5) but there are also examples of organisms that oxidize As(III) under anoxic conditions using an alternative electron acceptor (e.g. nitrate) (see Chapter 6) or anoxygenic photosynthesis (Budinhoff and Hollibaugh, 2008; Kulp *et al.*, 2008). Typically the arsenite oxidase that functions under anaerobic conditions (Arx) is distantly related to Aio (see Chapter 10).

4.3.2   *Arsenate respiration*

Discovered in *Sulfurospirillum arsenophilum* more than 15 years ago (Ahmann *et al.*, 1994), As(V) respiration was subsequently identified in several Chrenarchaeota and more than 20 genera of Bacteria (Stolz *et al.*, 2006). In all cases, As(V) serves as the terminal electron acceptor. These organisms are most often metabolically versatile, using organic (e.g., acetate, lactate) and inorganic (e.g., H_2, H_2S) electron donors as well as alternative electron acceptors (e.g., oxygen, nitrate, Fe(III), sulfate, and thiosulfate) (Laverman *et al.*, 1995; Macy *et al.*, 1996; Switzer Blum *et al.*, 1998; Hoeft *et al.*, 2004; Hollibaugh *et al.*, 2006; Baesman *et al.*, 2009). They all have in common the respiratory arsenate reductase (Arr) with molybdopterin containing large subunit (ArrA) and smaller 4Fe-4S cluster subunit (ArrB) (Krafft and Macy, 1998; Afkar *et al.*, 2003; Saltikov and Newman, 2003). Again, the proliferation of genome data has revealed that although the core enzyme (ArrAB) is highly conserved, there can be additional subunits, including an anchoring subunit, ArrC, and a putative chaperone protein, ArrD, as well as a variety of regulatory elements (Stolz *et al.*, 2006). Sequence analysis also suggests that horizontal gene transfer is common.

4.3.3   *Organoarsenicals in anaerobic respiration*

Although inorganic As (i-As) pesticides and herbicides were banned in the United States during the 1970s, organoarsenicals are still widely used. Methylated species are used in landscaping (for crab grass control) and microbial transformation has been demonstrated on golf courses (Feng *et al.*, 2005). Whether the organisms garner any energy from the process has not yet been established. Roxarsone, (3-nitro-4-hydroxybenzene arsonic acid), has been a common feed additive in the poultry industry. It prevents coccidiosis, and also accelerates chicken weight gain and improves pigmentation. Although little of the roxarsone accumulates in the birds, the compound ends up in the litter, where it degrades rapidly. Specifically, clostridial species, including an isolate from the Ohio River in Pittsburgh, *Alcaliphilus oremlandii*, reduce the nitro group, producing 3-amino-4-hydroxybenzene arsonic acid as an end product (Stolz *et al.*, 2007). When *A. oremlandii* is grown with lactate and roxarsone, cell yields are 10-fold higher than if cells are grown on lactate alone. The fact that *A. oremlandii* can also respire As(V) and thiosulfate, further supports the idea that this species can generate ATP through oxidative phosphorylation linked to roxarsone reduction. Indeed, a proteomic investigation comparing lactate-grown (fermentation) with roxarsone-grown cells revealed expression of the respiratory chain components under the latter growth condition (Chovanec *et al.*, 2010).

ACKNOWLEDGEMENTS

The author wishes to thank his students past and present, and colleagues (you know who you are) who have contributed to the deciphering of the microbial metabolism of As and its biogeochemical cycle.

REFERENCES

Achour, A.R., Bauda, P. & Billard, P.: Diversity of arsenite transporter genes from arsenic-resistant soil bacteria. *Res. Microbiol.* 158 (2007), pp. 128–137.

Ahmann, D., Roberts, A.L., Krumholz, L.R. & Morel, F.M.M.: Microbe grows by reducing arsenic. *Nature* 371 (1994), p. 750.

Afkar, E., Lisak, J., Saltikov, C., Basu, P., Oremland, R.S. & Stolz, J.F.: The respiratory arsenate reductase from *Bacillus selenitireducens* strain MLS10. *FEMS Microbiol. Letts.* 226 (2003), pp. 107–112.

Anderson, G.L., Williams, J. & Hille, R.: The purification and characterization of arsenite oxidase from *Alcaligenes faecalis*, a molybdenum-containing hydroxylase. *J. Biol. Chem.* 267 (1992), pp. 23674–23682.

Bachate, S.P., Cavalca, L. & Andreoni, V.: Arsenic-resistant bacteria isolated from agricultural soils of Bangladesh and characterization of arsenate-reducing strains. *J. Appl. Microbiol.* 107 (2009), pp. 145–156.

Baesman, S., Stolz, J.F. & Oremland, R.S.: Enrichment and isolation of *Bacillus beveridgei* sp. nov., a facultative anaerobic haloalkaliphile from Mono Lake California that respires oxyanions of tellurium, selenium, and arsenic. *Extremophile* 13 (2009), pp. 695–705.

Bentley, R. & Chasteen, T.G.: Microbial methylation of metalloids: arsenic, antimony, and bismuth. *Microbiol. Mol. Biol. Rev.* 66 (2002), pp. 250–271.

Budinhoff, C.R. & Hollibaugh, J.T.: Arsenite-dependent photoautotrophy by an *Ectothiorhodospira*-dominated consortium. *ISME J.* 2 (2008), pp. 340–344.

Cai, L., Liu, G.H., Rensing, C. & Wang, G.J.: Genes involved in arsenic transformation and resistance associated with different levels of arsenic-contaminated soils. *BMC Microbiol.* 9:4 (2009).

Chovanec, P., Stolz, J.F. & Basu, P.: A proteome investigation of roxarsone degradation by *Alkaliphilus oremlandii* strain OhILAs. *Metallomics* 2 (2010), pp. 133–139.

Cullen, W.R. & Bentley, R.: The toxicity of trimethylarsine: an urban myth. *J. Environ. Monit.* 7 (2005), pp. 11–15.

Dembitsky, V.M. & Levitsky, D.O.: Arsenolipids. *Prog. Lipid Res.* 43 (2004), pp. 403–448.

Edmonds, J.S. & Francesconi, K.A.: Arseno-sugars from brown kelp (*Ecklonia radiata*) as intermediates in cycling of arsenic in a marine ecosystem. *Nature* 289 (1981), pp. 602–604.

Feng, M., Shcrlau, J.E., Snyder, R., Snyder, G.H., Chen, M., Cisar, J.L. & Cai, Y.: Arsenic transport and transformation associated with MSMA application on a golf course green. *J. Agric. Food Chem.* 53 (2005), pp. 3556–3562.

Feldmann, J., John, K. & Pengprecha, P.: Arsenic metabolism in seaweed-eating sheep from Northern Scotland. *Fresenius J. Anal. Chem.* 368 (2000), pp. 116–121.

Garcia-Dominguez, E., Mumford, A., Rhine, E.D., Paschal, A. & Young, L.Y.: Novel autotrophic arsenite-oxidizing bacteria isolated from soil and sediments. *FEMS Microbiol. Ecol.* 66 (2008), pp. 401–410.

Green, H.H.: Description of a bacterium which oxidizes arsenite to arsenate, and one which reduces arsenate to arsenite, isolated from a cattle-dipping tank. *South Afr. J. Sci.* 14 (1918), pp. 465–467.

Hansen, H.R., Pickford, R., Thomas-Oates, J., Jaspars, M. & Feldmann, J.: 2-Dimethylarsinothioyl acetic acid identified in a biological sample: the first occurrence of a mammalian arsinothio(y)l metabolite. *Angewandte Chemie* 43 (2004), pp. 337–340.

Hoeft, S.E., Kulp, T.R., Stolz, J.F., Hollibaugh, J.T. & Oremland, R.S.: Dissimilatory arsenate reduction with sulfide as the electron acceptor: experiments with Mono Lake water and isolation of strain MLMS-1, a chemoautotrophic arsenate-respirer. *Appl. Environ. Microbiol.* 70 (2004), pp. 2741–2747.

Hoeft, S.E., Switzer Blum, J., Stolz, J.F., Tabita, F.R., Witte, B., King, G.M., Santini, J.M. & Oremland, R.S.: *Alkalilimnicola ehrlichii* sp. nov., a novel, arsenite-oxidizing haloalkaliphilic gammaproteobacterium capable of chemoautotrophic or heterotrophic growth with nitrate or oxygen as the electron acceptor. *Int. J. Syst. Evol. Microbiol.* 57 (2007), pp. 504–512.

Hollibaugh, J.T., Budinoff, C., Hollibaugh, R.A., Ransom, B. & Bano, N.: Sulfide oxidation coupled to arsenate reduction by a diverse microbial community in a soda lake. *Appl. Environ. Microbiol.* 72 (2006), pp. 2043–2049.

Kaur, S., Kamli, M.R. and Ali, A.: Diversity of Arsenate Reductase Genes (*arsC* Genes) from Arsenic-Resistant Environmental Isolates of *E. coli. Cur. Microbiol.* 59 (2009), pp. 288–294.

Krafft, T. & Macy, J.M.: Purification and characterization of the respiratory arsenate reductase of *Chrysiogenes arsenatis. Eur. J. Biochem.* 255 (1998), pp. 647–653.

Kulp. T.R., Hoeft, S.E., Asao, M., Madigan, M.T., Hollibaugh, J.T., Fisher, J.C., Stolz, J.F., Culbertson, C.W., Miller, L.G. & Oremland, R.S.: Arsenic(III) fuels anoxygenic photosynthesis in hot spring biofilms from Mono Lake, California. *Science* 321 (2008), pp. 967–970.

Laverman, A.M., Blum, J.S., Schaefer, J.K., Phillips, E.J.P., Lovley, D.R. & Oremland, R.S.: Growth of strain SES-3 with arsenate and other diverse electron acceptors. *Appl. Environ. Microbiol.* 61 (1995), pp. 3556–3561.

Macy, J.M., Nunan, K., Hagen, K.D., Dixon, D.R., Harbour, P.J., Cahill, M. & Sly, L.: *Chrysiogenes arsenatis*, gen. nov. sp. nov., a new arsenate-respiring bacterium isolated from gold mine wastewater. *Int. J. Syst. Bacteriol.* 46 (1996), pp. 1153–1157.

Mailloux, B. J., Alexandrova, E., Keimowitz, A., Wovkulich, K., Freyer, G. Herron, M., Stolz, J.F., Kenna, T., Pichler, T., Polizzotto, M., Dong, H., Bishop, M. & Knappett, P.: Microbial mineral weathering for nutrient acquisition releases arsenic. *Appl. Environ. Microbiol.* 75 (2009), pp. 2558–2565.

Mukhopadhyay, R., Rosen, B.P., Phung, L.T. & Silver, S.: Microbial arsenic: from geocycles to genes and enzymes. *FEMS Microbiol Revs.* 26 (2002), pp. 311–325.

Nakahara, H., Ishikawa, T., Sarai, Y. & Kondo, I.: Frequency of heavy-metal resistance in bacteria from inpatients in Japan. *Nature* 266 (1977), pp. 165–167.

Oremland, R.S., Hoeft, S.E., Santini, J.M., Bano, N., Hollibaugh, R.A. & Hollibaugh, J.T.: Anaerobic oxidation of arsenite in Mono Lake water and by a facultative, arsenite-oxidizing chemoautotroph, strain MLHE-1. *Appl. Environ. Microbiol.* 68 (2002), pp. 4795–4802.

Oremland, R.S. & Stolz, J.F.: The ecology of arsenic. *Science* 300 (2003), pp. 939–944

Oremland, R.S., Kulp, T.R., Switzer Blum, J., Hoeft, S.E., Baesman, S., Miller, L.G. & Stolz, J.F.: A microbial arsenic cycle in a salt-saturated, extreme environment: Searles Lake, California. *Science* 308 (2005), pp. 1305–1308.

Qin, J., Rosen, B.P., Zhang, Y., Wang, G., Franke, S. & Rensing, C.: Arsenic detoxification and evolution of trimethylarsine gas by a microbial arsenite S-adenosylmethionine methyltransferase. *Proc. Natl. Acad. Sci. USA* 103 (2006), pp. 2075–2080.

Qin, J., Lehr, C.R., Yuan, C., Le, X.C., McDermott, T.R. & Rosen, B.P.: Biotransformation of arsenic by a Yellowstone thermoacidophilic eukaryotic alga. *Proc. Natl. Acad. Sci. USA* 106 (2009), pp. 5213–5217.

Rhine, D.E., Garcia-Dominguez, E., Phelps, C.D. & Young, L.Y.: Environmental microbes can speciate and cycle arsenic. *Environ. Sci. Technol.* 39 (2005), pp. 9569–9573.

Rhine, E.D., Phelps, C.D. & Young, L.Y.: Anaerobic arsenite oxidation by novel denitrifying isolates. *Environ. Microbiol.* 8 (2006), pp. 899–908.

Richey, C., Chovanec, P., Hoeft, S.E., Oremland, R.S., Basu, P. & Stolz, J.F.: Respiratory arsenate reductase as a bidirectional enzyme. *Biochem. Biophys. Res. Comm.* 382 (2009), pp. 298–302.

Rosen, B.P.: Biochemistry of arsenic detoxification. *FEBS Lett.* 529 (2002), pp. 86–92.

Rosen, B.P., Ajees, A.A. & McDermott, T.R.: Life and death with arsenic. *Bioessays* 33 (2011), pp. 350–357.

Saltikov, C.W. & Olsen, B.H.: Homology of *Escherichia coli* R773 *arsA, arsB*, and *arsC* genes in arsenic-resistant bacteria isolated from raw sewage and arsenic-enriched creek waters. *Appl. Environ. Microbiol.* 68 (2002), pp. 280–288.

Saltikov, C.W. & Newman, D.K.: Genetic identification of a respiratory arsenate reductase. *Proc. Nat. Acad. Sci. USA* 100 (2003), pp. 10983–10988.

Silver, S. & Phung, L.T.: Genes and enzymes involved in bacterial oxidation and reduction of inorganic arsenic. *Appl. Environ. Microbiol.* 71 (2005), pp. 599–608.

Stolz, J.F., Perera, E., Kilonzo, B., Kail, B., Crable, B., Fisher, E., Ranganathan, M., Wormer, L. & Basu, P.: Biotransformation of 3-nitro-4-hydroxybenzene arsonic acid and release of inorganic arsenic by *Clostridium* species. *Environ. Sci. Tech.* 41 (2007), pp. 818–823.

Stolz, J.F., Basu, P., Santini, J.M. & Oremland, R.S.: Selenium and arsenic in microbial metabolism. *Annu. Rev. Microbiol.* 60 (2006), pp. 107–130.

Stolz, J.F. & Oremland, R.S.: *Microbial metal and metalloids metabolism: advances and applications.* ASM Press, (2011) Washington DC, p. 368.

Sun, W., Sierra-Alvarez, R., Milner, L. & Field, J.A.: Anaerobic oxidation of arsenite linked to chlorate reduction. *Appl. Environ. Microbiol.* 76 (2010 b), pp. 6804–6811.

Switzer Blum, J., Burns Bindi, A., Buzzelli, J., Stolz, J.F. & Oremland, R.S.: *Bacillus arsenicoselenatis sp. nov.*, and *Bacillus selenitireducens sp. nov.*: two haloalkaliphiles from Mono Lake, California which respire oxyanions of selenium and arsenic. *Arch. Microbiol.* 171 (1998), pp. 19–30.

Tsai, S.-L., Singh, S. & Chen, W.: Arsenic metabolism by microbes in nature and the impact on bioremediation. *Curr. Opinion Biotech.* 20 (2009), pp. 659–667.

Yin, X.-X., Chen, J., Qin, J., Sun, G.-X., Rosen, B.P. & Zhu, Y.-G.: Biotransformation and volatilization of arsenic by three photosynthetic cyanobacteria. *Plant Phys.* 156 (2011), pp. 1631–1638

Zargar, K., Hoeft, S., Oremland, R. & Saltikov, C.W.: Identification of a novel arsenite oxidase gene, *arxA*, in the haloalkaliphilic, arsenite-oxidizing bacterium *Alkalilimnicola ehrlichii* strain MLHE-1. *J. Bacteriol.* 192 (2010), pp. 3755–3762.

CHAPTER 5

Prokaryotic aerobic oxidation of arsenite

Thomas H. Osborne & Joanne M. Santini

5.1 INTRODUCTION

Arsenic (As) is a toxic metalloid found in inorganic and organic forms (Hughes *et al.*, 2002). The two soluble inorganic forms of As (i-As), arsenite [As(III)] and arsenate [As(V)], are widely, but not evenly, distributed in terrestrial and aquatic environments (Smedley and Kinniburgh, 2002). As(III) is more toxic and mobile than As(V), the mobility of which is constrained by its greater tendency to adsorb to minerals such as ferrihydrite and alumina (Smedley and Kinniburgh, 2002). Natural and anthropogenic sources of i-As have resulted in contaminated ecosystems and in some cases mass poisoning of human populations, a problem widely reported in countries such as West Bengal, Bangladesh, Chile, Vietnam and Cambodia (Ravenscroft *et al.*, 2009).

The speciation of i-As in the environment is greatly influenced by microbial transformation (Oremland and Stolz, 2003). Many microorganisms, for instance, possess an As-resistance mechanism, which permits survival and growth in environments with increased levels of i-As. The As-resistance mechanism (Ars) encoded by the *ars* operon reduces As(V) to As(III) using a cytoplasmic As(V) reductase (ArsC); the As(III) is expelled by a specific efflux pump (ArsB) at an energetic cost to the cell. Some organisms also detoxify i-As by methylation, producing organoarsenicals such as arsenobetaine and trimethylarsenate that are considerably less toxic than i-As (Hughes, 2002). Despite the toxicity of As(III) and As(V) some prokaryotes are capable of gaining energy from its oxidation and reduction, respectively, and are important in cycling of i-As in the environment (Rhine *et al.*, 2005). Organisms that can use As(V) as the terminal electron acceptor in anaerobic respiration, reducing it to As(III) using the enzyme As(V) reductase, have been isolated from anoxic environments and can use a variety of electron donors for growth (Stolz *et al.*, 2006). Some microorganisms can oxidize As(III), and do this aerobically and/or anaerobically using nitrate and chlorate as electron acceptors, or phototrophically. This chapter will focus on aerobic As(III) oxidation.

5.2 AEROBIC ARSENITE-OXIDIZING BACTERIA

Bacterial oxidation of As(III) to the less toxic and less mobile species, As(V), was first observed when a heterotrophic As(III) oxidizer, designated *Bacterium arsenoxydans*, was isolated from As-containing cattle-dipping fluid in Onderstepoort, South Africa (Green, 1918). Unfortunately *B. arsenoxydans* was lost and bacterial As(III) oxidation was not further investigated until Turner (1949, 1954) isolated several heterotrophic *Pseudomonas* spp. and a *Xanthomonas* sp., again from cattle-dipping fluids, in Queensland, Australia. Since these first discoveries, over 40 strains of aerobic As(III)-oxidizing bacteria have been isolated from diverse As-contaminated environments, such as hot springs (e.g., Gihring and Banfield, 2001), hydrothermal sediment (e.g., Handley *et al.*, 2009a), mine tailings and drainage water (e.g., Duquesne *et al.*, 2008), arsenopyrite-containing samples (e.g., Santini *et al.*, 2000), soil (e.g., Cai *et al.*, 2009a) and sewage (e.g., Philips and Taylor, 1976). The As(III)-oxidizing bacteria that have been isolated to date have diverse phenotypes and include acidophiles such as *Acidocaldus* sp. str. AO5 (D'Imperio *et al.*, 2007) and *Hydrogenobaculum acidophilum* str. H55 (Donahoe-Christiansen *et al.*, 2004); thermophiles, e.g. *Thermus* sp. str. HR13 (Gihring and Banfield, 2001) and *Thermus* sp. str. AO3C (Connon *et al.*, 2008); mesophiles such as NT-26 (Santini *et al.*, 2000), NT-14 (Santini et al., 2002) *Alcaligenes faecalis* (Osborne and Ehrlich, 1976; Phillips and Taylor, 1976) and *Herminiimonas arsenicoxydans* str. ULPAs1 (Muller *et al.*, 2003); and psychrophiles, for example GM1 (Osborne *et al.*, 2010). A list of the characterized aerobic As(III)-oxidizing bacteria is presented in Table 5.1, along with the environment they were isolated from and their mode of As(III) oxidation [i.e. chemolithoautrophic where CO_2 serves as the sole carbon source and As(III) as the electron donor, or heterotrophic where organic matter is required for As(III) oxidation].

The isolated aerobic As(III) oxidizers (Table 5.1), are also phylogenetically diverse, spanning the Proteobacteria, Aquificiae, Deinococcus-Thermus, Chloroflexi and Actinobacteria phyla of Bacteria, as demonstrated in the phylogenetic tree shown in Figure 5.1. In addition to the As(III) oxidizers presented in Table 5.1 and Figure 5.1, reports of many more strains of putative aerobic As(III) oxidizers, which lack published physiological data, are common in the literature. The organisms reported in some of these studies further increases the diversity of the already established As(III)-oxidizing taxa, such as the Alphaproteobacteria, Betaproteobacteria and Aquificiae (Hamamura *et al.*, 2009; Quemeneur *et al.*, 2008) but also indicates that aerobic As(III) oxidation is performed by members of the phyla Firmicutes and Bacteroidetes (Heinrich-Salmeron *et al.*, 2011).

At the time of writing, aerobic As(III) oxidation has been observed only in members of the Bacteria. As(III) oxidation by members of the Archaea is not yet established, as preliminary findings of As(III) oxidation by *Sulfolobus acidocaldarius* (Sehlin and Lindstrom, 1992) could not be reproduced (Santini unpublished data).

5.3   ARSENITE METABOLISM

Aerobic As(III) oxidation can occur both chemolithoautotrophically and heterotrophically. Chemolithoautotrophic As(III)-oxidizing organisms do not require an organic source of carbon for growth, as they are capable of fixing inorganic carbon in the form of CO_2 or HCO_3^-. In chemolithoautotrophic growth, organisms gain energy by using As(III) as the sole electron donor and coupling its oxidation to the reduction of O_2 to H_2O. The $\Delta G^{0'}$ value for the following equation demonstrates that energy is available from aerobic As(III) oxidation (Santini *et al.*, 2000):

$$2H_3AsO_3 + O_2 \rightarrow HAsO_4^{2-} + H_2AsO_4^- + 3H^+ \quad (\Delta G^{0'} = -175.8 kJ/mol)$$

The As(III)-oxidizing strains NT-26, isolated from an arsenopyrite-containing gold mine in the Northern Territory, Australia (Santini *et al.*, 2000), and *Thiomonas* sp. str. 3As from an acid-mine-drainage site, Carnoulès, France (Duquesne *et al.*, 2008), are the most extensively studied aerobic chemolithoautotrophic As(III) oxidizers. *Acidocaldus* sp. str. AO5, isolated from an acid-sulfate-chloride hot spring, is the only apparent obligate chemolithoautotrophic As(III) oxidizer currently known (D'Imperio *et al.*, 2007).

Heterotrophic As(III)-oxidizing bacteria either require an organic source of carbon for growth, or are auxotrophic for a particular nutrient (e.g., a specific amino-acid). Of the heterotrophic As(III) oxidizers isolated to date, *A. faecalis* has been the most extensively studied, in terms of its As(III) oxidase (Aio) [which catalyzes As(III) oxidation], while *H. arsenicoxydans* str. ULPAs1 has been the subject of genomic, transcriptomic and proteomic studies to investigate cellular responses to As(III) and the physiology of As(III) oxidation (Muller *et al.*, 2007; Weiss *et al.*, 2009; Cleiss-Arnold *et al.*, 2010; Koechler *et al.*, 2010). In heterotrophic As(III) oxidation, the oxidation of As(III) to As(V) has been shown to occur in the late-log phase of growth except in the psychrotolerant bacterium GM1 (Osborne *et al.*, 2010) and it was originally presumed that this process was a detoxification mechanism. However, it was observed that when grown in the presence of As(III), *A. faecalis* and NT-14 exhibited a decreased generation time and an increased final cell yield, respectively, indicating that these organisms can also gain energy from As(III) oxidation (Anderson *et al.*, 2003; vanden Hoven and Santini, 2004).

Table 5.1. Isolated aerobic As(III)-oxidizing bacteria.

Organism	Phylogenetic affiliation	Environment/sample isolated from	Metabolism	Reference
Acidicaldus sp. str. AO5	α	Acid-sulfate-chloride geothermal spring	fac chem	D'Imperio *et al.*, 2007
Agrobacterium albertimagni str. AOL15	α	Macrophyte material from geothermal waters	het	Salmassi *et al.*, 2002
Agrobacterium tumefaciens str. 5A	α	As-spiked soil	het	Macur *et al.*, 2004
Ancylobacter sp. str. OL-1	α	Heavy metal and hydrocarbon-contaminated soil	fac chem	Garcia-dominguez *et al.*, 2008
BEN-5	α	As-contaminated mine water	fac chem	Santini *et al.*, 2002
Bosea sp. str. WAO	α	Black shale	fac chem	Rhine *et al.*, 2008
NT-4	α	Arsenopyrite from gold mine	fac chem	Santini *et al.*, 2002
NT-26	α	Arsenopyrite from gold mine	fac chem	Santini *et al.*, 2000
Ochrobactrum tritici	α	Soil	het	Branco *et al.*, 2009
Sinorhizobium sp. str. M14	α	Gold mine waters	het	Drewniak *et al.*, 2008
Thiobacillus sp. str. S1	α	Heavy metal and hydrocarbon-contaminated soil	fac chem	Garcia-Dominguez *et al.*, 2008
Achromobacter sp. str. SY8	β	As-impacted soil	het	Cai *et al.*, 2009a; Cai *et al.*, 2009b
Alcaligenes faecalis	β	Sewage; soil	het	Osborne and Ehrlich, 1976; Phillips and Taylor, 1976
BEN-4	β	As-contaminated mine water	het	Santini *et al.*, 2002
Comamonas testosteroni str. W30E1a	β	As-contaminated aquifer	het	Sultana *et al.*, 2011
GM1	β	As-contaminated biofilm inside gold mine	het	Osborne *et al.*, 2010
Herminiimonas arsenicoxydans str. ULPAs1	β	Industrial sludge	het	Muller *et al.*, 2003; Muller *et al.*, 2006
Hydrogenophaga sp. str. YED6-4	β	Macrophyte material from geothermal waters	het	Salmassi *et al.*, 2006
Hydrogenophaga sp. str. YED1-8	β	Macrophyte material from geothermal waters	het	Salmassi *et al.*, 2006
Hydrogenophaga sp. str. YED6-21	β	Macrophyte material from geothermal waters	het	Salmassi *et al.*, 2006
Hydrogenophaga sp. str. CL-3	β	Heavy metal and hydrocarbon-contaminated soil	fac chem	Garcia-Dominguez *et al.*, 2008
Leptothrix sp. str. S1.1	β	Mine drainage water	het	Battaglia-Brunet *et al.*, 2006a

(*Continued*)

Table 5.1. Continued.

Organism	Phylogenetic affiliation	Environment/sample isolated from	Metabolism	Reference
NT-10	β	Arsenopyrite from gold mine	het	Santini et al., 2002
NT-14	β	Arsenopyrite from gold mine	het	Santini et al., 2002
Ralstonia sp. str. 22	β	Soil from mine site	het	Lieutaud et al., 2010
Thiomonas arsenivorans	β	Gold mine	fac chem	Battaglia-Brunet et al., 2006b
Thiomonas sp. str. 3As	β	Acid-mine-drainage	fac chem	Duquesne et al., 2008
Variovorax sp. str. 34	β	As-spiked soil	het	Macur et al., 2004
Variovorax paradoxus	β	Mine drainage water	het	Battaglia-Brunet et al., 2006a
Marinobacter santoriniensis	γ	Hydrothermal sediment	het	Handley et al., 2009a; Handley et al., 2009b
Pseudomonas sp. str. 31	γ	As-spiked soil	het	Macur et al., 2004
Pseudomonas sp. str. TS44	γ	As-impacted soil	het	Cai et al., 2009a; Cai et al., 2009b
Pseudomonas sp. str. VC-1	γ	Desert sediment	het	Campos et al., 2010
Arthrobacter sp. str. 15b	Actinobacteria	Sewage treatment site	fac chem*	Prasad et al., 2009
Microbacterium oxydans str. W702	Actinobacteria	As-contaminated aquifer	het	Sultana et al., 2011
Hydrogenobaculum acidophilum str. H55	Aquificae	Acid-sulfate-chloride geothermal spring	chem	Donahoe-Christiansen et al., 2004
Chloroflexus aurantiacus	Chloroflexi	Hot spring	het*	Lebrun et al., 2003
Thermus sp. str. A03C	Deinococcus-Thermus	Hot spring	het	Connon et al., 2008
Thermus sp. str. HR13	Deinococcus-Thermus	Hot spring	het	Gihring and Banfield, 2001
Thermus sp. str. YT-1	Deinococcus-Thermus	Yellowstone National Park	het	Gihring et al., 2001
Thermus thermophilus str. HB8	Deinococcus-Thermus	Yellowstone National Park	het	Gihring et al., 2001

chem: chemolithoautotrophic; fac: facultative; het: heterotrophic; α, β, γ: Alpha-, Beta-, Gammaproteobacteria, respectively. *Claimed or implied by authors although no physiological data has been published. Both of these organisms are included as they have homologues of the As(III) oxidase (Aio) that have been characterized. Isolated organisms that have homologues of the aio genes but have not been shown to oxidize As(III) are not included. Organisms which for which no 16S rRNA gene sequence is available are also not included.

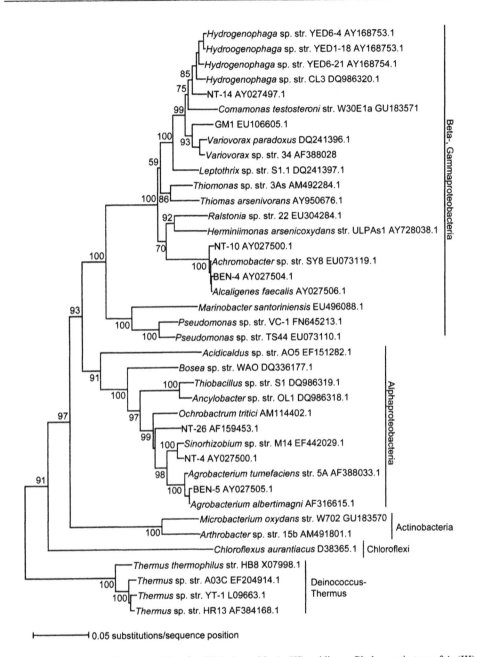

Figure 5.1. Phylogenetic relationship of published aerobic As(III) oxidizers. Phylogenetic tree of As(III) oxidizers based on the 16S rRNA gene. Tree is rooted with 16S rRNA gene of the archaeon *Pyrobaculum calidifontis* (not shown). Significant bootstrap values (per 100 trials) of major branch points are shown. Accession numbers are displayed next to organism name. *Hydrogenobaculum acidophilum* str. H55 and *Pseudomonas* sp. str. 31 are omitted due to incorrectly deposited 16S rRNA gene sequences.

5.3.1 *Arsenite oxidase and encoding genes*

Aerobic As(III) oxidation is catalyzed by Aio, which is phylogenetically distinct from Arx, the enzyme that catalyzes As(III) oxidation in the anaerobic As(III) oxidizer *Alkalilimnicola ehrlichii* str. MLHE-1 (Richey *et al.*, 2009). Aio is a member

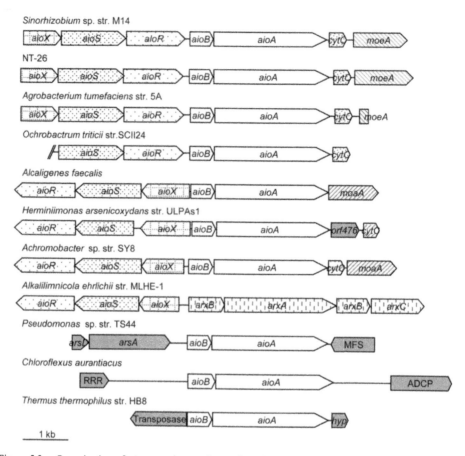

Figure 5.2.   Organization of *aio* gene clusters from selected As(III)-oxidizing bacteria. *aioS*, sensor histidine kinase gene; *aioR*, transcriptional regulator gene; *aioA*, As(III) oxidase large subunit gene; *aioB*, As(III) oxidase small subunit gene; *cytC*, cytochrome *c* gene; *moeA* and *moaA*, molybdenum cofactor biosynthesis genes; *arxB'ABC*, genes encoding for Arx As(III) oxidase; *aioX*, As(III)-binding protein gene; *arsA*, As(III)-pump ATPase gene; *arsD*, As resistance operon repressor gene; MFS, major facilitator superfamily protein gene; RRR, response-regulator receiver gene; ADCP, ATPase domain containing protein gene; *hyp*, hypothetical protein gene. Adapted from Sardiwal *et al.* (2010).

of the complex iron-sulfur molybdoenzyme family (Rothery *et al.*, 2008). It is a periplasmic heterodimer that consists of a large catalytic subunit (AioA; ~100 kDa) and a smaller subunit (AioB; ~15 kDa). AioA contains a single molybdenum atom (coordinated by *bis*-molybdopterin guanine dinucleotide) and a 3Fe-4S cluster, while the AioB contains a 2Fe-2S Rieske cluster (Ellis *et al.*, 2001). The AioA and AioB subunits are encoded by the *aioA* and *aioB* genes, respectively, which have been identified and sequenced in several aerobic As(III) oxidizers. Homologues of *aioA* and *aioB* have also been identified in the genomes of several prokaryotes that belong to taxa not represented among the characterized As(III) oxidizers, including two members of the phylum Chlorobia, a member of the Nitrospirae and three members of the domain Archaea: *Pyrobaculum calidifontis*, *Aeropyrum pernix* and *Sulfolobus tokodaii*.

In all known As(III)-oxidation gene clusters the *aioA* and *aioB* genes are in the same orientation, with *aioB* located upstream of *aioA* (Fig. 5.2). Several of the Proteobacterial As(III) oxidizers have other genes involved in As(III) oxidation located upstream and downstream of *aioB* and *aioA*, respectively (Fig. 5.2). Genes encoding a sensor histidine kinase (*aioS*) and a response regulator (*aioR*) are located upstream of *aioB* and comprise a two-component signal transduction system which regulates transcription of the *aioA* and *aioB* genes (Sardiwal *et al.*, 2010; see Chapter 9). Interestingly, the regulator/sensor pair is also present upstream of the genes encoding Arx in *A. ehrlichii* str. MLHE-1 (Richey *et al.*, 2009). A gene designated *aioX* is also conserved upstream of *aioB* in the Proteobacterial As(III) oxidizers and encodes a periplasmic As(III)-binding protein which is essential for up-regulation of the *aioA* and *aioB* genes in *Agrobacterium tumefaciens* str. 5A (Liu *et al.*, in press).

Genes encoding *c*-type cytochromes (*cytC*), are located downstream of the *aioA* gene in some *aio* gene clusters. The *cytC* gene in the NT-26 *aio* gene cluster encodes a physiological electron acceptor to the Aio (Santini *et al.*, 2007) while in *Ochrobactrum tritici* the co-expression of the *cytC* gene with the *aioA* and *aioB* genes was essential for an As(III)-oxidizing phenotype in a heterologous expression system (Branco *et al.*, 2009). The protein encoded by the *cytC* gene in *Ralstonia* sp. str. 22 has also been shown to act as an electron acceptor to the organism's Aio *in vitro* (Lieutaud *et al.*, 2010). Genes that encode proteins involved in molybdenum cofactor biosynthesis (*moeA* and *moaA*) are also present downstream of *aioA* in some As(III) oxidizers.

5.4   AEROBIC ARSENITE-OXIDIZING COMMUNITIES

5.4.1   *Bacterial communities associated with arsenite oxidation in geothermal springs*

As shown in Table 5.1, aerobic As(III)-oxidizing bacteria have been isolated from a diverse range of environments. In many of these environments As(III) oxidation has been observed *in situ* and attributed to the communities of microbes present, highlighting the role that these organisms play in the cycling of i-As in the environment. Several studies have investigated the microbial communities where As(III) oxidation occurs, using culture-dependent and -independent methods. These studies have been primarily done on biofilms in As-rich geothermal waters (Jackson *et al.*, 2001; Inskeep *et al.*, 2004; Salmassi *et al.*, 2006; Connon *et al.*, 2008). All of the studies on the outflow channels of geothermal springs identify a gradient of either increasing As(V) or declining As(III) concentration with increasing distance from the source of the As-rich water. They all link the As(III) oxidation in the channels to the microorganisms present in the environment, and analyze the microbial communities by targeting the 16S rRNA gene sequence.

A study of microbial mats in the outflow channel of a hot spring [58–62°C, pH 3.1, 33 μM total t-As, 31 μM As(III)] in Yellowstone National Park (YNP), used the culture-independent method of Denaturing Gradient Gel Electrophoresis (DGGE) of the 16S rRNA gene, to determine the dominant taxa in the mat (Jackson *et al.*, 2001). This study showed that the community became more phylogenetically diverse with distance from the water source. The investigators attributed the increase in complexity to a reduction in temperature, not in As(III) concentration (Jackson *et al.*, 2001). A separate study on the Alvord Hot Spring, Oregon [pH ~ 7, 69.5–78.2°C, 60 μM t-As, 56 μM As(III)], from which *Thermus* sp. str. A03C was isolated, used 16S rRNA gene libraries to determine the microbial diversity of upstream and downstream areas of an As(III)-oxidizing biofilm (Connon *et al.*, 2008). Like the study by Jackson *et al.* (2001), this study also found the downstream biofilm sample to have a greater phylogenetic diversity than the upstream sample. Microbial mats in another geothermal spring located in the Norris Basin of YNP [pH ~ 3, 53–64°C, 33 μM t-As, 24 μM As(III)] were also analyzed using DGGE of the 16S rRNA gene. The study identified many novel archaeal and bacterial species and also identified a change in community composition and phylogenetic diversity with distance from the spring source, as the As(III) concentration decreased (Inskeep *et al.*, 2004). The common phenomenon observed throughout these three studies is that there is an increase in the phylogenetic diversity of a community as conditions become less extreme [i.e., temperatures and/or As(III) concentrations decline]. The studies identified species of Bacteria and Archaea present in different areas of the outflow channels, but they could not give an indication as to what proportion, or subset, of the microbial population was capable of oxidizing As(III). A study of the biofilms associated with submerged macrophytes (aquatic plants) in Hot Creek, California, included a DGGE analysis of the 16S rRNA gene; the biofilm contained Alpha- and Betaproteobacteria and eukaryotes. The study also resulted in the isolation of three *Hydrogenophaga* spp. capable of oxidizing As(III) (Table 5.1) and included a most-probable number analysis of cultured heterotrophs [where the original sample was serially diluted in growth medium and checked for growth and As(III) oxidation], to estimate the proportion of As(III)-oxidizing bacteria in the community as 6–56%. The broad range may indicate either that the analysis used an insufficient number of replicates, that the authors did not account for the fact that samples were taken from different plants or areas of the creek, or another factor caused a large variation among their replicates (Salmassi *et al.*, 2006).

5.4.2   *Development of culture-independent methods to detect arsenite oxidizers*

The increasing number of *aioA* and *aioB* gene sequences available in GenBank has facilitated the development of a culture-independent method for detecting and analyzing As(III) oxidizers in environmental samples. The development of degenerate primers that when used in the polymerase chain reaction amplify a portion of the *aioA* gene, has shown that this gene can be used as a marker for the presence of As(III)-oxidizing bacteria (Inskeep *et al.*, 2007). Several degenerate primer pairs have been designed and shown to amplify portions of the *aioA* gene from several strains of bacteria (Inskeep *et al.*, 2007; Rhine *et al.*, 2007; Quemeneur *et al.*, 2008). The primers designed by Inskeep *et al.* (2007) were designed using *aioA* genes from phylogenetically diverse As(III) oxidizers, including members of the phyla Proteobacteria, Chlorobia and Deinococci-Thermus. The primer pair is also the most extensively tested, successfully amplifying *aioA* fragments from a diverse range of bacterial taxa from a wide range of environments including geothermal microbial mats, soil and lake sediments. The primer pair was subsequently adapted by Hamamura *et al.* (2009) to amplify and examine *aioA* sequences from Aquificiae strains, as these were found to be the dominant taxa in diverse geothermal spring systems across YNP, ranging from acidic to neutral pH and t-As concentrations from 3–126 µM. The primers designed by Rhine *et al.* (2007) and Quemeneur *et al.* (2008) did not use as diverse a range of sequences and were tested only on Proteobacteria, and environmental surveys with the Quemeneur *et al.* (2008) primer pair only returned Proteobacterial sequences. Recently, Heinrich-Salmeron *et al.* (2011) used another set of degenerate PCR primers, which successfully amplified a phylogenetically diverse range of *aioA* sequences from an As-impacted sediment near Gabes-Gottes mine, France [150 µg L^{-1} t-As, 14.6 µg L^{-1} As(III)]. Uniquely, this primer pair successfully amplified *aioA* sequences from members of the phyla Firmicutes, Bacteroidetes and Actinobacteria.

The Inskeep *et al.* (2007) primer pair was used to analyze the As(III)-oxidizing communities in an As-contaminated biofilm in a subarctic gold mine (Giant Mine; see Chapter 2) in Yellowknife, Northwest Territories, Canada, from which the psychrotolerant As(III) oxidizer GM1 was isolated (Osborne *et al.*, 2010). The study analyzed the diversity of As(III) oxidizers in samples from the top and bottom of the biofilm, which had As(III) concentrations of 691 mg L^{-1} and 450 mg L^{-1}, respectively. The diversity of As(III) oxidizers, assessed by phylogenetic analysis of amplified *aioA* genes, was similar in the two sub-samples, with 99% of sequences falling into three distinct clades. However, the composition of the three clades differed between the two samples, indicating that the As(III)-oxidizing community structures differed between the two samples.

Degenerate primers can also be used to accurately estimate the numbers of As(III) oxidizers in environmental samples and compare those with the overall bacterial population size. A recent study used Quantitative Real Time PCR and DGGE of *aioA* genes to assess and analyze As(III)-oxidizing communities in the As-impacted waters of the Upper Isle river basin, France (Quemeneur *et al.*, 2010). The study compared the copy number of *aioA* with that of the 16S rRNA gene to estimate the proportion of As(III) oxidizers in a series of samples that differed in As concentration; the As(III)-oxidizing community was largest in the samples that contained the most As.

5.4.3   *Detection of* aioA *expression with RT-PCR*

The studies described above have shown that *aioA* is a good indicator of the presence of As(III) oxidizers in the environment. However, it does not give an indication of whether bacterial As(III) oxidation is actually occurring. Inskeep *et al.* (2007) demonstrated that degenerate *aioA* primers can be used in conjunction with reverse transcriptase-PCR to amplify *aioA* cDNA from mRNA transcripts, to show whether *aioA* genes are being expressed in an As(III)-contaminated environment. By coupling the RT-PCR of *aioA* transcripts with a Quantitative Real Time approach, it should be possible to identify and quantify the active As(III) oxidizers in a particular environment, which may prove useful for the monitoring and control of As-impacted environments.

5.5 SUMMARY AND FUTURE DIRECTIONS

The development of degenerate *aioA* primers has enabled *aioA* to be used as a marker to detect prokaryotic As(III) oxidizers and As(III) oxidation. However, without *aioA* sequences from isolated As(III) oxidizers and genome assignments it is difficult to taxanomically assign As(III)-oxidizing communities accurately, as demonstrated in the studies by Heinrich-Salmeron *et al.* (2011), Inskeep *et al.* (2007) and Osborne *et al.* (2010). The As(III)-oxidizing members of a microbial community can only be identified using a metagenomic approach. Metagenomic (identification of the total DNA content from an environmental sample) and metaproteomic (identification of the total protein content from an environmental sample) approaches have been used recently to characterize the microbial communities in the As-impacted environment of the Reigous stream near the Carnoulès mine, France (Bertin *et al.*, 2011), and revealed the different metabolic processes occurring in the ecosystem. These approaches, along with metatranscriptomics (identification of the total mRNA from an environmental sample) would be extremely useful tools in the characterization of As(III)-oxidizing communities, allowing an accurate visualization of which organisms are active in the environment and how other metabolic processes and organisms affect aerobic As(III) oxidation.

The archaeal and several bacterial phyla are not currently represented among the characterized As(III) oxidizers. They should be examined in future work, in order to realize the diversity and understand the mechanisms of, aerobic As(III) oxidation, which in turn would lead to a greater understanding of how As(III) oxidizers influence the fate of As in the environment.

ACKNOWLEDGEMENTS

THO and JMS would like to acknowledge funding from the Natural Environment Research Council and M. Heath for advice on the manuscript.

REFERENCES

Anderson, G.L., Love, M. & Zeider, B.K.: Metabolic energy from arsenite oxidation in *Alcaligenes faecalis. J. Phys. IV France* 107 (2003), pp. 49–52.

Battaglia-Brunet, F., Yann, I., Garrido, F., Delorme, F., Crouzet, C., Greffie, C. & Joulian, C.: A simple biogeochemical process removing arsenic from mine drainage water. *Geomicrobiol. J.* 23 (2006a), pp. 201–211.

Battaglia-Brunet, F., Joulian, C., Garrido, F., Dictor, M.C., Morin, D., Coupland, K., Johnson, B., Hallberg, K.B. & Baranger, P.: Oxidation of arsenite by *Thiomonas* strains and characterization of *Thiomonas arsenivorans* sp. nov. *Antonie van Leeuwenhoek* 89 (2006b), pp. 99–108.

Branco, R., Francisco, R., Chung, A.P. & Morais, P.V.: Identification of an *aox* system that requires cytochrome *c* in the highly arsenic-resistant bacterium *Ochrobactrum tritici* SCII24. *Appl. Environ. Microbiol.* 75 (2009), pp. 5141–5147.

Bertin, P.N., Heinrich-Salmeron, A., Pelletier, E. *et al.*: Metabolic diversity among main microorganisms inside an arsenic-rich ecosystem revealed by meta- and proteo-genomics. *ISME J.* (2011).

Cai, L., Liu, G., Rensing, C. & Wang, G.: Genes involved in arsenic transformation and resistance associated with different levels of arsenic-contaminated soils. *BMC Microbiol.* 9:4 (2009a).

Cai, L., Rensing, C., Li, X.Y. & Wang, G.: Novel gene clusters involved in arsenite oxidation and resistance in two arsenite oxidizers: *Achromobacter* sp SY8 and Pseudomonas sp TS44. *Appl. Microbiol. Biotech.* 83 (2009b), pp. 715–725.

Campos, V.L., Valenzuela, C., Yarza, P., Kampfer, P., Vidal, R., Zaror, C., Mondaca, M.A., Lopez-Lopez, A. & Rossello-Mora, R.: *Pseudomonas arsenicoxydans* sp nov., an arsenite-oxidizing strain isolated from the Atacama desert. *Syst. Appl. Microbiol.* 33 (2010), pp. 193–197.

Cleiss-Arnold, J., Koechler, S., Proux, C., Fardeau, M.L., Dillies, M.A., Coppee, J.Y., Arsene-Ploetze, F. & Bertin P.N.: Temporal transcriptomic response during arsenic stress in *Herminiimonas arsenicoxydans*. *BMC Genomics* 11:709 (2010).

Connon, S.A., Koski, A.K., Neal, A.L., Wood, S.A. & Magnuson, T.S.: Ecophysiology and geochemistry of microbial arsenic oxidation within a high arsenic, circumneutral hot spring system of the Alvord Desert. *FEMS Microbiol. Ecol.* 64 (2008), pp. 117–128.

D'Imperio, S., Lehr, C.R., Breary, M. & McDermott, T.R.: Autecology of an arsenite chemolithotroph: sulfide constraints on function and distribution in a geothermal spring. *Appl. Environ. Microbiol.* 73 (2007), pp. 7067–7074.

Donahoe-Christiansen, J., D'Imperio, S., Jackson, C.R., Inskeep, W.P. & McDermott, T.R.: Arsenite-oxidizing *Hydrogenobaculum* strain isolated from an acid-sulphate-chloride geothermal spring in Yellowstone National Park. *Appl. Environ. Microbiol.* 70 (2004), pp. 1865–1868.

Drewniak, L., Matlakowska, R. & Sklodowska, A.: Arsenite and arsenate metabolism of *Sinorhizobium* sp. M14 living in the extreme environment of the Zloty Stok gold mine. *Geomicrobiol. J.* 25 (2008), pp. 363–370.

Duquesne, K., Lieautaud, A., Ratouchniak, J., Muller, D., Lett, M.C. & Bonnefoy, V.: Arsenite oxidation by a chemoautotrophic moderately acidophilic *Thiomonas* sp.: from the strain isolation to the gene study. *Environ. Microbiol.* 10 (2008), pp. 228–237.

Ellis, P.J., Conrads, T., Hille, R. & Kuhn, P.: Crystal structure of the 100 kDa arsenite oxidase from *Alcaligenes faecalis* in two crystal forms at 1.64 angstrom and 2.03 angstrom. *Structure* 9 (2001), pp. 125–132.

Garcia-Dominguez, E., Mumford, A., Rhine, E.D., Paschal, A. & Young, L.Y.: Novel autotrophic arsenite-oxidizing bacteria isolated from soil and sediments. *FEMS Microbiol. Lett.* 66 (2008), pp. 401–410.

Gihring, T.M. & Banfield, J.F.: Arsenite oxidation and arsenate respiration by a new *Thermus* isolate. *FEMS Microbiol. Lett.* 204 (2001), pp. 335–340.

Gihring, T.M., Druschel, G.K., McClesky, R.B., Hamers, R.J. & Banfield, J.F.: Rapid arsenite oxidation by *Thermus aquaticus* and *Thermus thermophilus*: field and laboratory investigations. *Environ. Sci. Technol.* 19 (2001), pp. 3857–3862.

Green, H.H.: Description of a bacterium which oxidizes arsenite to arsenate, and of one which reduces arsenate to arsenite, isolated from a cattle-dipping tank. *S. Afr. J. Sci.* 14 (1918), pp. 465–467.

Hamamura, N., Macur, R.E., Korf, S., Ackerman, G., Taylor, W.P., Kozubal, M., Reysenbach, A.L. & Inskeep, W.P.: Linking microbial oxidation of arsenic with detection and phylogenetic analysis of arsenite oxidase genes in diverse geothermal environments. *Environ. Microbiol.* 11 (2009), pp. 421–431.

Handley, K.M., Héry, M. & Lloyd, J.R.: Redox cycling of arsenic by the hydrothermal marine bacterium *Marinobacter santoriniensis*. *Environ. Microbiol.* 11 (2009a), pp. 1601–1611.

Handley, K.M., Héry, M. & Lloyd, J.R.: *Marinobacter santoriniensis* sp nov., an arsenate-respiring and arsenite-oxidizing bacterium isolated from hydrothermal sediment. *Int. J. Syst. Microbiol.* 59 (2009b), pp. 886–892.

Heinrich-Salmeron, A., Cordi, A., Brochier-Armanet, C., Halter, D., Pagnout, C., Abbaszadeh-fard, E., Montaut, D., Seby, F., Bertin, P.N., Bauda, P. & Asène-Ploetze, F.: Unsuspected diversity of Arsenite-oxidizing bacteria as revealed by widespread distribution of the *aoxB* gene in prokaryotes. *Appl. Environ. Microbiol.* 77 (2011), pp. 4685–4692.

Hughes, M.F.: Arsenic toxicity and potential mechanisms of action. *Toxicol. Lett.* 133 (2002), pp. 1–16.

Inskeep, W.P., Macur, R.E., Harrison, G., Bostick, B.C. & Fendorf, S.: Biomineralization of As(V)-hydrous ferric oxyhydroxide in microbial mats of an acid-sulphate-chloride geothermal spring, Yellowstone National Park. *Geochim. Cosmochim. Acta* 68 (2004), pp. 3141–3155.

Inskeep, W.P., Macur, R.E., Hamamura, N., Warelow, T.P., Ward, S.A. & Santini, J.M.: Detection, diversity and expression of aerobic bacterial arsenite oxidase genes. *Environ. Microbiol.* 9 (2007), pp. 934–943.

Jackson, C.R., Langner, H.W., Donahoe-Christiansen, J., Inskeep, W.P. & McDermott, T.R.: Molecular analysis of microbial community structure in an arsenic-oxidizing acidic thermal spring. *Environ. Microbiol.* 3 (2001), pp. 532–542.

Koechler, S., Cleiss-Arnold, J., Proux, C., Sismeiro, O., Dillies, M.A., Goulhen-Chollet, F., Hommais, F., Lievremont, D., Arsène-Ploetze, F., Coppée, J.Y. & Bertin, P.: Multiple controls affect arsenite oxidase gene expression in *Herminiimonas arsenicoxydans. BMC Microbiol.* 10:53 (2010).

Lebrun, E., Brugna, M., Baymann, F., Muller, D., Lievremont, D., Lett, M.C. & Nitschke, W.: Arsenite oxidase, an ancient bioenergetic enzyme. *Mol. Biol. Evol.* 20 (2003), pp. 686–693.

Lieutaud, A., van Lis, R., Duval, S., Capowiez, L., Muller, D., Lebrun, R., Lignon, S., Fardue, M.L., Lett, M.C., Nitschke, W. & Schoepp-Cothenet, B.: Arsenite Oxidase from Ralstonia sp. 22: characterization of the enzyme and its interaction with soluble cytochromes. *J. Biol. Chem.* 285 (2010), pp. 20442–20451.

Liu, G., Liu, M., Kim, E.-H., Matty, W., Bothner, B., Lei, B., Rensing, C., Wang, G. & McDermott, T.R.: *Environ. Microbiol.* Dec 19. doi: 10.1111/j.1462-2920.2011.02672.x. [Epub ahead of print]

Macur, R.E., Jackson, C.R., Botero, L.M., McDermott, T.R. & Inskeep, W.P.: Bacterial populations associated with the oxidation and reduction of arsenic in unsaturated soil. *Environ. Sci. Technol.* 38 (2004), pp. 104–111.

Muller, D., Simenova, D.D., Riegel, P., Mangenot, S., Koecheler, S., Lievremont, D., Bertin, P.N. & Lett, M.C.: *Herminiimonas arsenicoxydans* sp. nov., a metalloresistant bacterium. *Int. J. Syst. Bacteriol.* 56 (2006), pp. 1765–1769.

Muller, D., Medigue, C., Koechler, S. *et al.*: A tale of two oxidation states: bacterial colonization of arsenic-rich environments. *PloS Genet.* 4:e53 (2007).

Oremland, R.S. & Stolz, J.F.: The ecology of arsenic. *Science* 300 (2007), pp. 939–944.

Osborne, F.H. & Ehrlich, H.L.: Oxidation of arsenite by a soil isolate of *Alcaligenes. J. App. Bacteriol.* 41 (1976), pp. 295–305.

Osborne, T.H., Jamieson, H.E., Hudson-Edwards, K.A., Nordstrom, D.K., Walker, S.R., Ward, S.A. & Santini, J.M.: Microbial oxidation of arsenite in a subarctic environment: diversity of arsenite oxidase genes and identification of a psychrotolerant arsenite oxidizer. *BMC Microbiol.* 10:205 (2010).

Philips, S.E. & Taylor, M.L. Oxidation of arsenite to arsenate by *Alcaligenes faecalis. Appl. Environ. Microbiol.* 32 (1976), pp. 392–399.

Prasad, K.S., Subramanian, V. & Paul, J.: Purification and characterization of arsenite oxidase from *Arthrobacter* sp. *Biometals* 22 (2009), pp. 711–721.

Quemeneur, M., Heinrich-Salmeron, A., Muller, D., Lievremont, D., Jauzein, M., Bertin, P.N., Garrido, F. & Joulian, C.: Diversity surveys and evolutionary relationships of aoxB genes in aerobic arsenite-oxidizing bacteria. *App. Environ. Microbiol.* 74 (2008), pp. 4567–4573.

Quemeneur, M., Cebron, A., Billard, P., Battaglia-Brunet, F., Garrido, F., Leyval, C. & Joulian, C.: Population structure and abundance of arsenite-oxidizing bacteria along an arsenic pollution gradient in waters of the upper isle river basin, France. *App. Environ. Microbiol.* 76 (2010), pp. 4566–4570.

Ravenscroft, P., Brammer, H. & Richards, K.: *Arsenic pollution a global synthesis.* Wiley-Blackwell, UK, 2009.

Rhine, E.D., Garcia-Dominguez, E., Phelps, C.D. & Young, L.Y.: Environmental microbes can speciate arsenic. *Environ. Sci. Technol.* 39 (2005), pp. 9569–9573.

Rhine, E.D., Onesios, K.M., Serfes, M.E., Reinfelder, J.R. & Young, L.Y.: Arsenic transformation and mobilization from minerals by the arsenite oxidizing strain WAO. *Environ. Sci. Technol.* 42 (2008), pp. 1423–1429.

Richey, C., Chovanec, P., Hoeft, S.E., Oremland, R.S., Basu, P. & Stolz, J.F.: Respiratory arsenate reductase as a bidirectional enzyme. *Biochem. Biophys. Res. Comm.* 382 (2009), pp. 298–302.

Rothery, R.A., Workun, G.J. & Weiner, J.H.: The prokaryotic complex iron-sufur molybdoenzymes family. *BBA-Biomembranes* 1178 (2008), pp. 1897–1929.

Salmassi, T.M., Venkateswaren, K., Satomi, M., Newman, D.K. & Hering, J.G.: Oxidation of arsenite by *Agrobacterium albertimagni*, AOL15, sp. nov., isolated form Hot Creek California. *Geomicrobiol. J.* 19 (2002), pp. 53–66.

Salmassi, T.M., Walker, J.J., Newman, D.K., Leadbetter, J.R., Pace, N.R. & Hering, J.G.: Community and cultivation analysis of arsenite oxidizing biofilms at Hot Creek. *Environ. Microbiol.* 8 (2006), pp. 50–59.

Santini, J.M., Sly, L.I., Schnagl, R.D. & Macy, J.M.: A new chemolithoautotrophic arsenite-oxidizing bacterium isolated from a gold mine: phylogenetic, physiological, and preliminary biochemical studies. *Appl. Environ. Microbiol.* 66 (2000), pp. 92–97.

Santini, J.M., Sly, L.I., Wen, A., Comrie, D., De Wulf-Durand, P. & Macy, J.M.: New arsenite-oxidizing bacteria isolated from Australian gold-mining environments–phylogenetic relationships. *Geomicrobiol. J.* 19 (2002), pp. 67–76.

Santini, J.M., Kappler, U., Ward, S.A., Honeychurch, M.J., vanden Hoven, R.N. & Bernhardt, P.V.: The NT-26 cytochrome c_{552} and its role in arsenite oxidation. *Biochim. Biophys. Acta* 1767 (2007), pp. 189–196.

Sardiwal, S., Santini, J.M., Osborne, T.H. & Djordjevic, S.: Characterisation of a two-component signal transduction system that controls arsenite oxidation in the chemolithoautotroph NT-26. *FEMS Microbiol. Lett.* 313 (2010), pp. 20–28.

Sehlin, H.M. & Lindstrom, E.B.: Oxidation and reduction of arsenic by *Sulfolobus acidocaldarius* strain BC. *FEMS Microbiol. Lett.* 93 (1992), pp. 87–92.

Smedley, P.L. & Kinniburgh, D.G.: A review of the source, behaviour and distribution of arsenic in natural waters. *Appl. Geochem.* 17 (2002), pp. 517–568.

Stolz, J.F., Basu, P., Santini, J.M. & Oremland, R.S.: Arsenic and selenium in microbial metabolism. *Annu. Rev. Microbiol.* 60 (2006), pp. 107–130.

Sultana, M., Härtig, C., Planer-Friedrich, B., Seifert, J. & Schlömann, M.: Bacterial communities in Bangladesh aquifers differing in aqueous arsenic concentration. *Geomicrobiol. J.* 28:3 (2011), pp. 198–211.

Turner, A.W.: Bacterial oxidation of arsenite. *Nature* 164 (1949), pp. 76–77.

Turner, A.W.: Bacterial oxidation of arsenite I. Description of bacteria isolated from arsenical cattle-dipping fluids. *Aust. J. Biol. Sci.* 7 (1954), pp. 452–478.

vanden Hoven, R.N. & Santini, J.M.: Arsenite oxidation by the heterotroph *Hydrogenophaga* sp. str. NT-14: the arsenite oxidase and its physiological electron acceptor. *Biochem. Biophys. Acta* 1656 (2004), pp. 148–155.

Weiss, S., Carapito, C., Cleiss, J., Koechler, S., Turlin, E., Coppee, J.Y., Heymann, M., Kugler, V., Stauffert, M., Cruveiller, S., Medigue, C., Van Dorsselaer, A., Bertin, P.N. & Arsene-Ploetze, F.: Enhanced structural and functional genome elucidation of the arsenite-oxidizing strain *Herminiimonas arsenicoxydans* by proteomic data. *Biochimie* 91 (2009), pp. 192–203.

CHAPTER 6

Anaerobic oxidation of arsenite by autotrophic bacteria: the view from Mono Lake, California

Ronald S. Oremland, John F. Stolz & Chad W. Saltikov

6.1 INTRODUCTION

The phenomenon of arsenite [As(III)] oxidation by aerobic bacteria was first reported by Green (1918), and the many subsequent discoveries made in this realm, most occurring over the past three decades, are the primary focus of this book. In contrast, the fact that select anaerobes can also achieve this feat was an entirely serendipitous discovery. As often occurs in science, the intended path leading towards a stated goal can take an unexpected turn, ultimately leading to greater rewards than those originally anticipated. The intellectual freedom to meander such a path of curiosity-driven research is a great gift especially when one arrives at an unexpected revelation. It is perhaps the most rewarding aspect of a scientist's career. Such was the case when we first uncovered the phenomenon of anaerobic As(III) oxidation.

Our arsenic-related field work focused on Mono Lake, California because of its exceptionally high levels of dissolved inorganic arsenic ($\sim$200 μM), and the fact that we had previously isolated two novel species of arsenate [As(V)]-respiring bacteria, *Bacillus arseniciselenatis* and *B. selenitireducens* from its bottom sediments (Switzer Blum *et al.*, 1998). Radiotracer investigations employing [73]As(V) measured high As(V) reductase activity in the anoxic water column of the lake, yielding an estimate that this electron sink could mineralize approximately 8–14% of annual phytoplankton productivity (Oremland *et al.*, 2000), a value confirmed independently on the basis of mass balance considerations (Hollibaugh *et al.*, 2005). In both studies both groups also used cultivation-based methods (Most-Probable-Numbers) to estimate the densities of As(V)-respiring bacteria in the anoxic water column, and arrived at similar low but detectable values (e.g., 10^2–10^3 ml^{-1}). The next goal was to determine what taxa of As(V)-respiring prokaryotes were involved in these water-column transformations, using culture-independent analyses (Denaturing Gradient Gel Electrophoresis) of As(V)-amended anoxic bottom water.

We had expected to find 16S rRNA gene amplicon sequences similar to those from the bacilli we isolated from the sediments, but instead found a few rather unremarkable amplicons in the Epsilon, Gamma and Delta proteobacteria; yet these incubations showed a complete reduction of the added As(V), caused by sulfide-linked oxidation by resident chemoautotrophs of the Deltaproteobacteria (Hoeft *et al.*, 2004; Hollibaugh *et al.*, 2006). This As(V) reductase activity was inhibited by nitrate (Fig. 6.1A), while addition of As(III) to nitrate-amended waters resulted in the formation of As(V) (Fig. 6.1B). This observation led us to conclude that there was anaerobic biological oxidation of As(III) to As(V), linked to the provided nitrate ions (Hoeft *et al.*, 2002).

6.2 NITRATE-RESPIRING ARSENITE-OXIDIZERS

The results described above led us to conduct enrichment experiments with nitrate plus As(III)-amended Mono Lake anoxic bottom water. These ultimately resulted in a stable consortium that was transferred into artificial medium and then purified by serial dilution (Oremland *et al.*, 2002). The isolate, strain MLHE-1, was identified as a member of the Gammaproteobacteria, phylogenetically distant from all previously described species of aerobic As(III) oxidizers, which clustered in the Alpha and Betaproteobacteria. Strain MLHE-1 grew in mineral salts medium

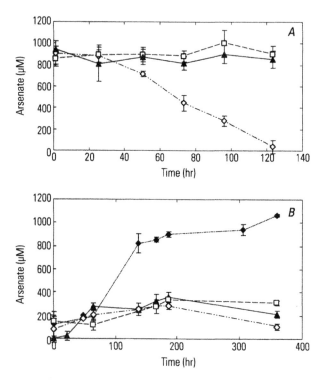

Figure 6.1. Arsenic oxyanion metabolism by incubated anoxic Mono Lake water. A): Reductive removal of As(V). Symbols: ◇, no additions; ▲, + nitrate; □, filter sterilized. B): Oxidation of As(III) to As(V). Symbols: ◇, no additions; ■, + nitrate; ▲, filter sterilized + nitrate; □, filter sterilized. Symbols represent the mean of 3 samples and vertical bars indicate ± 1 std. dev. Adapted from Hoeft *et al.* (2002) with permission from Geomicrobiology Journal.

with As(III) as the only electron donor and nitrate as the only electron acceptor (Fig. 6.2); no reduction beyond nitrite was observed. It also grew chemoautotrophically with nitrate when either sulfide or hydrogen was the electron donor. Strain MLHE-1 proved to be flexible and facultative in regard to chemoautotrophy. It was capable of growth as a heterotroph with acetate either under anaerobic (nitrate) conditions or in a fully aerated medium. After further phylogenetic analysis, and physiological and morphological characterization, the strain was given the name *Alkalilimnicola ehrlichii* (Hoeft et al., 2007).

Anaerobic As(III) oxidation was not confined to unusual extremophiles from Mono Lake. Rhine *et al.* (2008) reported the isolation of two new strains, DAO-1 and DAO-10 from As-contaminated soils, the former related to the *Azoarcus* clade in the Betaproteobacteria while the latter aligned with a *Sinorhizobium* of the Alphaproteobacteria. Both strains could grow chemoautotrophically by oxidizing As(III) with nitrate, and both contained RuBisCo Type II genes (*cbbM*) with which to fix CO_2 into cell material. They differed from *A. ehrlichii* in that they were capable of conducting a complete denitrification of nitrate to N_2 and had a viable and active N_2O reductase (*nosZ*). When compared with other aerobic autotrophic As(III) oxidizers (Garcia-Dominguez et al., 2008), their As(III) oxidase (i.e. AioA and AioB) amino acid sequences were conserved; but these differed sufficiently from those of heterotrophic As(III) oxidases to form a separate phylogenetic clade (Rhine *et al.*, 2007). One tested strain (DAO-1) failed to amplify the *aioAB* genes, suggesting the existence of an alternative mechanism as occurs in *A. ehrlichii* (see below).

Nitrate-linked oxidation of As(III) has been observed to occur in freshwater lake sediments (Senn and Hemond, 2002); as well in laboratory experiments with uncontaminated sediments and sewerage sludge (Sun *et al.*, 2008), by continuous flow sand columns inoculated with stable

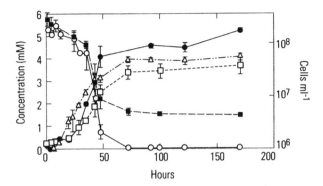

Figure 6.2. Growth of strain MLHE-1 on As(III) with nitrate. Cell densities are indicated on the right Y-axis. Symbols: ○, As(III); ●, As(V); ■, nitrate; □, nitrite; △, cells. Symbols represent the mean of 3 cultures and vertical bars indicate ± 1 std. dev. Adapted from Oremland *et al.* (2002) with permission from Applied and Environmental Microbiology.

enrichment cultures from the same experiment (Sun *et al.*, 2009) and in bioreactors (Sun *et al.*, 2010a). In all cases it was necessary to invoke a complete denitrification pathway in order to achieve a mass balance of reactants and products, which included the evolved N_2 gas. These workers went a step further and reported that As(III) oxidation could also be achieved with both "heterotrophic" and "autotrophic" enrichment cultures with chlorate as an electron acceptor (Sun *et al.*, 2010b). Chlorate was reduced to chloride, but it was not entirely clear if this was a strictly anaerobic process linked directly to chlorate: since As(III) oxidation was noted to also occur under microaerophilic conditions, it is possible that reductive dissimilation of chlorate releases molecular oxygen, making the system cryptically aerobic. The authors were able to PCR-amplify fragments of a conventional As(III) oxidase (*aioA*) gene, suggesting at least the possibility of oxygen-linked oxidation. In this context it is worth mentioning that the nitrate-linked methane oxidation by *Methylomirabalis oxyfera* has been shown recently to be a cryptically aerobic phenomenon (Ettwig *et al.*, 2010). *M. oxyfera* actually carries out a conventional aerobic pathway of C_1 oxidation involving methane monooxygenase, which is all made possible by the dismutation of the intermediate NO to equimolar N_2 and O_2. Another form of anaerobic As(III) oxidation was shown to be linked to selenate (Fisher and Hollibaugh, 2008). In this case, the isolate came from Mono Lake and contained a respiratory As(V) reductase gene (*arrA*) that aligned closely with those of two other arsenate/selenate respirers isolated from that system, namely *Bacillus selenitireducens* and *B. arseniciselenatis* (Switzer Blum *et al.*, 1998). Collectively, these results indicated that oxidation of As(III) with electron acceptors other than oxygen may be a widespread microbiological phenomenon that merits more attention.

6.3   AN ANNOTATED ARSENATE REDUCTASE THAT RUNS IN REVERSE

Several surprising observations were made when the genome of *A. ehrlichii* strain MLHE-1 was sequenced to completion and annotated. Pathways for the Calvin Benson Cycle (e.g., RuBisCo), CO oxidation, and sulfide oxidation were complete (Table 6.1). However, other pathways were either incomplete or absent altogether. Complete operons encoding nitric oxide reductase (Nor) and nitrous oxide reductase (Nos) were found, suggesting that *A. ehrlichii* might be capable of denitrification; yet growth experiments indicated that nitrite was the end product of nitrate reduction (Hoeft *et al.*, 2004). The solution lay in the fact that neither of the nitrite reductases that produce nitric oxide (NirS or NirK) was present in the genome. Another, more puzzling, discovery was that although this organism was isolated under anoxic conditions (using a mineral salts medium with As(III) as the electron donor and nitrate as the electron acceptor), and although it clearly grew by oxidizing As(III) (Fig. 6.2), it had no As(III) oxidase (e.g., *aioA*). Instead, it

Table 6.1.   Annotated gene homologs in *A. ehrlichii* reconciled with their observed phenotypic expression.

Gene	Annotated	Physiology Confirmed
Arsenite Oxidase, *aioBA*	No	Yes
Arsenate Reductase, *arrAB*	Yes	No
arrBA		
Arsenate Resistance, *arsC*	Yes	Yes
RuBisCo, *cbbl*-form q	Yes	Yes
Reverse TCA Cycle	Yes	Not Tested
Serine Pathway	Yes	Not Tested
Methane monooxygenase, *mmoC*	Yes	No
CO oxidation, *coxLMS*	Yes	Yes
Sulfide Oxidation Pathway	Yes	Yes
Hydrogenase, *hyaB*	Yes	Yes

had two distinct homologs of respiratory As(V) reductases (*arrA*) (Table 6.1). However, no clear As(V) reductase activity was observed.

The absence in the genome of an obvious homolog for As(III) oxidation when combined with the presence of two homologs for a physiologically non-functioning As(V) reductase led us to hypothesize that these latter enzymes, if expressed, could be running in reverse *in vivo*. The respiratory As(V) reductases and As(III) oxidases are both members of the broad molybendenum-family of dimethylsulfoxide (DMSO) reductase enzymes, (Fig. 6.4) but they have very different heterodimeric structures (Silver and Phung, 2005). Although other members of this family, such as DMSO reductase, had been shown to function reversibly (Shultz *et al.*, 1995), the conventional As(III) oxidase, with its high-potential Rieske-type iron sulfur cluster AioB, dictates a one-way flow of electrons, meaning that the enzyme complex AioBA could function only as an electron donor. Thus we first set out to determine if either of the Arr homologs in *A. ehrlichii* could couple the oxidation of As(III) to the reduction of an artificial electron acceptor such as 2,6-dichlorophenolindolephenol (DCPIP) or viologen dye (e.g., methyl or benzyl viologen). Indeed, protein extracts not only exhibited As(III) oxidase activity *in vitro*, but also reduced As(V) when the assays were run in reverse, using reduced DCPIP or viologen dyes. A protein complex that exhibited both these activities was then separated using "clear native" preparative polyacrylamide gels and identified as the "mlg_0216" homolog (Richey *et al.*, 2009). Subsequent investigations employing the "knockout" mutant approach confirmed that the As(V) reductase of MLHE-1 functions *in vivo* as a *de facto* As(III) oxidase (Zargar *et al.*, 2010). Of the two annotated genes, mutating *mlg_0216* eliminated the organism's ability to oxidize As(III) and to reduce nitrate, while the second *arrA* homolog gene, *mlg_2426*, had no such critical function. Therefore, *mlg_0216* has been designated as *arxA*, because it has a greater homology to *arrA* than to *aioA*, yet it warrants a new classification as a novel gene encoding a protein involved in arsenite oxidation.

Closer inspection of the genomic region surrounding *arxA* (genome locus tag: *mlg_0216*) reveals an operon structure that is both *aio-* and *arr*-like (Fig. 6.3). In contrast to *arr*, the first gene of an *aio* operon usually encodes a small Fe-S subunit called *aioB*. In MLHE-1 this subunit contains a Cys-X-X-Cys-His motif, an amino-acid motif common to all *c*-type hemes. The second gene (*mlg_0216*) encodes ArxA, the catalytic subunit similar to AioA (and ArrA). The third gene of the MLHE-1 *arx* operon encodes a putative Fe-S subunit with greater similarity to ArrB of the arsenate respiratory reductases than those of the arsenite oxidases (AioB/A). The fourth gene, *arxC*, is predicted to encode a membrane protein similar to ArrC of the arsenate reductases. This protein is thought to function as a quinol oxidoreductase that transfers electrons from the reduced quinone pool to ArrAB and arsenate. This would be reversed for Arx pathways. The last gene of the MLHE-1 *arx* operon, *arxD*, encodes a putative chaperone protein that may aid in molybdenum cofactor incorporation into ArxA.

Based on phylogenetic analysis, the ArxA of MLHE-1 is clearly a member of the DMSO reductase family of molybdenum-containing enzymes. It is also clear that ArxA is of a different

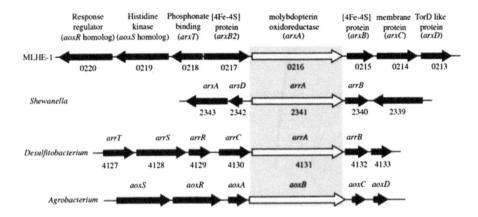

Figure 6.3. The *arx* containing genomic region of *Alkalilimnicola erhlichii* MLHE-1 in relationship to several bacteria that have *arr* and *aio* (designated *aox* in this figure) gene clusters. The numbers below the genes (arrows) are the corresponding genome locus tag numbers for that organism. The grey shaded region emphasizes the location of the main catalytic subunit for either the arsenite oxidase, employing the obsolete terms for *aio* genes (*aoxS* = *aioS*, *aoxR* = *aioR*, *aoxA* = *aioB*, *aoxB* = *aioA*; Lett *et al.*, 2011) the anaerobic arsenite oxidase (ArxA) or the arsenate respiratory reductase (ArrA).

evolutionary lineage than the AioA type of arsenite oxidases. ArxA is deeply branching within the ArxA clade, implying a shared evolutionary origin with the ArrA arsenate respiratory reductases. Based on phylogeny alone, one would conclude that ArxA is merely a distant ArrA homolog. However, genetic and biochemical studies have shown that ArxA functions as an As(III) oxidase but has some residual As(V) reductase activity *in vitro*.

Work on the regulation of *arx* operon in MLHE-1 is in the early stages. Gene expression studies showed that *arxA* is highly expressed under anaerobic conditions with As(III) (Zargar *et al.*, 2010). However, cultures grown anaerobically in the presence of As(V) or aerobically in the presence of As(III) exhibited low to undetectable *arxA* expression. These results suggest that the *arx* operon is regulated on several levels: one that is As(III)-dependent and another involving aerobic/anaerobic sensing. As(III)-dependent regulation may be carried out by a mechanism similar to that described in *aio*-containing organisms like *Agrobacterium tumefacians* (Kashyap *et al.*, 2006) and NT-26 (Sardiwal *et al.*, 2010). In these organisms, AioS and AioR regulate the As(III) oxidase expression (see Chapter 9). AioS encodes a membrane-bound sensory histidine kinase that phosphorylates the response regulator AioR. The phosphorylated form of AioR is predicted to bind the promoter region of the *aio* operon and induce its expression. As with AioSR, MLHE-1 contains a gene cluster (ArxTSR) near *arxB2ABCD* (Fig. 6.3). The putative ArxS and ArxR proteins are similar to sensory histidine kinase and response regulator family proteins. Moreover, AioS and AioR are close homologs to ArxS and ArxR. Based on the preliminary gene expression work of *arxA* in MLHE-1 and genomic comparisons of *arx* to *aio* operons, the regulatory cascade leading to *arx* gene expression may involve sensing As(III) in the periplasm (*via* ArxT), phosphorylation of the membrane-bound ArxS, a phospho group transfer to ArxR in the cytoplasm, and finally activation of *arx* operon expression.

6.4 ANOXYGENIC PHOTOSYNTHESIS FUELED BY ARSENITE

Photosynthetic bacteria have long been known to be capable of achieving anoxygenic photo-synthesis using low-electrochemical-potential compounds as sources of the electrons needed to reduce CO_2 into their cellular material. While sulfide, H_2, and certain organics (e.g., acetate) were known to fulfill this role, more recently both Fe(II) and nitrite have been implicated as well (Widdel *et al.*, 1993; Griffin *et al.*, 2007). It was suggested that As(III) could also serve in

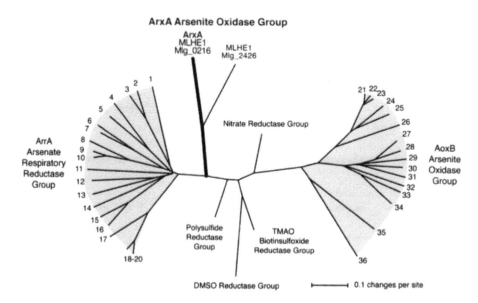

Figure 6.4.   Phylogenetic analysis of arsenate respiratory reductases (ArrA), AoxB-type Arsenite Oxidases (now referred to as AioA), and the ArxA-type arsenite oxidase of MLHE-1. The unrooted tree was constructed using a Neighbor-Joining method; gaps were ignored in the final phylogeny. The numbering refers to representative amino acid sequences ArrA and AioA as described below: Arsenate respiratory reductase Group (ArrA): 1, *Chrysiogenes* AAU11839*; 2 *Geobacter lovleyi* ZP 01593421; 3, *Geobacter uraniireducens* Rf4 ZP 01140714; 4, *Bacillus selenitireducens* AAQ19491*; 5, *Bacillus arsenicoselenatis* AAU11841*; 6, *Sulfurospirillum barnesii* AAU11840*; 7, *Wolinella succinogenes* NP 906980*; 8, *Desulfosporosinus* ABB02056*; 9–10, *Desulfitobacterium* YP 520364 & ZP 01372404*; 11, MLMS-1 ZP 01288668*; 12, *Natranaerobius thermophilus* YP_001916826; 13, *Halarsenatibacter silvermanii* SLAS-1 ACF74513*; 14, *Desulfonatronospira thiodismutans* ASO3-1 ZP_03737819; 15, *Alkaliphilus metalliredigens* ZP 00800578; 16, *Alkaliphilus oremlandii* OhILAs ZP 01360543*; 17, *Shewanella piezotolerans* (WP3, YP_002311519); 18–20, *Shewanella* group (*AAQ01672, ZP_01704274, 1YP_964317); Arsenite Oxidase AioA Group: 21, NT26 AAR05656**; 22, *Agrobacterium tumefaciens* ABB51928**; 23, *Ochrobactrum tritici* ACK38267**; 24, *Xanthobacter autotrophicus* Py2 ZP 01198801; 25, *Nitrobacter hamburgensis* YP_571843; 26, *Roseovarius* sp. 217 ZP 01034989; 27, *Ralstonia* sp. 22 ACX69823; 28, *Alcaligenes* AAQ19838**; 29, *Herminiimonas arsenoxydans* AAN05581**; 30, *Burkholderia multivorans* ZP 0157266830; 31, *Rhodoferax ferrireducens* YP 524325; 32, *Thiomonas* sp. 3As CAM58792**; 33, *Pseudomonas* sp. TS44 ACB05943; 34, *Halomonas* sp. HAL1 ACF77048; 35, *Chloroflexus aurantiacus* ZP 00356; 36, *Thermus thermophilus* YP 145366**. Symbols,* and** indicate the organism is known to respire arsenate or oxidize arsenite, respectively. The figure was reproduced from Zargar *et al.* (2010) with permission from the Journal of Bacteriology.

this capacity, and an *Ectothiorhodospira*-dominated enrichment culture established from light-incubated Mono Lake sediments was shown to oxidize As(III) to As(V) in the light, but not the dark (Budinhoff and Hollibaugh, 2008). In a separate study using red biofilms taken from Mono Lake hot springs, As(III) oxidation was also noted in light-incubated but not dark-incubated samples (Kulp *et al.*, 2008). From this biofilm-derived enrichment, a pure culture was established of *Ectothiorhodospira* strain PHS-1, which grew anaerobically in the light using As(III) as its electron donor (Fig. 6.5). Light-associated As(III) oxidation was also noted in green-pigmented biofilms from a hot spring dominated by *Oscillatoria*-like cyanobacteria (Kulp *et al.*, 2008).

Attempts to amplify As(III) oxidase gene fragments using primers designed for *aioA* were not successful, implying that strain PHS-1 possessed another means of As(III) oxidation. Successful amplification of an As(V)-reductase-like (*arrA*) gene fragment, however, was achieved. Alignment of the gene sequences revealed a close sequence identity (68%) of the *arxA* gene (the "reverse running" As(V) reductase) to that of *A. ehrlichii* (Kulp *et al.*, 2008). The full-length

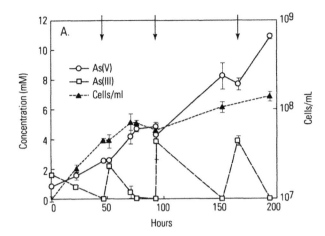

Figure 6.5. Anoxygenic photosynthesis and growth linked to As(III) oxidation by strain PHS-1. Arrows at the top indicate approximate times of pulsed additions of more As(III) after depletion was noted (modified from Kulp *et al.*, 2008). Reproduced from Science with permission.

arxA and surrounding DNA region has been sequenced (unpublished data). Inspection of this region reveals an *arx* operon arrangement that is nearly identical to the *arx* operon structure of MLHE-1. These results suggest that ArxA-type of As(III) oxidases may be widespread in nature, although this generalization is limited by the fact that the gene has been observed in only two Gammaproteobacteria isolated from Mono Lake, and possibly in an *Azoarcus*-like isolate DAO1 from soil (Rhine *et al.*, 2007). Subsequent investigations of the Mono Lake hot-spring red biofilm have demonstrated a capacity for both anaerobic As(III) oxidation (in the light) and As(V) reduction (in the dark) (Hoeft *et al.*, 2010). The former process was probably carried out by *Ectothiorhodospira*-like bacteria that contained the *arxA* gene (23 clones were identified) and the latter by other anaerobes containing homologs of respiratory arsenate reductase (*arrA*) (13 clones identified). It is clear from the above that two new species of anaerobic As(III)-oxidizing bacteria of the Gammaproteobacteria, one chemoautotrophic and one photoautotrophic can conserve energy for growth using a novel mechanism of As(III) oxidation. Whether this phenomenon is restricted to these microbes, is unique to the microbial ecosystem of Mono Lake, or has a broader significance and geographic range remains for future research to determine.

REFERENCES

Budinhoff, C.R. & Hollibaugh, J.T.: Arsenite-dependent photoautotrophy by an *Ectothiorhodospira* dominated consortium. *ISME J.* 2 (2008), pp. 340–344.

Ettwig, K.F., Butler, M.K., Le Paslier, D., Pelletier, E., Mangenot, S., Kuyper, M.M.M., Schrieber, F., *et al.*: Nitrite-driven anaerobic methane oxidation by oxygenic bacteria. *Nature* 464 (2010), pp. 543–548.

Fisher, J.C. & Hollibaugh, J.T.: Selenate-dependent anaerobic arsenite oxidation by a bacterium from Mono Lake, California. *Appl. Environ. Microbiol.* 74 (2008), pp. 2588–2594.

Garcia-Dominguez, E., Mumford, A., Rhine, E.D., Paschal, A. & Young, L.Y.: Novel autotrophic arsenite-oxidizing bacteria isolated from soil and sediments. *FEMS Microbiol. Ecol.* 66 (2008), pp. 401–410.

Green, H.H.: Description of a bacterium which oxidizes arsenite to arsenate, and one which reduces arsenate to arsenite, isolated from a cattle-dipping tank. *So. Afr. J. Sci.* 14 (1918), pp. 465–467.

Griffin, B.M., Schott, J. & Schink, B.: Nitrite, an electron donor for anoxygenic photosynthesis. *Science* 316 (2007), pp. 1870.

Hoeft, S.E., Lucas, F., Hollibaugh, J.T. & Oremland, R.S.: Characterization of microbial arsenate reduction in the anoxic bottom waters of Mono Lake, California. *Geomicrobiol. J.* 19 (2002), pp. 23–40.

Hoeft, S.E., Kulp, T.R., Stolz, J.F., Hollibaugh, J.T. & Oremland, R.S.: Dissimilatory arsenate reduction with sulfide as the electron acceptor: experiments with Mono Lake water and isolation of strain MLMS-1, a chemoautotrophic arsenate-respirer. *Appl. Environ. Microbiol.* 70 (2004), pp. 2741–2747.

Hoeft, S.E., Switzer Blum, J., Stolz, J.F., Tabita, F.R., Witte, B., King, G.M., Santini, J.M. & Oremland, R.S.: *Alkalilimnicola ehrlichii* sp. nov., a novel, arsenite-oxidizing haloalkaliphilic gammaproteobacterium capable of chemoautotrophic or heterotrophic growth with nitrate or oxygen as the electron acceptor. *Int. J. Syst. Evol. Microbiol.* 57 (2007), pp. 504–512.

Hoeft, S.E., Kulp, T.R., Han, S., Lanoil, B. & Oremland, R.S.: Coupled arsenotrophy in a hot spring photosynthetic biofilm at Mono Lake, California. *Appl. Environ. Microbiol.* 76 (2010), pp. 4633–4639.

Hollibaugh, J.T., Carini, S., Gürleyük, H., Jellison, R., Joye, S.B., LeCleir, G., Meile, C., Vasquez, L. & Wallschläger, D.: Arsenic speciation in Mono Lake, California: response to seasonal stratification and anoxia. *Geochim. Cosmochim. Acta* 69 (2005), pp. 1927–1937.

Hollibaugh, J.T., Budinoff, C., Hollibaugh, R.A., Ransom, B. & Bano, N.: Sulfide oxidation coupled to arsenate reduction by a diverse microbial community in a soda lake. *Appl. Environ. Microbiol.* 72 (2006), pp. 2043–2049.

Kashyap, D.R., Botero, L.M., Franck, W.L., Hassett, D.J. & McDermott, T.R.: Complex regulation of arsenite oxidation in *Agrobacterium tumefaciens. J. Bacteriol.* 188 (2006), pp. 1081–1088.

Kulp. T.R., Hoeft, S.E., Asao, M., Madigan, M.T., Hollibaugh, J.T., Fisher, J.C., Stolz, J.F., Culbertson, C.W., Miller, L.G. & Oremland, R.S.: Arsenic(III) fuels anoxygenic photosynthesis in hot spring biofilms from Mono Lake, California. *Science* 321 (2008), pp. 967–970.

Lett, M.C, Muller, D., Liévremont, D., Silver, S. & Santini, J.: Unified nomenclature for genes involved in prokaryotic aerobic arsenite oxidation. *J. Bacteriol.* Nov 4. [Epub ahead of print]

Oremland, R.S., Dowdle, P.R., Hoeft, S., Sharp, J.O., Schaefer, J.K., Miller, L.G., Switzer Blum, J., Smith, R.L., Bloom, N.S. & Wallschlaeger, D.: Bacterial dissimilatory reduction of arsenate and sulfate in meromictic Mono Lake, California. *Geochim. Cosmochim. Acta* 64 (2000), pp. 3073–3084.

Oremland, R.S., Hoeft, S.E., Santini, J.M., Bano, N., Hollibaugh, R.A. & Hollibaugh, J.T.: Anaerobic oxidation of arsenite in Mono Lake water and by a facultative, arsenite-oxidizing chemoautotroph, strain MLHE-1. *Appl. Environ. Microbiol.* 68 (2002), pp. 4795–4802.

Rhine, E.D., Phelps, C.D. & Young, L.Y.: Anaerobic arsenite oxidation by novel denitrifying isolates. *Environ. Microbiol.* 8 (2006), pp. 899–908.

Rhine, E.D., Chadhain, N., Zylstra, G.J. & Young, L.Y.: The arsenite oxidase genes (aroAB) in novel chemoautotrophic arsenite oxidizers. *Biochem. Biophys. Res. Comm.* 354 (2007), pp. 662–667.

Richey, C., Chovanec, P., Hoeft, S.E., Oremland, R.S., Basu, P. & Stolz, J.F.: Respiratory arsenate reductase as a bidirectional enzyme. *Biochem. Biophys. Res. Comm.* 382 (2009), pp. 298–302.

Sardiwal, S., Santini, J.M., Osborne, T.H. & Djordjevic, S.: Characterization of a two-component signal transduction system that controls arsenite oxidation in the chemolithoautotroph NT-26. *FEMS Lett.* 313 (2010), pp. 20–28.

Schultz, B.E., Hille, R. & Holme, R.H.: Direct oxygen atom transfer in the mechanism of action in *Rhodobacter sphaeroides* dimethyl sulfoxide reductase. *J. Am. Chem. Soc.* 117 (1995), pp. 827–828.

Senn, D.B. & Hemond, H.F.: Nitrate controls on iron and arsenic in an urban lake. *Science* 296 (2002), pp. 2372–2376.

Silver, S. & Phung, L.T.: Genes and enzymes involved in bacterial oxidation and reduction of inorganic arsenic. *Appl. Environ. Microbiol.* 71 (2005), pp. 599–608.

Sun, W., Sierra, R. & Field, J.A.: Anoxic oxidation of arsenite linked to denitrification in sludges and sediments. *Water Res.* 42 (2008), pp. 4569–4577.

Sun, W., Sierra-Alvarez, R., Milner, L., Oremland, R. & Field, J.A.: Arsenite and ferrous iron oxidation linked to chemolithotrophic denitrification for the immobilization of arsenic in anoxic environments. *Environ. Sci. Technol.* 43 (2009), pp. 6585–6591.

Sun, W., Sierra-Alvarez, R. & Field, J.A.: The role of denitrification on arsenite oxidation and arsenic mobility in an anoxic sediment column model with activated alumina. *Biotechnol. Bioeng.* 107 (2010a), pp. 786–794.

Sun, W., Sierra-Alvarez, R., Milner, L. & Field, J.A.: Anaerobic oxidation of arsenite linked to chlorate reduction. *Appl. Environ. Microbiol.* 76 (2010b), pp. 6804–6811.

Switzer Blum, J., Burns Bindi, A., Buzzelli, J., Stolz, J.F. & Oremland, R.S.: *Bacillus arsenicoselenatis* sp. nov., and *Bacillus selenitireducens* sp. nov.: two haloalkaliphiles from Mono Lake, California which respire oxyanions of selenium and arsenic. *Arch. Microbiol.* 171 (1998), pp. 19–30.

Widdel, F., Schnell, S., Heising, S., Ehrenreich, A., Assmus, B. & Schink, B.: Ferrous iron oxidation by anoxygenic phototrophic bacteria. *Nature* 362 (1993), pp. 834–836.

Zargar, K., Hoeft, S., Oremland, R. & Saltikov, C.W.: Identification of a novel arsenite oxidase gene, *arxA*, in the haloalkaliphilic, arsenite-oxidizing bacterium *Alkalilimnicola ehrlichii* strain MLHE-1. *J. Bacteriol.* 192 (2010), pp. 3755–3762.

CHAPTER 7

Arsenite oxidase

Matthew D. Heath, Barbara Schoepp-Cothenet, Thomas H. Osborne &
Joanne M. Santini

7.1 INTRODUCTION

The two enzymes known to carry out bioenergetic arsenite [As(III)] oxidation are As(III)
oxidases – Aio and Arx (see also Chapters 6 and 10 for detailed description of Arx). To date,
only Aio has been purified and characterized.

Aio oxidizes As(III) to arsenate As(V) according to the following reaction:

$$As^{III}O_2^- + 2H_2O \rightarrow As^V O_4^{3-} + 4H^+ + 2e^-$$

The reaction sequence is completed once the electrons generated are passed to a physiological
electron acceptor, which has been found to be a c-type cytochrome or azurin (Anderson et al.,
1992; Santini et al., 2007 and vanden Hoven and Santini, 2004). Aio activity has also been liked
to chlorate reduction (Sun et al., 2010) and photosynthesis (Lebrun et al., 2003; Duval et al.,
2008) in some organisms.

Arx is a variant of arsenate [As(V)] reductase (Arr), but may act as an As(III) oxidase in
anaerobic respiration of either nitrate (Hoeft et al., 2007) or selenate (Fisher and Hollibaugh,
2008) or in anoxygenic photosynthesis (Kulp et al., 2008). The reversible nature of Arx [i.e.
functioning both as an As(III) oxidase and as an As(V) reductase (though only in vitro)] was an
intriguing discovery; possibly similar to what is observed for succinate dehydrogenase/fumarate
reductase and dimethylsulfoxide (DMSO) reductase, for example (Maklashina and Cecchini,
1999; Shultz et al., 1995).

Aio has been isolated from a variety of organisms which include: Rhizobium sp. str. NT-26,
Hydrogenophaga sp. str. NT-14, Alcaligenes faecalis, Arthrobacter sp. str. 15b and Ralstonia sp.
str. 22 (Santini and vanden Hoven, 2004; vanden Hoven and Santini, 2004; Anderson et al., 1992;
Prasad et al., 2009; Lieutaud et al., 2010). NT-26 is a chemolithoautotrophic As(III) oxidizer
and is a member of the Alphaproteobacteria. NT-14, A. faecalis and 22 are all members of the
Betaproteobacteria and oxidize As(III) heterotrophically. Table 7.1 summarizes the properties of
the purified enzymes, and their structural and functional characteristics, which are the main focus
of this chapter.

Aio consists of two heterologous (α-AioA and β-AioB) subunits; a large catalytic α-subunit
contains a molybdenum atom coordinated by two pterin molecules, and a [3Fe-4S] cluster.
The small β-subunit contains a Rieske [2Fe-2S] cluster. Since Aio possesses two pyranopterin
molecules coordinating the molybdenum atom it can be assigned to the DMSO reductase family
of molybdoenzymes.

An unexplained observation is the subunit conformation of the Aio; the NT-26 enzyme is a
$\alpha_2\beta_2$, NT-14 a $\alpha_3\beta_3$, while A. faecalis and 22 are a $\alpha_1\beta_1$. These enzymes all catalyse the oxidation
of As(III) to As(V) and have slight variations in their turnover and affinity for their substrate
(Table 7.1).

Aio is transported from the cytoplasm to the periplasm as a fully folded protein complex via the
Twin Arginine Translocation (Tat) pathway. Transport via the Tat pathway requires possession of
a conserved twin arginine N-terminal signal peptide sequence (S/T)-R-R-X-F-L-K), recognized

Table 7.1.   Comparison of purified As(III) oxidases.

	NT-26	NT-14	A. faecalis	Ralstonia sp. 22	Arthrobacter sp. str. 15b
Location	Periplasm	Periplasm	'Membrane'	'Membrane'	'Membrane'
Native molecular mass (kDa)	219	309	100	110	100
Subunit composition	α 98	α 86	α 85	α 97	α 85
	β14	β16	β15	β16	β14
Oligomeric state	$\alpha_2\beta_2$	$\alpha_3\beta_3$	$\alpha_1\beta_1$	$\alpha_1\beta_1$	$\alpha_1\beta_1$
Cofactors	Mo, Fe, S	Mo, Fe, S	Mo, Fe, S	Mo, Fe, S	Mo, Fe, S
K_m (μM)	61[#]	35[#]	8[†]	7[⊥], 13[⊥]	26[#]
V_{max} (μmol min^{-1} mg^{-1})	2.4[#]	6.1[#]	2.88[†]	5.7[#], 140[⊥], 140[⊥]	2.45[#]

'Membrane' and periplasm indicates where Aio is at its highest abundance: in membrane fractions (i.e. outer surface of inner membrane) or soluble periplasmic fractions, respectively; this may vary depending on the method of cell disruption; [#]calculated using the artificial electron acceptor 2,6-dichlorophenolindophenol (DCPIP); [†]calculated using the native co-purified azurin; [⊥]Calculated using cytochrome c_{552} and c_{554}, respectively from Ralstonia sp. str. 22; N.D. not determined (Santini and vanden Hoven, 2004; vanden Hoven and Santini, 2004; Anderson et al., 1992; Lieutaud et al., 2010; Prasad et al., 2009)

by the translocase machinery (Berks et al., 2005). An N-terminal Tat leader sequence is found on the small Rieske subunit of all arsenite oxidases. The Tat pathway has been most studied in the model organism Escherichia coli; here, many of the Tat substrates/chaperones assist in folding of the apo-protein and bind cofactors to coordinate their insertion. It is currently unknown whether there are specific groups of chaperones required for specific sub-sets of cofactors (Berks et al., 2005). It has been suggested that chaperone-mediated 'proof-reading' steps may exist to allow the Tat system to identify the correct folded state of the protein, prior to its export (Berks et al., 2005). For complex substrates like Aio, which have multiple cofactors and exist in heterodimeric configurations; such quality-control steps may occur at different stages before interaction with the translocase machinery.

The N-terminal Tat leader sequence found on AioB is also thought to be responsible for its membrane attachment (anchor) in some circumstances (Lieutaud et al., 2010). Examples of other Rieske proteins that are anchored to the membrane via the Tat signal sequence include the Rieske subunit of the cytochrome bc_1 complex (Bachmann et al., 2006). As mentioned in Table 7.1, the relative abundance of Aio in soluble periplasmic fractions and membrane fractions varies depending on the method of cell disruption. In examples where the signal peptide is thought to be cleaved, Aio may be attached to the membrane via another protein.

In this chapter, we outline some of the main characterization studies of Aio, to provide an insight into the structure and function of this bioenergetic metalloenzyme.

7.2   CHARACTERISTICS OF THE ARSENITE OXIDASE

7.2.1   X-ray crystallographic structure of Aio

7.2.1.1   Overview

To date, one crystal structure has been determined to 1.64 and 2.03 Å resolution, from the betaproteobacterium A. faecalis (NCIB 8687) (Ellis et al., 2001). The Aio from this mesophile is a heterodimeric ($\alpha_1\beta_1$) 100-kDa Mo- and FeS-containing protein, shown in Figure 7.1. The large α-subunit (825 amino-acid residues) contains the Mo center and a [3Fe-4S] cluster. It is associated with a small β-subunit (134 amino-acid residues): a Rieske protein containing a [2Fe-2S] cluster. The heterodimeric structure contains a network of hydrogen bonds at the interface

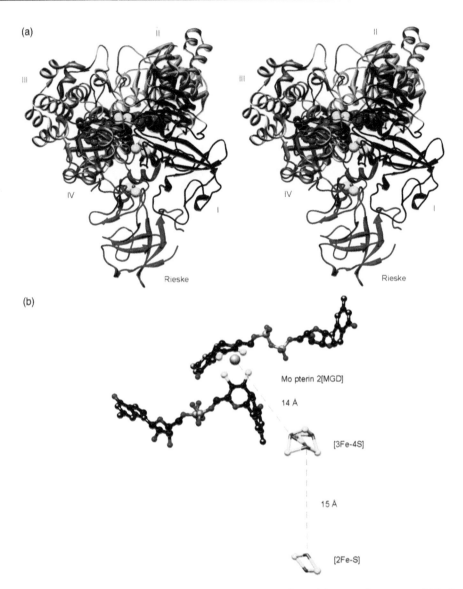

Figure 7.1.   Aio crystal structure from *A. faecalis*. (a) Stereo view of the crystal structure of Aio from *A. faecalis* and schematic representation of domain structure with respect to its primary sequence. The large α-subunit is shown in blue (domain I), green (domain II), orange (domain III) and pink (domain IV), the small β-subunit in red. (b) Arrangement of the metal centers involved in electron transfer (Mo → [3Fe-4S] → [2Fe-2S]), the electrons are then transferred to a periplasmic electron acceptor (azurin or *c*-type cytochrome) (Anderson *et al.*, 1992).

between the two subunits, and adopts an overall dimension of 75 Å × 75 Å × 50 Å (Ellis *et al.*, 2001). Since no published genome sequence of *A. faecalis* was available at the time of this structure's publication, the amino-acid sequences of the two subunits of the enzyme were inferred from the electron density maps (the genes were later sequenced and submitted in 2004: accession numbers AAQ19838 and AAQ19839).

There are many unique properties associated with Aio, compared with other members of the DMSO reductase family with known structures. Most significant is the absence of any covalent

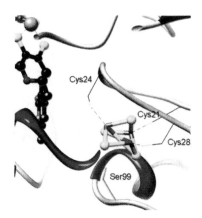

Figure 7.2.   Ribbon diagram of the large subunit around the [3Fe-4S] cluster. The amino-acid side chains coordinating the Fe are shown. The Mo center (top) is situated approximately 14 Å from the [3Fe-4S] cluster. The [3Fe-4S] cluster is coordinated by residues Cys21-X2-Cys24-X$_3$-Cys28-Ser99. The serine retains the fourth position and orientation of the fourth cysteine and is thought to play a role in electron transfer to the Rieske [2Fe-2S] center.

bond between the protein and the Mo atom, which is found in all other members of the DMSO reductase family (Ellis *et al.*, 2001; Romao, 2009). Other distinguishing features of Aio include: (1) the Fe-S center is a [3Fe-4S] instead of a 4Fe-4S; and (2) AioB is a Rieske subunit, rather than the more common 4[4Fe-4S] subunit. In fact, Aio is the only known enzyme with a Rieske subunit associated with a molybdopterin subunit. We will now describe the significance of these findings, and the crystal structure of Aio.

7.2.1.2   *Large subunit and molybdenum center*
A total of four domains make up the large subunit of Aio (Ellis *et al.*, 2001) (Fig. 7.1). Each domain presents a characteristic alpha-beta-alpha helix-sheet-helix sandwich topology and exhibits significant structural homology to other iron-sulfur-possessing members of this family. Domain I, consisting of three antiparallel beta-sheets and six helices, binds the [3Fe-4S] cluster coordinated by the motif Cys21-X$_2$-Cys24-X$_3$-Cys28–Ser99 near the interface of domains III and IV (Ellis *et al.*, 2001).

Domains II and III are related to one another in the protein structure, by a pseudo-two-fold axis of symmetry and both possess homologous dinucleotide-binding folds (Fig. 7.1). A seven-stranded, mainly parallel, beta-sheet makes up domain II; with five helices on one side of the beta-sheet, and seven helices and a small, two-stranded, antiparallel beta sheet on the other side (Ellis *et al.*, 2001). Domain III has a similar alpha-beta-alpha (helix, sheet, helix) sandwich topology, consisting of a fully parallel, five-stranded, beta-sheet with six helices on one side and nine on the other, with a small two-stranded parallel beta-sheet. Domain IV consists of a six-stranded beta-barrel flanked by alpha-helices (Fig. 7.1) (Ellis *et al.*, 2001).

A highly polar solvent-access tunnel leads from the top of the structure through to the center bound by domains I, II and III (Fig. 7.3) (Ellis *et al.*, 2001). The surface of this tunnel is almost entirely formed of serine, aspartate, asparagine, glutamate, lysine, histidine, arginine, and tyrosine residues. At the base of this tunnel lie a number of hydrophilic residues (His195, Glu203, Arg419, and His423) that have been implicated in binding As(III) (Ellis *et al.*, 2001) (Fig. 7.3). The manual docking of As(III) as As(OH)$_3$ shows possible hydrogen bonding with these four residues; suggesting an optimal position of As(III) for the interaction with the Mo center. In addition to this, experimental evidence has already implied that the involvement of at least one histidine in As(III)

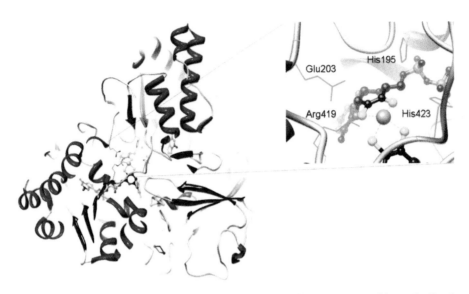

Figure 7.3.   Top-down ribbon diagram of the Aio. A highly polar, solvent-access tunnel is seen leading from the top through to the centre of the structure. The amino-acid side chains of His195, Glu203, Arg419, and His423 are shown at the base of this tunnel, thought to be involved in binding As(III) (main diagram and inset).

binding is essential for enzyme activity (McNellis and Anderson, 1998). To determine/confirm amino-acids involved in As(III) binding, a substrate-bound structure or mutagenesis work is required.

It is the base of this tunnel that holds the Mo center, which forms the active site. In the space between the four domains of the large subunit, two antiparallel molybdopterin guanine dinucleotide (MGD) centers are positioned, stabilized by a complex network of hydrogen bonding and salt bridges (Ellis *et al.*, 2001). The Mo atom is coordinated in a similar manner to that seen in other members of the DMSO reductase family of Mo enzymes: *via* two molybdopterin molecules, each providing a *cis*-dithiolene moiety. Electron density maps reveal a fifth ligand, modelled as a single O atom at 1.6Å from the Mo – suggesting a Mo=O bond; this completes the five-coordinate geometry of the Mo. This five-coordinate structure is thought to represent the reduced form of the Mo center; suggesting that the protein has become reduced in the course of crystallographic analysis (Ellis *et al.*, 2001). The crystal structure reveals a lack of a protein ligand to the metal and thus offers a unique coordination chemistry seen at this site, compared with other enzymes of this group, to be discussed in section 7.2.1.5.

7.2.1.3   *Rieske subunit*

The small Rieske subunit consists of two sub-domains, principally containing beta-sheets (Fig. 7.1) (Ellis *et al.*, 2001). The overall fold is similar to other known Rieske-containing subunits such as cytochromes $b_6 f$ and bc_1, complexes and dioxygenases (Kurisu *et al.*, 2003; Xia *et al.*, 1997; Ferraro *et al.*, 2005). A common Cys-X-His-$X_{15\text{-}17}$-Cys-X_2-His sequence motif (i.e. Cys60-X-His62-X_{15}-Cys78-X_2-His81), observed for other Rieske proteins, binds the Rieske [2Fe-2S] cluster. The two loops (Fig. 7.4), containing Cys60, His62 and Cys78, His81, coordinate the cluster and are held together by a disulfide bond between Cys60 and Cys80. The imidazol rings of the histidine residues coordinate the Fe nearer the surface of the subunit, with His81 exposed to solvent but His62 buried within the αβ-subunit interface. The second iron and its two coordinating cysteine ligands are positioned within the protein and are not exposed (Ellis *et al.*, 2001).

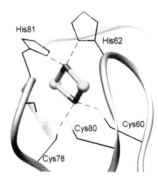

Figure 7.4.   Ribbon diagram of the small subunit around the Rieske [2Fe-2S] center. The amino-acid side chains coordinating the Fe are shown. The [2Fe-2S] cluster is coordinated by residues Cys60-X-His62-X_{15}-Cys78-X_2-His81.

7.2.1.4   *Electron transport pathway*

The oxidation of As(III) releases two electrons that are thought to be transferred from the Mo center as follows: Mo → [3Fe-4S] → [2Fe-2S] (see Fig. 7.1). From the Rieske center the electrons are then transferred to a physiological electron acceptor, either azurin or cytochrome *c* (Anderson *et al.*, 1992; vanden Hoven and Santini, 2004). Approximate distances between redox centers are outlined in Figure 7.1b; the [3Fe-4S] cluster lies ∼14 Å from the Mo atom while the two Fe-S clusters are also separated by a distance of ∼15 Å (Ellis *et al.*, 2001).

Electron transfer is thought to be mediated by a complex network of covalent- and hydrogen-bonding interactions, between the pyrazine ring of the pterin cofactor and the Fe-S clusters. In the shortest path considered, electrons pass through Ser99 (the residue whose natural change from cysteine is responsible for the formation of a [3Fe-4S] rather than a [4Fe-4S] cluster in the protein) and His62 of the small subunit (which lies at the interface between the two subunits) (Ellis *et al.*, 2001).

His81 has been implicated in electron transfer out of Aio to its physiological oxidants; it is the sole ligand exposed to solvent, being positioned near the surface of the Rieske subunit. Other residues in this vicinity are well-suited for the interaction/binding of small globular proteins (Ellis *et al.*, 2001). However, to date no site-directed mutagenesis has been performed to study these interactions.

7.2.1.5   *Comparison with other molybdenum enzyme structures*

There is relatively little (∼26%) sequence identity between Aio from *A. faecalis* and other non-Aio members of the DMSO reductase family (Ellis *et al.*, 2001). However, the overall topologies of Aio's individual domains of the α-subunit are similar to those in the other members of the DMSO reductase family. A BLAST search has identified the large Aio subunit to be most similar to the assimilatory nitrate reductase (Nas) from *Pseudomonas syringae pv. tabaci* (26%*) and to the formate dehydrogenase (Fdh) from *Methanosaeta harundinacea* (25%*) [*topological identities calculated from uniprot (default settings)].

The structural relationships of the two subunits of the Aio are mutually exclusive in that no other enzyme with a Rieske subunit has so far been shown to contain a molybdopterin subunit. Phylogenetic analyses reveal a very ancient origin of Aio, pre-LUCA, i.e. existing before the split of the Bacteria and Archaea (Lebrun *et al.*, 2003) (see Chapter 10). Aio is considered a good example of an enzyme assembled using redox domains from a restricted set of redox-protein building blocks that would have existed during the early stages of evolution (i.e., pre-LUCA) (Lebrun *et al.*, 2003).

As previously mentioned, the large subunit possesses a serine residue instead of a cysteine in the last cluster-binding motif – appearing to play a role in electron transfer. This natural change is responsible for the formation of a high-potential [3Fe-4S] cluster, rather than a [4Fe-4S] cluster

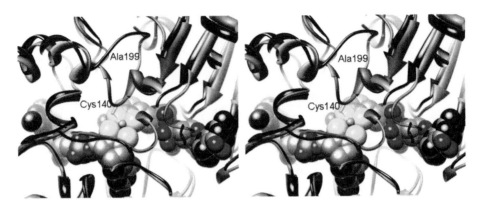

Figure 7.5. Stereo ribbon diagram of the active site of the Aio. Aio is shown in blue, superimposed on the active site of the periplasmic nitrate reductase (Nap) (from *Desulfovibrio desulfuricans*), shown in green. The cysteine side chain coordinating the Mo atom in Nap lies near the metal atom. The corresponding alanine residue in the Aio is far from the active site.

seen, for example, in formate dehydrogenase (Boyington *et al.*, 1997) and other DMSO reductase family members with cysteines at this position. However, the most significant difference between Aio and other Mo enzymes is the lack of an additional ligand, seen in other members of the DMSO reductase family, in the form of a covalent bond with an amino-acid side chain from cysteine, selenocysteine, aspartic acid or serine (Romao, 2009). The amino-acid residue corresponding to one of these four residues is Ala199 in the *A. faecalis* Aio; this forms part of a conserved amino-acid substitution (Phe in some cases), compared with other known Aio's. As a result, the active site becomes more exposed than in other members of the DMSO reductase family; the structure folds the loop containing this residue away from the Mo site, which is unable to form any covalent linkage (Fig. 7.5) (Ellis *et al.*, 2001). Consequently, Aio represents a fifth subgroup of the DMSO reductase family, which possesses two equivalents of the organic pterin molecule coordinating the Mo; with the other four subgroups containing serine, cysteine, aspartic acid or selenocysteine as a protein ligand to the metal. In fact, the overall coordination more closely resembles that of tungsten-containing enzymes, which also possess two equivalents of the pterin cofactor, no polypeptide ligand, and a proposed dioxo-coordination in the oxidized enzyme.

7.2.2 *Spectroscopic features of the redox-active centers*

As introduced above, the three redox-active centers in Aio are a Mo center, a [3Fe-4S] cluster and a Rieske [2Fe-2S] cluster. The mechanism of As(III) oxidation may be clarified by characterizing the spectral and redox properties of these three centers. These properties can be determined by using techniques such as Visible, Electron Paramagnetic Resonance (EPR) and X-ray spectroscopies.

7.2.2.1 *Visible absorption spectroscopy on arsenite oxidase*

The three centers of Aio (Fig. 7.6) do not contribute to intense optical absorption. The optical contributions from both Fe-S centers, ranging typically from 400 nm to 600 nm, are overlapping and are therefore difficult to distinguish by visible spectroscopy. In one study (Hoke *et al.*, 2004) the [2Fe-2S] and [3Fe-4S] centers were titrated. Following the decrease of their absorption caused by their reduction, redox potentials (E_m) were +130 mV and +260 mV for the [2Fe-2S] and [3Fe-4S] centers, respectively. The Mo center presents a visible contribution somewhat distinct from the two first centers, typically around 600 nm–700 nm. The optical properties of the Mo center have been characterized in-depth in DMSO reductase (Bastian *et al.*, 1991; Benson *et al.*, 1992 and Finnegan *et al.*, 1993) but there is presently no published in-depth study of Aio. However, observations of Anderson *et al.* (1992) and Hoke *et al.* (2004) during the course of a

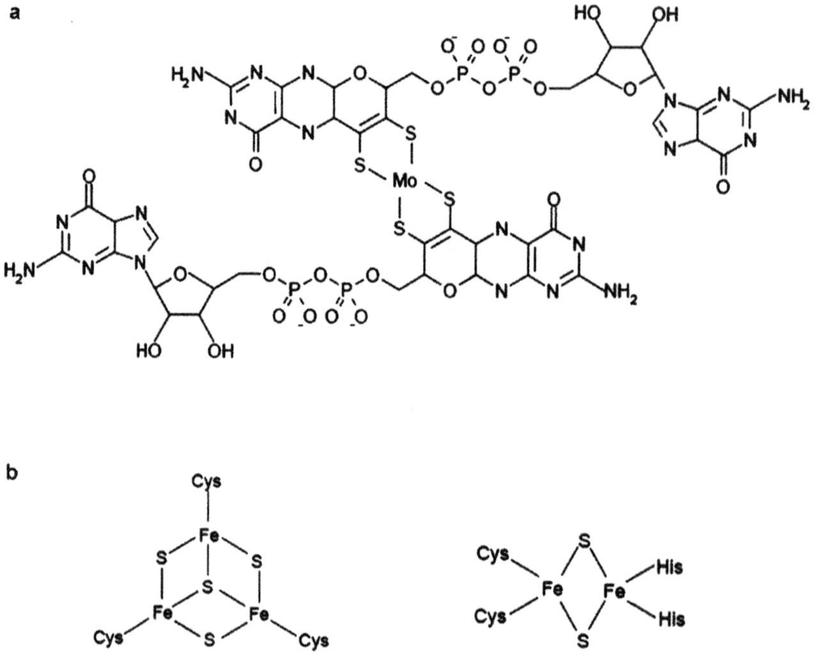

a

b

Figure 7.6.   Redox centers of the Aio. (a) Molybdopterin cofactor in the form of two antiparallel molyb-dopterin guanine dinucleotide (MGD) molecules, coordinating the molybdenum atom via two cis-dithiolene moieties in antiparallel fashion. (b) Iron-sulfur clusters of the Aio ([3Fe-4S] and Rieske [2Fe-2S] cluster, respectively).

reductive optical/EPR titration, suggest the contribution at 682 nm to correspond to the Mo-center. The preliminary EPR results obtained on Aio (see section 7.2.2.2) suggest this technique to be promising for the detection and characterization of Aio centers.

7.2.2.2   Continuous-wave electron paramagnetic resonance spectroscopy on arsenite oxidase

Since the three centers of Aio do not feature intense optical absorption contributions, EPR spectroscopy has been a particularly useful tool for characterizing the redox properties of Aio. Continuous-Wave (CW) EPR relies on the absorption of electromagnetic radiation in the microwave region by paramagnetic systems. In Aio, the three cofactors possess paramagnetic states, so CW EPR can be used to determine their redox properties. The oxidized form of the enzyme exhibits a rhombic signal, observable between 6 K and 20 K, which has g-values of 2.03, 2.0 and 1.99 arising from the oxidized [3Fe-4S] cluster (Fig. 7.7; upper curve). Upon reduction with As(III), ascorbate or dithionite, this signal disappears and is replaced by a second rhombic signal, observable between 15 K and 40 K; this has g-values of 2.02, 1.88 and 1.77 arising from the reduced Rieske [2Fe-2S] center (Fig. 7.7; lower curve). The g_{av} at 1.88, characteristic of Rieske centers, is due to the two His ligands of the cluster. This differs from 'classical' ferredoxins where only cysteines are cluster ligands (Duval et al., 2010).

The increase in the size of the [2Fe-2S] EPR signal during reduction of the enzyme allows determination of this cluster's redox potential and the dependence of this potential on pH. The [2Fe-2S] centers in Aio from NT-26, 22 and the putative arsenite oxidizer Chloroflexus aurantiacus (a thermophilic, green non-sulfur bacterium) have constant E_m values up to pH 8 at ~+210 mV. Above this pH value, the E_m values of the centers are pH-dependent, as is observed for the Rieske/cytb complexes (Fig. 7.8; Duval et al., 2010).

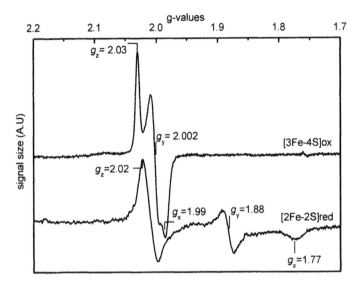

Figure 7.7. EPR spectra recorded on the Aio from NT-26. Spectrum of the [3Fe-4S]ox center was recorded on ferricyanide oxidized enzyme. EPR parameters: 12 K, 1-milliwat microwave power, 9.40-GHz microwave frequency, 10-gauss modulation frequency. Spectrum of the [2Fe-2S]red center was recorded on ascorbate reduced enzyme [Aio oxidase reduced by As(III) or dithionite shows identical spectrum]. EPR parameters: 12 K, 6.4-milliwat microwave power, 9.40-GHz microwave frequency, 16-gauss modulation frequency.

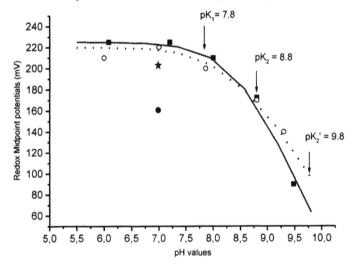

Figure 7.8. The reduction potential of the Rieske center from Aio as a function of pH. The open circles and closed squares represent data obtained with the 22 and NT-26 enzymes, respectively. The data obtained on the 22 Rieske cluster are fitted by assuming two ionization equilibria with pK_a-values of 8 and 9.8 and E_m (low pH) of $+210$ mV whereas the data obtained on NT-26 Rieske are fitted by assuming two ionization equilibria with pK_a-values of 8 and 8.8 and E_m (low pH) of $+220$ mV. The closed star and closed circle represent the E_m pH 7 value obtained with the *C. aurantiacus* ($+200$ mV) and *A. faecalis* ($+155$ mV) enzymes, respectively. Adapted from Duval *et al.* (2010).

As detailed above, AioB is homologous to the Rieske protein from the Rieske/cyt*b* complex. Extensive EPR characterization of wild-type and mutant Rieske/cyt*b* complexes has allowed the development of a structural model that predicts the redox and spectral properties of the cluster (Link, 1999; Schoepp *et al.*, 1999). The redox properties of the NT-26, 22 and *C. aurantiacus*

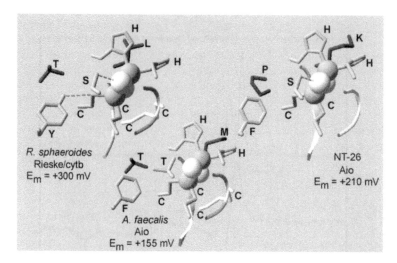

Figure 7.9.   Correlation between redox potential and structural determinants surrounding the Rieske cluster. In the Rieske/cyt*b* complex from *Rhodobacter sphaeroides*, cumulative effects of a disulfide bridge and hydrogen bonds from Tyr and Ser residues are responsible for a high redox potential at +300 mV. In *A. faecalis,* the cumulative effects of the absence of Ser and Tyr are responsible for a low redox potential at +155 mV. In NT-26, the presence of the Ser but absence of the Tyr, are responsible for an intermediate redox potential without any effect of the absence of the disulfide bridge. Adapted from Duval *et al.* (2010).

Aio-Rieskes, together with the low E_m value (+130 mV obtained by optical spectroscopy and +155 mV obtained by EPR spectroscopy) of the *A. faecalis* Rieske, suggest that Ser and Tyr residues surrounding the Fe-S cluster play the same roles in determining the [2Fe-2S] cluster redox properties in Aio as they do in Rieske/cyt*b* complexes. Figure 7.9 illustrates the cumulative effects of a disulfide bridge and hydrogen bonds from Tyr and Ser residues that are thought to be responsible for a high redox potential (+300 mV) in *Rhodobacter sphaeroides*. While the absence of these residues surrounding the Fe-S cluster in *A. faecalis* are responsible for a low redox potential (+155 mV); an intermediate redox potential is seen in NT-26 where the presence of Ser but absence of Tyr is apparent. However, two cysteine residues strictly conserved in all Rieske/cyt*b*-Rieske proteins, and considered to be crucial for its redox properties and function, are not fully conserved in the AioB counterpart and appear to influence neither the absolute value of redox potential nor its dependence on pH (Duval *et al.*, 2010).

   We have investigated the [3Fe-4S] cluster's redox potential and its dependence on pH. Redox titrations of the enzyme at pH 9 yielded a midpoint potential of +120 mV. Aio is exceptional among representatives of the DMSO reductase family in possessing a high-potential [3Fe-4S] cluster rather than the usual low-redox-potential [4Fe-4S] center, as a proximal Fe-S center in the catalytic subunit (Duval *et al.*, 2009). Redox titrations of the enzyme at pH values below pH 9 yielded interesting results (Duval *et al.*, 2009). At pH 7 and 8 the signature of the oxidized [3Fe-4S]$^+$ cluster disappears in a specific window of ambient redox potentials (Fig. 7.10 left panel). And at pH 6 the oxidation state of the [3Fe-4S]$^+$ cluster was no longer stable even if stored at cryogenic temperatures (70K) (Fig 7.10 right panel). This instability is eliminated by the addition of sulfite, known to be a potent inhibitor of the enzyme (Lieutaud *et al.*, 2010; see also section 7.2.3). The redox potential determined in the presence of sulfite, +270 mV, is in line with the value obtained by optical spectroscopy on the *Alcaligenes* enzyme (Hoke *et al.*, 2004). The results obtained in the absence of sulfite may be explained by electron transfer between the [3Fe-4S]$^+$ cluster and the Mo center at low temperature, but may also be explained by a strong paramagnetic interaction between the two cofactors.

   There are only limited data on FS0 (the Fe-S center of the catalytic subunit) from other members of the DMSO reductase family that can be compared with the results obtained on the [3Fe-4S]

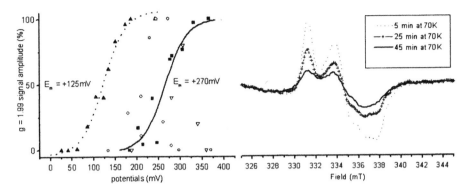

Figure 7.10. Redox-controlled EPR study of the [3Fe-4S] center. Redox titrations were performed at various pH values. Left panel, Only at pH 9 (closed triangles) could a conventional redox titration of the cluster be obtained; it yielded E_m pH 9 at +120 mV. At pH 8 (open circles), 7 (open triangles) or 6 (closed squares) the signal is lost within a certain potential window and/or unstable Right panel, even at low temperature. At pH 6, a stable signal is obtained after adding sulfite to the sample. A conventional redox titration thus determines the E_m pH 6 at +270 mV. Adapted from Duval *et al.*, (2009).

center from Aio. There are only two enzymes for which the signal of the FS0, a $[4\text{Fe-4S}]^+$ cluster, has been detected, i.e. a dissimilatory nitrate reductase (Nar) and an assimilatory nitrate reductase (Nas) (Jepson *et al.*, 2004; Rothery *et al.*, 2004). Using redox titrations the redox potentials of the FS0 center in Nar and Nas have been determined to be -55 mV and -190 mV, respectively.

The Mo center is the best-studied (by EPR) redox center in the DMSO reductase family of molybdenum enzymes. In addition to being the detection technique for redox titrations, CW EPR spectroscopy has been extensively applied to the Mo center to resolve weak nuclear hyperfine interactions from nearby atoms. Observation of such ligand hyperfine interactions has been used to probe the ligation geometries of Mo(V) (the only EPR observable redox state of the Mo center) sites within enzymes. Even when a crystallographic structure is available, CW EPR remains a useful complementary technique since atomic structures obtained by crystallography do not always reflect the functional states of the enzymes.

Aio is a very unusual member of the DMSO reductase family, in that despite considerable effort no Mo(V) signal has been detected (Anderson *et al.*, 1992; Duval *et al.*, 2009). This suggests that during the progressive reduction of the enzyme, the only stable redox states of this center are Mo(IV) and Mo(VI), which are both EPR-silent. Interpreting the reduction of Mo(VI) to Mo(IV) in *A. faecalis* Aio, during cyclic voltammetry experiments, as a cooperative 2-electron process (Hoke *et al.*, 2004) favors this hypothesis. However, this view was challenged in an independent study on Aio from NT-26 by cyclic voltammetry, in which the shape of the electrochemical signal obtained under turnover conditions was interpreted as supporting a one-electron rate-limiting step during turnover (Bernhardt and Santini, 2006). Another hypothesis could thus be that the EPR signature of Mo(V) is not detected because it strongly interacts with the proximal $[3\text{Fe-4S}]^+$ center. In support of the latter hypothesis, EPR experiments suggest that the oxidized $[3\text{Fe-4S}]^+$ cluster in Aio may be responsible for cancelling out the Mo(V) EPR signature. The signature of the oxidized $[3\text{Fe-4S}]^+$ cluster disappears (Fig. 7.10) in a specific potential window, and it can be speculated that this results from magnetic coupling with Mo(V). One possibility is therefore that the redox transition of Mo(VI) to Mo(V), in the course of the titration, occurs in the same redox potential range as the redox transition of $[3\text{Fe-4S}]^{+1}$ to $[3\text{Fe-4S}]^0$ and that reduced $[3\text{Fe-4S}]^0$ is able to transfer its electron to Mo(V) even at cryogenic temperatures, producing Mo(IV) and becoming re-oxidized again.

7.2.2.3 *Extended X-ray absorption fine structure (EXAFS) spectroscopy*
As(III) is known to be a potent inhibitor of some Mo-containing enzymes. Studies of As(III)-inhibited enzymes, such as xanthine oxidase, reveal a thiolate bridging of a S atom between

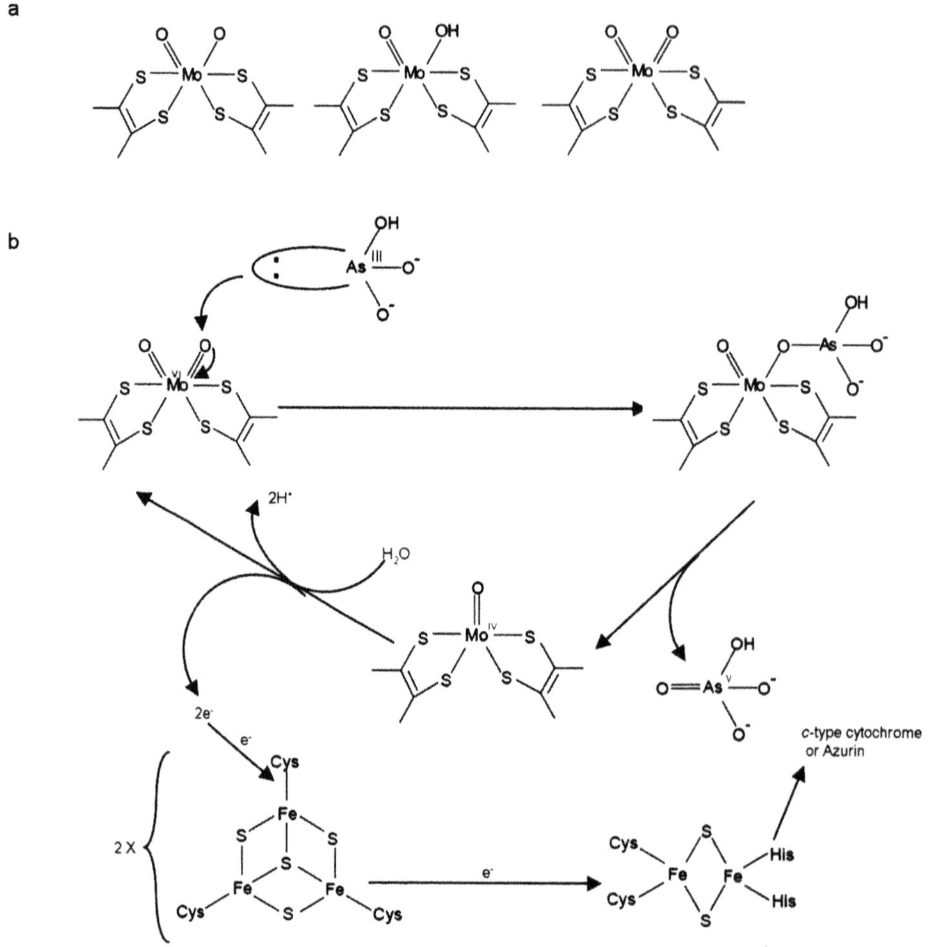

Figure 7.11.   Possible configurations of the Aio active site and proposed reaction mechanism. (a) Possible configurations of the Aio active site. The Mo is coordinated by four *cis*-dithiolene sulfur atoms of the two MGD molecules. The dioxo configuration is considered the most likely based on protein film voltammetry and EXAFS data (Hoke *et al.*, 2004 and Conrads *et al.*, 2002). (b) Proposed reaction mechanism of Aio. The oxo group with the longer bond is subject to nucleophilic attack by the lone electron pair. As(III) is oxidized to As(V), reducing Mo from a VI to IV. Hydrolysis restores the Mo to the VI state and the two electrons are passed to the 3Fe-4S cluster, on to the Rieske cluster, and to a soluble electron acceptor such as a *c*-type cytochrome or azurin. Adapted from Ellis *et al.* (2001).

As(III) and the Mo cofactor; giving rise to a Mo-S-As moiety (Cao *et al.*, 2011). Here, As(III) blocks the active site and inhibits electron transfer. However, Aio does not contain a sulfido group coordinated to the Mo (Fig. 7.11). The Mo active site of *A. faecalis*, in both the oxidized and reduced forms, was investigated using Extended X-ray Absorption Fine Structure (EXAFS), providing data for both the oxidized [Mo(VI)] and the reduced [Mo(IV)] forms of the enzyme (Conrads *et al.*, 2002). The analysis revealed four Mo-S interactions at 2.47 Å and 2.37 Å for the oxidized and reduced forms, respectively. This distance falls in line with the *bis*-enedithiolate coordination of the Mo atom seen in the crystal structure, while confirming the Mo, in *bis*-MGD, as having no additional fifth ligand in the form of an amino-acid side chain such as cysteine, selenocysteine, aspartic acid or serine, observed in other members of the DMSO reductase family.

Apart from four sulfur ligands, EXAFS analysis revealed that Mo possesses an =O group at 1.70 Å, which is consistent with a five-coordinate square-pyramidal coordination present in the crystal structure. The oxidized enzyme also possesses a second oxo-ligand which is thought to be present as $-O$, $-OH$ or a stretched =O group at 1.83 Å (Fig. 7.11). Given the limits in resolution of EXAFS data, the exact difference in bond lengths between the oxo at 1.70 Å and the second Mo-O was difficult to determine. However, a normal Mo-O bond length would be approximately 2 Å. While resonance Raman spectra of as-isolated, redox-cycled, O-labeled and reduced Aio strongly suggest a weaker, elongated Mo=O ligand, of oxidized Aio. In addition, Hoke *et al.* (2004) protein film voltammetry (PFV) studies support the assignment of the active site in the oxidized state as an asymmetric dioxo center, L_2MoO_2, rather than (a protonated) $L_2MoO(OH)$. Whether the additional atom ligand in the oxidized form of the enzyme is a distended Mo=O or Mo$-$OH, remains unclear (Conrads *et al.*, 2002).

7.2.3   Steady-state kinetics

7.2.3.1   Overview
A limited amount of data has, so far, been reported on the enzymology of Aio (Table 7.1). The turnover of Aio occurs *via* double displacement ('ping-pong') kinetics, in which the enzyme cycles between oxidized and reduced forms; this is common in oxido-reduction enzymes. As(V) is released before azurin (or c-type cytochrome) binds to the reduced enzyme.

Since As(III) is a two-electron donating substrate, the catalytic turnover is assumed to begin with the oxidation of As(III) at the Mo center (which can accept up to two electrons). The electrons are thought to be transferred one at a time to the [3Fe-4S] center and subsequently to the [2Fe-2S] center, which reduces a soluble electron-carrier protein (vanden Hoven and Santini, 2004). A few proteinaceaous electron carriers have been examined, but most of the enzymatic data currently available on Aio have been obtained using 2,4 dichlorophenolindolphenol (DCPIP), a non-physiological chemical electron acceptor (Anderson *et al.*, 1992; Santini and vanden Hoven, 2004; vanden Hoven and Santini, 2004). Enzymatic properties of Aio deduced from these studies, therefore, do not necessarily reflect the physiological reaction. The use of DCPIP may explain the very high K_m measured with NT-26 and NT-14 enzymes, but cyclic voltammetry confirmed this value for NT-26 (Bernhardt and Santini, 2006). The V_{max} measured with the NT-26, NT-14 and *A. faecalis* enzymes and DCPIP are also very low compared with those measured using the cytochromes from 22, c_{552} and c_{554} (Table 7.1).

The systematic use of physiological electron acceptors should perhaps be considered for future kinetic analysis of Aio. The fact that c-type cytochromes have been co-purified with the enzymes, that cytochrome c-encoding genes are often present in the *aio* gene clusters and that the As(III)-oxidation process in *Ochrobactrum tritici* requires the cytochrome c encoded by the *aio* operon (Branco *et al.*, 2009) indicate that the catalytic cycle of As(III) oxidation results in the reduction of a soluble cytochrome (Lieutaud *et al.*, 2010) in Proteobacteria. Another proteinaceaous electron carrier, azurin, has been shown to accept electrons from Aio (Anderson *et al.*, 1992; Lieutaud *et al.*, 2010). This protein, however, has not been shown to be a physiological electron acceptor to the enzyme. The only in-depth work performed with c-type cytochromes has been published by Lieutaud *et al.* (2010) on the 22 enzyme. Their data challenged not only the calculated K_m but also the determined pH optimum obtained with DCPIP (pH 6). With c_{552} and c_{554}, co-purified with 22 Aio, the proposed pH optimum value is $\sim$8. However, the kinetics of the enzyme from NT-26 has been investigated using PFV. This technique provides a unique advantage over traditional kinetic assays since the redox processes occurring are not masked by the redox potential of an artificial electron-transfer mediator. Instead, the enzyme's redox chemistry can be measured directly. Using cyclic voltammetry (independent of any electron acceptor), Hoke *et al.* (2004) and Bernhardt and Santini (2006) determined a pH optimum value (pH 6) and an electrochemical K_m value (46 μM) similar to those reported for the enzyme in solution using DCPIP.

Enzyme-inhibitory studies can provide useful information about the enzyme's active site, but little is known about potential inhibitors of Aio. Prasad *et al.* (2009) determined that 1 mM

Table 7.2.   Selectivity of Aio towards cytochromes.

	Bovine heart c	A. aeolicus c_{555}	R. NT-26 c_{552}	R. 22 c_{554}	P. aeruginosa azurin	DCPIP
Rhizobium NT-26 Aio[1]	++	−	++	−	−	+
Ralstonia 22 Aio[2]	−	++	−	++	++	+
H. arsenicoxydans Aio[3]	−	++	−	++	++	+
A. faecalis Aio[4]	−	++	−	++	++	+

+ indicates detected activity using DCPIP, ++ indicates detected activity using a corresponding cytochrome, − denotes lack of activity. [1]+/++ = Santini and vanden Hoven, 2004. − = Schoepp-Cothenet, 2011 unpublished data. [2]Lieutaud et al., 2010. [3]Schoepp-Cothenet, 2011 unpublished data. [4]Schoepp-Cothenet, 2011 unpublished data.

Co^{2+} and Zn^{2+} inhibited the enzyme. Because H_2S was reported to inhibit As(III) oxidation by *Hydrogenobaculum* whole cells, sulfide and sulfite were tested (Donahoe-Christiansen et al., 2004; Lieutaud et al., 2010). Whereas sulfide showed an i_{50} (sulfide concentration yielding 50% inhibition) of about 70 μM, sulfite appeared to strongly inhibit the Aio with an i_{50} of 10 μM with a 'mixed mode of inhibition', i.e. with an effect not only on affinity (K_m) but also on catalysis (K_m/V_{max}).

While other inhibitors are being investigated, kinetics of the enzyme is being investigated using various electron acceptors (*c*-type cytochromes or azurin) (Anderson et al., 1992, vanden Hoven and Santini, 2004; Santini et al., 2007) that have been co-purified. And since cytochrome *c*-encoding genes are often present in the *aio* gene clusters – systematic use of natural electron acceptors should be considered for future kinetic studies. In addition, Aio displays strong selectivity toward its electron-transfer partners which is described in more detail below.

7.2.3.2   Arsenite oxidase displays strong selectivity toward its electron-transfer partners

The large majority of known proteobacterial *aio* clusters contain a gene encoding a *c*-type cytochrome (Muller, 2004; Kashyap et al., 2006; Santini et al., 2007; Slyemi et al., 2008; Banco et al., 2009; Cai et al., 2009; Lieutaud et al., 2010) but this gene is not present in all proteobacterial clusters, and does not occur in the clusters of strains in any other phylogenetic lineages. Kinetic studies of the Proteobacteria have shown that Aio displays strong selectivity among the electron transfer partners listed in Table 7.2 (Anderson et al., 1992; Santini et al., 2007; Lieutaud et al., 2010).

Since 22, *H. arsenicoxydans* and *A. faecalis* belong to the Betaproteobacteria whereas NT-26 belongs to the Alphaproteobacteria, these kinetics results suggest that the selectivity of Aio towards cytochromes could be related to phylogeny. BLAST results reinforce this hypothesis, establishing that each of the betaproteobacterial cytochromes detected in *aio* clusters or isolated together with the enzymes displays higher sequence similarity with other betaproteobacterial cytochromes than with alphaproteobacterial or mitochondrial cytochromes (vanden Hoven and Santini, 2004; Santini et al., 2007). Consequently, we might predict that the NT-14 cytochrome, a betaproteobacterial one, would interact with the As(III) oxidases from 22, *H. arsenicoxydans* and *A. faecalis* but not with the NT-26 enzyme. This is, however, in conflict with the published data on NT-14 (vanden Hoven and Santini, 2004), since its partially characterized physiological electron acceptor, c_{551}, could accept electrons from NT-26 Aio. The observed selectivity is particularly striking since it does not appear to be related to the redox potential of the carrier [all the above cited carriers have E_mpH 8 values of around +240 mV (Santini et al., 2007; Lieutaud et al., 2010)]. The structural basis for this selectivity, observed in all examined As(III) oxidases remains to be determined.

7.2.4 *Conclusions*

Aio, the enzyme that catalyzes the oxidation of As(III) to As(V), is distantly related to Arx found in some anaerobic arsenite oxidizers (see Chapters 6 and 10). Aio has been purified from representatives of the Alpha- and Betaproteobacteria (Table 7.1) and is a heterodimer consisting of a large α subunit which contains a Mo atom at its active site and a [3Fe-4S] cluster and a small β subunit which contains a Rieske [2Fe-2S] cluster. The enzyme is normally located in the periplasm and is exported by the Tat pathway with the leader sequence on the β subunit directing the export of the fully folded αβ complex. Aio is a member of the DMSO reductase family of molybdenum-containing enzymes with the large subunit sharing similarities to the α subunit of the formate dehydrogenase and assimilatory nitrate reductase. The small Aio subunit however shares similarities to the Rieske subunit of the cytochrome bc_1 complex. The crystal structure of the *A. faecalis* Aio shows that the Mo is coordinated by four *cis*-dithiolene sulfur atoms of the two MGD molecules but unlike other members of the DMSO reductase family there is no covalent bond between the Mo and the protein. This finding makes the overall coordination more similar to that of W-containing enzymes, which also possess two equivalents of the pterin cofactor and no polypeptide ligand to the metal. The other similarity between the Aio and W-containing enzymes is that the oxidized enzyme contains a dioxo coordination to the Mo (not yet proven in the Aio but the most likely). Although no W-containing versions of the Aio have been identified to date, one can speculate that the ancestral (perhaps pre-LUCA) version of the enzyme contained W instead of Mo.

A variety of spectroscopic techniques have been used to elucidate the role of the redox centers in electron transfer. As(III) oxidation to As(V) results in the transfer of two electrons, the reaction occurs at the Mo active site and results in the reduction of Mo(VI) to Mo(IV). From the active site one electron at a time is transferred from the [3Fe-4S] to the Rieske [2Fe-2S] cluster and then on to a soluble electron carrier (e.g., *c*-type cytochrome and azurin). The high potential [3Fe-4S] cluster is not usually found in molybdenum-containing enzymes and instead the lower [4Fe-4S] cluster is present. Likewise Aio is the only example of a molybdenum-containing enzyme with a Rieske [2Fe-2S] cluster. Many attempts have been made to visualize the Mo(V) signal by EPR with no success and thus no information is available on the redox potential of the Mo center. Efforts will be made to study the enzyme structure and As(III) oxidation mechanism using a combination of crystallography, spectroscopy and mutagenesis.

Mo- and W-containing enzymes represent an important group of enzymes in biology. Study of Aio in different model systems will not only contribute significantly to the understanding of enzymology within this field but will provide the framework for industrial applications such as the use of these enzymes for bioremediation or biosensors. The ability to engineer enzymes for biotechnological purposes requires the detailed understanding of the enzyme mechanism and the ability to express the enzyme in a suitable host such as *E. coli*. We have recently achieved the latter and will now endeavor to use this to not only improve our understanding of enzyme function but to engineer the enzyme to improve its suitability as a biosensor.

ACKNOWLEDGEMENTS

MDH is funded by the Wellcome Trust PhD Interdisciplinary Program in Structural, Chemical and Computational Biology. THO is funded by the National Environment Research Council. We would like to thank Russ Hille for his constructive comments on the manuscript.

REFERENCES

Anderson, G., Williams, J. & Hille, R.: The purification and characterization of arsenite oxidase from *Alcaligenes faecalis*, a molybdenum-containing hydroxylase. *J. Biol. Chem.* 267 (1992), pp. 23674–23682.

Bachmann, J., Bauer, B., Zwicker, K., Ludwig, B. & Anderka, O.: The Rieske protein from *Paracoccus denitrificans* is inserted into the cytoplasmic membrane by the twin-arginine translocase. *FEBS J.* 273:21 (2006), pp. 4817–4830.

Bastian, N.R., Kay, C.J., Barber, M.J. & Rajagopalan, K.V.: Spectroscopic studies of the molybdenum-containing dimethyl sulfoxide reductase from *Rhodobacter sphaeroides* f. sp. *denitrificans*. *J. Biol. Chem.* 266:1 (1991), pp. 45–51.

Benson, N., Farrar, J.A., McEwan, A.G. & Thomson, A.J.: Detection of the optical bands of molybdenum(V) in DMSO reductase (*Rhodobacter capsulatus*) by low-temperature MCD spectroscopy. *FEBS Lett.* 307:2 (1992), pp. 169–172.

Bernhardt, P.V. & Santini, J.M.: Protein film voltammetry of arsenite oxidase from the chemolithoautotrophic arsenite-oxidizing bacterium NT-26. *Biochemistry* 45 (2006), pp. 2804–2809.

Berks, B.C., Palmer, T. & Sargent, F.: Protein targeting by the bacterial twin-arginine translocation (Tat) Pathway. *Curr. Opin. Microbiol.* 8:2 (2005), pp. 174–181.

Boyington, J.C., Gladyshev, V.N., Khangulov, S.V., Stadtman, T.C. & Sun, P.D.: Crystal structure of formate dehydrogenase H: catalysis involving Mo, molybdopterin, selenocysteine, and an Fe4s4 cluster. *Science* 275:5304 (1997), pp. 1305–1308.

Branco, R., Francisco, R., Chung, A.P. & Morais, P.V.: Identification of an *aox* system that requires cytochrome *c* in the highly arsenic-resistant bacterium *Ochrobactrum tritici* SCII24. *Appl. Environ. Microbiol.* 75:15 (2009), pp. 5141–5147.

Cai, L., Liu, G., Rensing, C. & Wang, G.: Genes involved in arsenic transformation and resistance associated with different levels of arsenic-contaminated soils. *BMC Microbiol.* 9:4 (2009).

Cao, H., Hall, J. & Hille, R.: X-Ray crystal structure of arsenite-inhibited xanthine oxidase: M-sulfido, M-oxo double bridge between molybdenum and arsenic in the active site. *J. Am. Chem. Soc.* (2011).

Conrads, T., Hemann, C., George, G.N., Pickering, I.J., Prince, R.C. & Hille, R.: The active site of arsenite oxidase from *Alcaligenes faecalis*. *J. Am. Chem. Soc.* 124:38 (2002), pp. 11276–11277.

Donahoe-Christiansen, J., D'Imperio, S., Jackson, C.R., Inskeep, W.P. & McDermott, T.R.: Arsenite-oxidizing *Hydrogenobaculum* strain isolated from an acid-sulfate-chloride geothermal spring in Yellowstone National Park. *Appl. Environ. Microbiol.* 70:3 (2004), pp. 1865–1868.

Duval, S., Santini, J.M., Nitschke, W. & Schoepp-Cothenet, B.: Spectroscopic study of arsenite oxidase. *6th Gordon Research Conference on Molybdenum and Tungsten Enzymes*, Lucca, 2008.

Duval, S., Santini, J.M., Nitschke, W., Hille, R. & Schoepp-Cothenet, B.: The small subunit AroB of arsenite oxidase: lessons on the [2Fe-2S]-Rieske protein superfamily. *J. Biol. Chem.* 285:27 (2010), pp. 20442–20451.

Ellis, P.J., Conrads, T., Hille, R. & Kuhn, P.: Crystal structure of the 100 kDa arsenite oxidase from *Alcaligenes faecalis* in two crystal forms at 1.64 Å and 2.03 Å. *Structure* 9 (2001), pp. 125–132.

Finnegan, M.G., Hilton, J., Rajagopalan, K.V. & Johnson, M.K.: Optical transitions of molybdenum(V) in glycerol-inhibited DMSO reductase from *Rhodobacter sphaeroides*. *Inorg. Chem.* 32 (1993), pp. 2616–2617.

Ferraro, D.J., Gakhar, L. & Ramaswamy, S.: Rieske business: Structure-function of Rieske non-heme oxygenases. *Biochem. Biophys. Res. Comms.* 338:1 (2005), pp. 175–190.

Fisher, J.C. & Hollibaugh J.T.: Selenate-dependent anaerobic arsenite oxidation by a bacterium from Mono lake, California. *Appl. Environ. Micriobiol.* 74 (2008), pp. 2588–2594.

Hoeft, S.E., Blum, J.S., Stolz, J.F., Tabita, F.R., Witte, B., King, G.M,. Santini, J.M. & Oremland, R.S.: *Alkalilimnicola ehrlichii* sp. nov., a novel, arsenite-oxidizing haloalkaliphilic gammaproteobacterium capable of chemoautotrophic or heterotrophic growth with nitrate or oxygen as the electron acceptor. *Int. J. Sys. Evol. Microbiol.* 57 (2007), pp. 504–512.

Hoke, K.R., Cobb, N., Armstrong, F.A. & Hille, R.: Electrochemical studies of arsenite oxidase: an unusual example of a highly cooperative two-electron molybdenum center. *Biochemistry* 43 (2004), pp. 1667–1674.

Jepson, B.J.N., Anderson, L.J., Rubio, L.M., Taylor, C.J., Butler, C.S., Flores, E., Herrero, A., Butt, J.N. & Richardson, D.J.: Tuning a nitrate reductase for function. The first spectropotentiometric characterization of a bacterial assimilatory nitrate reductase reveals novel redox properties. *J. Biol. Chem.* 279:31 (2004), pp. 3212–3218.

Kashyap, D.R., Botero, L.M., Franck, W.L., Hasset, D.J. & McDermott, T.R.: Complex regulation of arsenite oxidation in *Agrobacterium tumefaciens*. *J. Bacteriol.* 188:3 (2006), pp. 1081–1088.

Kulp, T.R., Hoeft, S.E., Aroa, M., Madigan, M.T., Hollibaugh, J.T., Fisher, J.C., Stolz, J.F., Culbertson, C.W., Miller, L.G. & Oremland, R.S.: Arsenic (III) fuels anoxygenic photosynthesis in hot spring biofilms from Mono Lake, California. *Science* 321 (2008), pp. 967–970.

Lebrun, E., Brugna, M., Baymann, F., Muller, D., Lièvremont, P., Lett, M.-C. & Nitschke, W.: Arsenite oxidase, an ancient bioenergetic enzyme. *Mol. Biol. Evol.* 20 (2003), pp. 686–693.

Lieutaud, A., van Lis, R., Duval, S., Capowiez, L., Muller, D., Lebrun, R., Lignon, S., Fardeau, M.L., Lett, M.C., Nitschke, W. & Schoepp-Cothenet, B.: Arsenite oxidase from *Ralstonia* sp. 22: characterization of the enzyme and its interaction with soluble cytochromes. *J. Biol. Chem.* 285:27 (2010), pp. 20433–20441.

Link, T.A.: The structures of Rieske and Rieske-type proteins. In: A.G. Sykes & R. Cammack (eds): *Advances in inorganic Chemistry. Iron-sulfur proteins.* Academic Press, San Diego, CA, USA, 47 (1999), pp. 83–157.

Maklashina, E. & Cecchini, G.: Comparison of catalytic activity and inhibitors of quinone reactions of succinate dehydrogenase (Succinate-ubiquinone oxidoreductase) and fumarate reductase (Menaquinol-fumarate oxidoreductase) from *Escherichia coli. Arch. Biochem. Biophys.* 369 (1999), pp. 223–232.

McNellis, L. & Anderson G.L.: Redox-state dependent chemical inactivation of arsenite oxidase. *J. Inorg. Biochem.* 69:4 (1998), pp. 253–257.

Muller, D.: *Analyse génétique et moléculaire du stress arsenic de souches bactériennes isolées d'environnements contaminés par l'arsenic.* PhD Thesis, Université de Strasbourg, France, 2004.

Prasad, K.S., Subramanian, V. & Paul, J.: Purification and characterization of arsenite oxidase from *Arthrobacter* sp. *Biometals* 22:5 (2009), pp. 711–721.

Romao, M.J.: Molybdenum and tungsten enzymes: A crystallographic and mechanistic overview. *Dalton Trans.* 21 (2009), pp. 4053–4068.

Rothery, R.A., Bertero, M.G., Cammack, R., Palk, M., Blasco, F., Strynadka, N.C.J. & Weiner, J.H.: The catalytic subunit of *Escherichia coli* nitrate reductase A contains a novel [4Fe-4S] cluster with a high-spin ground state. *Biochemistry* 43 (2004), pp. 5324–5333.

Rothery, R.A., Workun, G.J. & Weiner, J.H.: The prokaryotic complex iron-sulfur molybdoenzyme family. *Biochim. Biophys. Acta* 1778:9 (2008), pp. 1897–1929.

Santini, J.M. & vanden Hoven, R.N.: Molybdenum-containing arsenite oxidase of the chemolithoautotropic arsenite oxidizer NT-26. *J. Bacteriol.* 186:6 (2004), pp. 1614–1619.

Santini, J.M., Kappler, U., Ward, S.A., Honeychurch, M.J., vanden Hoven, R.N. & Bernhardt, P.V.: The NT-26 cytochrome c_{552} and its role in arsenite oxidation. *Biochim. Biophys. Acta* 1767:2 (2007), pp.189–196.

Schoepp, B., Brugna, M., Lebrun, E. & Nitschke, W.: Iron-sulfur centers involved in photosynthetic light reactions. In: A.G. Sykes & R. Cammack (eds): *Advances in inorganic chemistry.* Academic Press, San Diego, CA, USA, 47 (1999), pp. 335–360.

Schultz, B.E., Hille, R. & Holm, R. H.: Direct oxygen atom transfer in the mechanism of action of *Rhodobacter sphaeroides* dimethyl sulfoxide reductase. *J. Am. Chem. Soc.* 117:2 (1995), pp. 827–828.

Slyemi, D., Ratouchniak, J. & Bonnefoy, V.: Regulation of the arsenic oxidation encoding genes of a moderately acidophilic, facultative chemolithoautotrophic *Thiomonas* sp. *Adv. Mat. Res.* 20–21 (2008), 427–430.

Sun, W., Sierra-Alvarez, R., Milner, L. & Field, J.A.: Anaerobic oxidation of arsenite linked to chlorate reduction. *Appl. Environ. Microbiol.* (2010), pp. 6804–6811.

vanden Hoven, R.N. & Santini, J.M.: Arsenite oxidation by the heterotrophy *Hydrogenophaga* sp. str. NT-14: the arsenite oxidase and its physiological electron acceptor. *Biochim. Biophys. Acta* 1656 (2004), pp. 148–155.

Xia, D., Yu, C.A., Kim, H., Xia, J.Z., Kachurin, A.M., Zhang, Li., Yu, L. & Deisenhofer, J.: Crystal structure of the cytochrome Bc1 complex from bovine heart mitochondria. *Science* 277:5322 (1997), pp. 60–66.

CHAPTER 8

Microbial arsenic response and metabolism in the genomics era

Philippe N. Bertin, Lucie Geist, David Halter, Sandrine Koechler,
Marie Marchal & Florence Arsène-Ploetze

8.1 INTRODUCTION

By their ability to transfer elements through biotic and abiotic compartments of Earth, microorganisms are key players in biogeochemical cycles. Those living in heavily polluted environments have various means of dealing with high concentrations of toxins, and thus detoxifying their environment. The processes involved include not only physical changes – such as precipitation or solubilization, and adsorption or desorption (Borch *et al.*, 2010) – but also redox reactions (Gadd, 2010). Indeed, most metallic elements can play an important role in microbial physiology. Some are essential components of metalloproteins, while microorganisms can use others as electron donors or acceptors, in the reduction or oxidation of organic or inorganic substrates. Although, unlike many other metals, arsenic (As) cannot be regarded as essential for life, it does have an important role in the metabolism of various prokaryotes. On primordial Earth, As-based metabolism may have been important in the early stages of life in mineral-rich waters (Kulp *et al.*, 2008).

Until recently, the study of microbial metabolism required the isolation and culturing of specific microorganisms from samples collected in the field. The advent of genomics has revolutionized the field of microbial ecology, as it allows us to examine the genes present in whole microbial communities, rather than focusing on the description and classification of single organisms (Bertin *et al.*, 2008). This development is of great importance because the response of microbial communities to environmental factors is usually multifactorial, depending on the coordinated expression of sets of genes in different organisms. Currently, the most studied genomes belong to three bacterial lineages, i.e. Proteobacteria, Firmicutes and Actinobacteria, which represent the majority of known bacteria. However, recent sequencing efforts concern not only the genomics of microorganisms isolated from various environments and cultivated in laboratory conditions but also inventories of microbial communities containing uncultured microbes potentially expressing novel functions (http://www.genomeonline.org). The "omics" approaches, which include genome sequencing and comparison, as well as transcriptome or proteome profiling, should therefore improve the understanding of microbial metabolism, its diversity and its impact *in situ* on the functioning of ecosystems (Bertin *et al.*, 2008; Holmes *et al.*, 2009; Wilkins *et al.*, 2009).

In this review, we will focus on recent advances in descriptive and functional genomics, with reference to the metabolism of As and aspects of adaptive responses to its presence in the environment.

8.2 DESCRIPTIVE AND COMPARATIVE GENOMICS

The genomes of several As-metabolizing isolates from various environments have been or are currently being sequenced. They belong to different taxonomic groups, and have different carbon and energy metabolism (Table 8.1). In addition to their resistance to As, these microorganisms also express functions involved in other As transformations such as oxidation or methylation.

Resistance to As depends mainly on reductases associated with an efflux pump and is widespread in the microbial world (Silver and Phung, 2005). Even in the eukaryote *Saccharomyces cerevisiae*, the reductase genes *ACR* are functionally similar to the prokaryotic *arsRDABC* resistance genes (Mukhopadhyay *et al.*, 2002). Their low sequence and structural homology, however, suggests evolutionary convergence, highlighting the effectiveness of such systems against As stress (Mukhopadhyay *et al.*, 2002; Páez-Espino *et al.*, 2009). In addition, a second family of transporters of As compounds has been identified in *S. cerevisiae*, and includes the YCF1 gene, encoding an ATP-dependent transporter of As into vacuoles (Ghosh *et al.*, 1999).

As(III) oxidation serves principally as a detoxification mechanism in heterotrophic bacteria like *Herminiimonas arsenicoxydans* (Muller *et al.*, 2007). In contrast, in autotrophic bacteria, such as *Thiomonas* sp. or *Rhizobium/Agrobacterium* sp. NT-26, As(III) oxidation is involved in energy metabolism: these strains grow chemolithoautotrophically by expressing oxidation genes in the presence of As(III) (Santini and vanden Hoven, 2004; Bryan *et al.*, 2009). In eukaryotes, only the red alga *Cyanidioschyzon* sp. is thought to oxidize As(III) to As(V), but it has not been shown that this process requires an enzymatically catalyzed reaction (Qin *et al.*, 2009). Whether it does or not, the oxidation may be a side-effect of oxygen production by photosynthesis. What is clear is that no eukaryotic gene conferring the ability to oxidize As(III) has yet been isolated.

Finally, As(III) methylation depends on methyltransferases. As(III) methyltransferase is known to successively transfer three methyl groups from S-adenosylmethionine to As(III), leading to the production of various volatile methylated species, e.g. monomethylarsine (MMA), dimethylarsine (DMA) or trimethylarsine oxide (TMAO). The heterologous expression of eukaryotic arsenite methyltransferases in bacteria confers increased tolerance to As (Qin *et al.*, 2009; Yuan *et al.*, 2008), suggesting that methylation might be a detoxification mechanism. However, while As methylation is widespread in eukaryotes, arsenite methyltransferase genes are rarely found in bacterial genomes.

8.2.1   *Genome exploration of cultured microorganisms*

Herminiimonas arsenicoxydans is the first microorganism involved in As redox reactions whose genome was sequenced (Table 8.1). This Betaproteobacterium uses organic compounds as an electron donor, oxidizes As(III) and can resist up to 6 mM As(III) and 200 mM As(V) (Muller *et al.*, 2007). These high levels of resistance may be explained, at least partly, by the fact that *H. arsenicoxydans* has three As gene clusters containing arsenate reductase (*arsC*) and efflux pump (*arsB*) encoding genes (Table 8.2). Unusually, the three clusters are on its single chromosome, whereas the existence of multiple copies of resistance genes is usually associated with extrachromosomal DNA, e.g. the megaplasmids in *Cupriavidus metallidurans* (Janssen *et al.*, 2010; Mergeay *et al.*, 2003).

One of these clusters is flanked by insertion sequences, and contains both *aio* genes involved in As(III) oxidation and *pst* genes involved in high-specificity phosphate transport. Given that genes with related functions are often co-localized on the genome and expressed together (Ettema *et al.*, 2005), this suggests a possible link between As(III) oxidation and phosphate transport, perhaps related to the structural similarity between the two elements: the *pst* operon may help to maintain a sufficient intracellular phosphate concentration despite high levels of As(V) produced by As(III) oxidation (Muller *et al.*, 2007; Cleiss-Arnold *et al.*, 2010). A second *ars* locus is found in a putative genomic island (GEI), together with genes involved in heavy metal resistance. This GEI may thus have a general role in adaptive responses to stressful environments, as do many GEIs found in environment-specific strains (Dufresne *et al.*, 2008; Qin *et al.*, 2011), and may be considered as similar to the pathogenicity-associated islands in pathogenic strains. Such genomic islands may group genes regulating cell motility and/or biofilm formation and thus play an important role in the ecology of microorganisms (Juhas *et al.*, 2009).

The association of As metabolism with transposons or GEI raises the possibility that these genes may have been acquired by horizontal gene transfer (HGT). This association is not restricted to *H. arsenicoxydans*: *Thiomonas* strains, which are widespread in As-rich acid mine drainage (AMD)

Table 6.1 Sequenced genomes of prokaryotes metabolizing As.

Organism	Origin	General metabolism[b]	Genome available	Genome sequencing center or link	GenBank	Strain or genome reference
Herminiimonas arsenicoxydans	Activated sludge of an industrial water treatment plant, Germany	Chemoorganotroph heterotroph	Yes	https://www.genoscope.cns.fr/agc/microscope/arsenoscope	CU207211	Muller et al., 2007
Thiomonas arsenitoxydans 3As	AMD[a] Carnoulès, Gard, France	Chemolithoautotroph mixotroph	Yes	https://www.genoscope.cns.fr/agc/microscope/arsenoscope	FP475956-FP475957	Arsène-Ploetze et al., 2010
Thiomonas intermedia K12	Sewer, Germany	Chemolithoautotroph mixotroph	Yes	DOE[d], Joint Genome Institute	CP002021	Milde et al., 1983
Rhizobium sp. NT-26	Gold mine, Australia	Facultative Chemolithoautotroph	Finished	https://www.genoscope.cns.fr/agc/microscope/arsenoscope	–	Santini et al., 2000
Thermus thermophilus HB8	Hot spring, Japan	Heterotroph	Yes	DOE, Joint Genome Institute	AP008226	Henne et al., 2004
Alkalilimnicola ehrlichei MLHE-1	Mono Lake, CA, USA	Chemolithoautotroph or heterotroph	Yes	DOE, Joint Genome Institute	CP000453	Oremland et al., 2002
Alkaliphilus oremlandii OhILAs	Ohio River sediments, Pittsburgh PA, USA	Chemoorganotroph heterotroph	Yes	DOE, Joint Genome Institute	CP000853	Fisher et al., 2008
Bacillus selenitireducens MLS10	Anoxic muds of Mono Lake, CA, USA	Chemoorganotroph	Yes	DOE, Joint Genome Institute	CP001791	Switzer Blum et al., 1998
Sulfurospirillum barnesii SES-3	Selenate contaminated freshwater marsh in western Nevada, USA	Chemoorganotroph	Finished	DOE, Joint Genome Institute	–	Oremland et al., 1994
Wolinella succinogenes DSM 1740	Bovine rumen fluid	Chemoorganotroph	Yes	Max Planck Institute	BX571656	Baar et al., 2003
Pyrobaculum arsenaticum DSM 13514	Hot spring at Pisciarelli Solfatara, Naples, Italy	Chemolithoautotroph or organotroph	Yes	DOE, Joint Genome Institute	CP000660	Huber et al., 2000
Shewanella sp. ANA-3	Wooden pier within a brackish estuarine environment at Woods Hole, MA, USA	Chemoorganotroph heteroptroph	Yes	DOE, Joint Genome Institute	CP000469	Saltikov et al., 2003
Chrysiogenes arsenatis DSM 11915	Gold mine wastewater from the Ballarat Goldfields, Australia	Chemoorganotroph heterotroph	Finished	J. Craig Venter Institute	–	Macy et al., 1996
δ-proteobacterium MLMS-1	Mono Lake, CA, USA	Chemolithoautotroph	Abandoned	DOE, Joint Genome Institute	AAQF01000000	Hoeft et al., 2004

[a] AMD: Acid Mine Drainage; [b] organotroph: uses organic compounds as an electron donor; heterotroph: uses organic compounds as a carbon source; mixotroph: can use a mix of different energy and carbon sources; [c] Yes: genome is available in public databases; Finished: genome is sequenced but not available in public databases; [d] DOE: U.S. Department of Energy.

Table 8.2.   Genes involved in arsenic metabolism and/or resistance in the sequenced genomes of known As-metabolizing prokaryotes.

Organism	aioBA[a]	arsBC or acr3: number of copies[b]	arrAB[c]	arxA[d]	arsM[e]	As metabolism gene on GEI or plasmid (P)[f]	Strain or genome reference
Herminiimonas arsenicoxydans	+	3	NF	NF	NF	GEI	Muller et al., 2007
Thiomonas arsenitoxydans 3As	+	2	NF	NF	NF	GEI	Arsène-Ploetze et al., 2010
Thiomonas intermedia K12	+	1	NF	NF	NF	No	Milde et al., 1983
Rhizobium sp. NT-26	+	2	NF	NF	NF	P	Santini et al., 2000
Thermus thermophilus HB8	+	0	NF	NF	NF	P	Henne et al., 2004
Alkalilimnicola ehrlichei MLHE-1	NF	2	+	+	NF	ND	Oremland et al., 2002
Alkaliphilus oremlandii OhILAs	NF	1 *arsC*	+	NF	NF	–	Fisher et al., 2008
Bacillus selenitireducens MLS10	NF	1	+	NF	NF	ND	Switzer Blum et al., 1998
Sulfurospirillum barnesii SES-3	NF	NF	+	NF	NF	ND	Oremland et al., 1994
Wolinella succinogenes DSM 1740	NF	1	+	+	NF	–	Baar et al., 2003
Pyrobaculum arsenaticum DSM 13514	NF	2	+	NF	NF	–	Huber et al., 2000
Shewanella sp. ANA-3	NF	3	+	NF	NF	ND	Saltikov et al., 2003
Chrysiogenes arsenatis DSM 11915	ND	ND	+	ND	ND	ND	Macy et al., 1996
δ- proteobacterium MLMS-1	ND	ND	+	ND	ND	ND	Hoeft et al., 2004

[a] *aioBA* (also called *aroBA*, *asoBA* or *aoxAB*) encode the arsenite oxidase; [b] *arsBC* encode an As(III) extrusion pump, as *acr3*, and an arsenate reductase, respectively; [c] *arrAB* encode the respiratory arsenate reductase; [d] *arxA* encode a new arsenite oxidase divergent from *aioAB*; [e] *arsM* encodes an arsenite methyltransferase; [f] GEI: Genomic Island, and P: Plasmid. ND: not demonstrated; NF: no homologous gene was found in the genome.

ecosystems, also contain specific genomic islands. Recent CGH chip experiments have found genes involved in resistance to As and heavy metals clustered together in genomic islands (Arsène-Ploetze *et al.*, 2010): these GEI contain genes involved in resistance to As (*aio* and *ars* genes), Cd, Co, Zn, Cu, Ag, and Hg, together with others regulating motility and biofilm formation. Such GEI thus confer a suite of adaptations to specific environments, and their acquisition may have a fundamental role in the plasticity and the evolution of bacterial genomes (Juhas *et al.*, 2009), including those of As-metabolizing strains.

Other As-metabolizing bacteria may have multiple copies of the genes involved in As(III) resistance and oxidation. In some cases, these are thought to occur in genomic islands or plasmids, and may thus have been acquired by HGT processes (Table 8.2). Recently, arsenite oxidase (*aio*) genes were found in regions predicted to correspond to genomic islands, in the putative arsenite oxidizers *Vibrio splendidus* and *Pyrobaculum calidifontis*, or on (mega)plasmids in *Nitrobacter hamburgensis*, *Thermus thermophilus*, or *Rhizobium* sp. NT-26 (Heinrich-Salmeron *et al.*, 2011), but the ability of several of them to oxidize As(III) remains to be determined. So far, more than 200 bacterial or archaeal *aioA*-like genes have been characterized (Inskeep *et al.* 2007; Quéméneur *et al.*, 2008), and phylogenetic analysis suggests that HGT has had an important role: analysis of the phylogeny of *aioA* has revealed that it was probably transferred from Alpha- to Betaproteobacteria and Actinobacteria and from Betaproteobacteria to Gammaproteobacteria, Chlorobi/Bacteroidetes, Firmicutes and Actinobacteria (Heinrich-Salmeron *et al.*, 2011). Indeed, comparisons between 16S rRNA and AioA phylogenies showed the existence of some striking inconsistencies between organism and gene evolutionary histories. The existence of HGT events represents the most likely hypothesis to explain the presence of identical (or nearly identical) AioA sequences in some distant lineages. For example, some *Pseudomonas*, *Marinobacter*, *Halomonas*, and also Firmicutes and Chlorobi/Bacteroidetes AioA sequences were intermixed with those from Betaproteobacteria, suggesting that the former acquired their *aioA* gene from the latter by HGT.

Other genes involved in As metabolism have been characterized in bacterial genomes. The *arxA* gene, which encodes a novel oxidase that couples As(III) oxidation to nitrate reduction in the Gammaproteobacterium *Alkalilimnicola ehrlichii* MLHE-1 (Zargar *et al.*, 2010), was also found in the genome of *Wolinella succinogenes*, a bacterium belonging to the Delta/Epsilon subdivision. Moreover, an additional arsenate reductase associated with As respiration (ArrA) has been identified in *Shewanella* sp. strain ANA-3 (Saltikov and Newman, 2003) and in other phylogenetically distant bacteria (Duval *et al.*, 2008). These observations demonstrate that the diversity of As-metabolizing bacteria is larger than expected, and far from being fully explored.

As the data on gene phylogeny and the distribution of As-metabolism genes suggest: while ancient HGT between phylogenetically distant groups has been important in determining large-scale patterns, the more recent acquisition of As-metabolism genes on plasmids or in GEI may have been a major contributor to the diversity of ecotypes in As-contaminated ecosystems (Arsène-Ploetze *et al.*, 2010; Bertin *et al.*, 2011).

8.2.2 *Metagenomics: culture-independent characterization of arsenic-metabolizing microbial communities*

Unlike model organisms that can be easily cultivated in laboratory conditions, many micro-organisms are difficult or impossible to culture, so the study of microbial communities has been severely hindered. Recently, however, it has become possible to screen whole-community DNA libraries to find specific genes of interest, including those involved in As resistance (Chauhan *et al.*, 2009), or even to sequence the whole genomic DNA content (Bertin *et al.*, 2011). To date, more than 300 metagenome projects are referenced in specialized databases, including those focusing on the archaeal community of the Richmond mine and the bacterial commu-nity of the Carnoulès mine (http://www.genomeonline.org). Despite the difficulties in analyzing, interpreting and comparing metagenomic data (Foerstner *et al.*, 2006), and depending on the microbial diversity in the environments under study, it may still be possible to reconstruct some whole-community genomes. Questions such as which organisms are present and what metabolic functions are involved in biotransformations, including those of As, can now be addressed at a molecular level.

Regarding ecosystems contaminated by metallic elements, two studies have so far been published in environmental genomics: the metagenomes of an AMD biofilm (Tyson *et al.*, 2004), and an As-rich stream (Bertin *et al.*, 2011). The former study yielded two dominant strains: a bacterium, *Leptospirillum*, and an archaean, *Ferroplasma*. *Leptospirillum* was represented by several conspecific strains differing in nucleotide sequence by less than 1%. This limited genetic diversity may however lead to functional diversity, as the strains may play different roles. The different ecotypes may indeed express different metabolic pathways, so competition is weakened and coexistence permitted (Denef *et al.*, 2010a; Denef *et al.*, 2010b; Simmons *et al.*, 2008; Wilmes *et al.* 2010).

The second environmental genomics study (Bertin *et al.*, 2011) yielded complete descriptions of several microorganisms possibly involved in the natural attenuation of the Carnoulès ecosystem. These microbes significantly reduced the concentration of As by oxidation and co-precipitation with iron and sulfur. The analysis of metagenomic data led to the identification of the genes, i.e. *aio* in *Thiomonas* sp. and *rus* in *Acidithiobacillus* sp., involved in As(III)- and iron-oxidation, respectively. This analysis also allowed the characterization of genetic variants of a novel phylum, "*Candidatus* Fodinabacter comunificans", which may assist the growth of other microorganisms such as *Thiomonas* sp. and *Acidithiobacillus* sp. by recycling or transforming organic matter.

8.3   HIGH-THROUGHPUT GENOMICS REVEAL THE FUNCTIONING OF MICROORGANISMS

Knowing the complete genome of a microorganism might conceivably allow us to determine its responses to its environment. In practice, of course, genome sequencing has revealed a large percentage of identified genes that are unrelated to any known function (Bertin *et al.*, 2008). This difficulty has led to the development of functional genomic approaches, which reveal the functioning of an organism rather than its potential: what it actually does, rather than what its genome might enable it to do. Here, the aim is to infer gene expression, or the accumulation of the resulting products, from analyses of the cell's whole mRNA profile ("transcriptome"), entire protein content ("proteome"), or metabolites ("metabolome").

Gene expression could be studied using classical approaches derived from PCR technology (RT-PCR), but DNA chips (microarrays) are a more powerful tool that allows us to analyze the impact of environmental factors on the transcriptome. In future, the recently developed RNA-sequencing approach based on high-throughput sequencing (Coppée, 2008; Wang *et al.*, 2009) should greatly improve expression profiling.

The cell's protein profile depends not only on transcription rates but also, potentially, on post-translational modification or proteolysis, so proteomics may yield a description of cell functioning that is different from that obtained from transcriptomics. An organism's protein synthesis is usually analyzed by establishing a protein profile after two-dimensional electrophoresis. After visualization of proteins by coloration or fluorescent labeling, polypeptides are identified by mass spectrometry (MS) (Bertin *et al.*, 2008). Currently proteome maps can be obtained directly using MS coupled to liquid chromatography (LC-MS). The effect of a stress on protein synthesis or stability can then be analyzed by differential proteomics, i.e. by comparing the protein profiles obtained from cells grown in different conditions. The recent use of fluorescent labeling (DIGE) greatly facilitates such an approach (Arsène-Ploetze *et al.*, 2011).

Finally, many of the genes expressed by microbial cells code for proteins with enzymatic activity, which can generate a large number of metabolites. Although a full analysis of the metabolome is still impossible, the current rapid progress in techniques of analytical chemistry, such as nuclear magnetic resonance (NMR) or high-resolution liquid chromatography (HPLC), should, in the foreseeable future, allow identification of the various categories of metabolites present in a cell (Weckwerth, 2010; Rubakhin *et al.*, 2011).

These functional genomic methods have recently allowed the complex adaptive responses to As in various microorganisms to be deciphered, at least in part.

8.3.1 *Arsenic response in* Herminiimonas arsenicoxydans

The response to As has been extensively studied in *H. arsenicoxydans*. Parallel studies of changes in its proteome and transcriptome have revealed a range of physiological responses to the presence of As(III). This causes transcriptional up-regulation of genes involved in both As resistance and As(III) oxidation (Weiss *et al.*, 2009). As(III) also induces several other processes (Fig. 8.1): e.g. oxidative stress resistance, DNA repair, flagellum biosynthesis and phosphate transport (Carapito *et al.*, 2006; Muller *et al.*, 2007; Koechler *et al.*, 2010; Cleiss-Arnold *et al.*, 2010).

Detailed transcriptomic analyses have shown that this microbial adaptation to As relies on a two-step response. First, cells resist the toxic effects of As, by the induction of genes involved in oxidative stress resistance and glutathione (GSH) metabolism, and of others required for an efficient efflux of As(III). Next, their response to the presence of toxic metabolites results in an increase in motility and As(III) oxidation. Flagellum synthesis, cell motility and biofilm formation are thus all associated with arsenite oxidase activity in *H. arsenicoxydans* (Muller *et al.*, 2007; Cleiss-Arnold *et al.*, 2010; Marchal *et al.*, 2010).

Biofilm formation is usually associated with the synthesis of exopolysaccharides (EPS), which are known to trap antibacterial compounds – not only charged antibiotics but also metallic ions (Comte *et al.*, 2008; Guibaud *et al.*, 2009; Shigeta *et al.*, 1997; Wei *et al.*, 2011). Electron microscopy and quantitative mRNA analysis have shown that *H. arsenicoxydans* forms a biofilm in the presence of As and sequesters it in an EPS matrix (Muller *et al.*, 2007; Marchal *et al.*, 2010). Consequently, as has also been described for pathogenic strains exposed to antibiotics and metallic ions (Hoffman *et al.*, 2005; Majtán *et al.*, 2008; Ordax *et al.*, 2010), the enhanced exopolysaccharide synthesis observed in the presence of As may correspond to a general stress-resistance mechanism in *H. arsenicoxydans*. The presence of As(III) prolongs this initial stage of biofilm formation (i.e. EPS synthesis), and cell adhesion begins only when all the As(III) has been oxidized to As(V) (Marchal *et al.*, 2010): detoxification of the medium precedes the development of a biofilm, and may thus contribute to the establishment of other microorganisms in the community.

8.3.2 *Arsenic response in other bacteria*

As in *H. arsenicoxydans* (Koechler *et al.*, 2010), As(III) oxidation seems to be associated with biofilm development in *Rhizobium/Agrobacterium* sp. NT-26. Indeed, in the presence of As(III), this strain has been shown to synthesize flagella (Santini *et al.*, 2000), which are known to play a key role in various steps of biofilm formation (Verstraeten *et al.*, 2008). The arsenite oxidase is encoded by a cluster of genes on a large plasmid (Table 8.2), while genes for the synthesis of flagella lie on the chromosome; this suggests a possible cross-talk between clusters present on these two genetic elements.

Analysis of the As response suggests the widespread existence of common behavioral features in As-metabolizing microorganisms. For example, analysis of the As(V)-specific transcriptome of *C. metallidurans* CH34 has shown the induction of genes involved in As resistance, phosphate metabolism, and resistance to oxidative and general stress (Zhang *et al.*, 2009). As(V) induces overexpression of phosphate transporters in *Comamonas* sp. CNB-1, *Leptospirillum ferriphilum* and *Pseudomonas* sp. As-1 (Li *et al.*, 2010; Patel *et al.*, 2006; Zhang *et al.*, 2007). Finally, proteomic analysis of the As(III) response in *L. ferriphilum* and *Thiomonas* sp. (Fig. 8.2) has revealed a pattern similar to that observed in *H. arsenicoxydans*, i.e. an induction of proteins encoded by *ars* operons, and proteins involved in GSH metabolism, DNA repair and protection against oxidative stress (Li *et al.*, 2010; Bryan *et al.*, 2009).

Further distinct As-specific responses are known from other prokaryotes. For example, metabolic differences have been observed between various *Thiomonas* strains. In the presence of As(III), proteins involved in carbon dioxide fixation are preferentially accumulated in

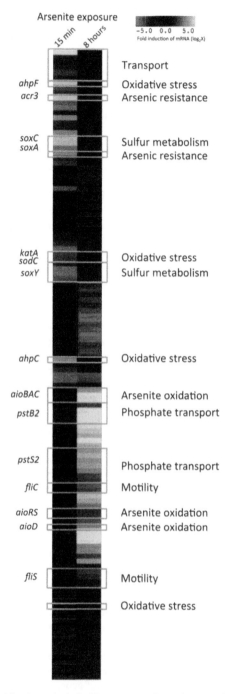

Figure 8.1. Transcriptome kinetic analysis in *H. arsenicoxydans* after 15 min (left column) and 8 hours (right column) exposure to As(III). Yellow and blue colors represent increased and reduced gene expression, respectively. The adaptation of *H. arsenicoxydans* to As relies on a two-step response: the early and late responses are illustrated by genes surrounded by orange and blue boxes respectively. Early response (orange boxes), which is as a general stress response, refers to genes whose expression gradually increases or declines, starting soon after exposure to arsenite. These genes are involved in various processes, including: oxidative stress defense, e.g. *ahpF* and *ahpC* (peroxiredoxin), *katA* (catalase) and *sodC* (superoxide dismutase); sulfur metabolism, e.g. *soxA, sox* and *soxY* (sulfur oxidation); and arsenate resistance, e.g. *acr3* (arsenite efflux pump). Late response (blue) – a specific As response – refers to genes whose expression increases or declines only after prolonged exposure to arsenite. They are implicated in mechanisms such as: arsenite oxidation, e.g. *aioABCD* (arsenite oxidase) and *aioRS* (*aioABCD* regulation); specific phosphate transport, e.g. *pstB2*, and *pstS2*; and motility, e.g. *fliC* and *fliS* (flagellar system) (Cleiss-Arnold *et al.*, 2010).

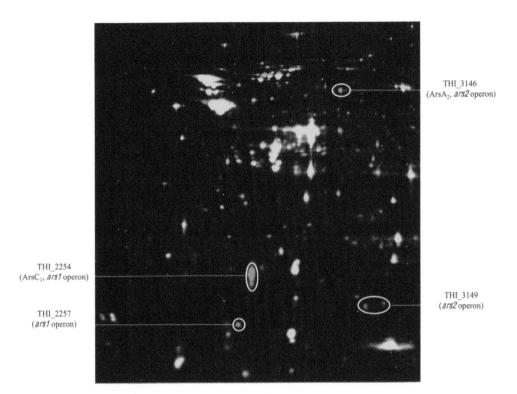

Figure 8.2.　Differential proteomic analysis of the *Thiomonas arsenitoxydans* response to As(III). On the gel are extracts obtained from bacteria cultivated in the absence (labeled with Cy5) or the presence (labeled with Cy3) of As(III). Spots correspond to proteins that were induced (green spots) or repressed (red) by As(III). Spots that are circled correspond to proteins involved in As metabolism.

T. arsenivorans, while in *T. arsenitoxydans,* levels of enzymes involved in organic carbon metabolism (glycolysis and gluconeogenesis) decline. These observations support the hypothesis that, unlike *T. arsenitoxydans*, *T. arsenivorans* is capable of optimal autotrophic growth in the presence of As(III), which is used as the inorganic electron donor (Bryan *et al.*, 2009). In other organisms, similar studies were performed to analyze the response to As(V). Transcriptomic profiling of the archaean *Pyrobaculum aerophilum* led to the identification of genes possibly involved in As(V) respiration. It revealed up-regulation of genes coding for a molybdopterin oxidoreductase, an iron sulfur protein and a membrane-anchoring subunit – the three subunits of the arsenate reductase. This observation suggests the induction in this organism of an operon encoding a membrane-bound respiratory arsenate reductase (Cozen *et al.*, 2009).

8.3.3　*Arsenic response in eukaryotes*

Until now, the most complete study has been performed in *Saccharomyces cerevisiae*. Although not known to be particularly tolerant to As, this yeast does have specific defense mechanisms. These were revealed by a "deletome" study. Here, each coding DNA sequence (CDS) longer than 100 nucleotides was systematically deleted, leading to the generation of a yeast collection composed of about 4700 strains. Screening of this library revealed several genes that are possibly involved, directly or indirectly, in As(III) tolerance. They were shown to be implicated in several physiological processes, e.g. sulfur metabolism, glutathione biosynthesis, vacuolar and endosomal transport and sorting, cell-cycle regulation, and lipid- and fatty acid-metabolism (Thorsen *et al.*, 2009). The screening also showed that genes involved in glutathione biosynthesis are essential for cells exposed to monomethylarsine but not to As(III); this is consistent with the fact that glutathione content increases significantly under MMA exposure but not with As(III) (Jo *et al.*, 2009). In contrast, other genes involved in DNA double-strand break repair, chromatin modification and tubulin folding seem important for exposure to both arsenical species, suggesting that the response to MMA has components in common with the response to As(III). In addition, the As(III)-specific transcriptome of *S. cerevisiae* showed an induction of genes implicated in GSH biosynthesis, sulfur assimilation, As detoxification, oxidative stress defense, redox maintenance and proteolytic activity, while genes encoding protein biosynthesis were down-regulated (Thorsen *et al.*, 2007). Several of these reactions – e.g. the regulation of GSH metabolism, oxidative stress, and As detoxification – are similar to the responses in *H. arsenicoxydans*, already discussed in section 8.3.1 (Cleiss-Arnold *et al.*, 2010). In *S. cerevisiae*, these results of transcriptome analysis are supported by metabolite profiling, which has revealed the accumulation of glutathione precursors (cysteine and gamma-glutamyl cysteine) (Thorsen *et al.*, 2007). Comparative transcriptomics has been used to determine the profile of genes specifically induced under As(III) stress; this has been compared with the transcriptome profiles observed under other heavy metal exposure, i.e. Ag, Cu, Cd, Hg, Zn and Cr (Jin *et al.*, 2008). Some genes are induced or repressed whatever the metal stress, e.g. genes involved in metal ion transport and homeostasis or detoxification of reactive oxygen species, and in polysaccharide biosynthesis and protein targeting and transport, respectively. In contrast, others are specific to As(III), e.g. genes involved in sulfur metabolism, protein folding, and energy production. Similar responses have been found in prokaryotes under heavy-metal or arsenic exposure (Moore *et al.*, 2005; Cleiss-Arnold *et al.*, 2010).

In the protozoan *Leishmania amazonia* and two of its As(III)-resistant variants, comparative transcriptomic studies under As(III) exposure revealed an up-regulation of genes involved in resistance to oxidative stress, drug resistance, DNA synthesis, electron transport chain, carbohydrate and fatty acid metabolism, and flagellum biosynthesis (Lin *et al.*, 2008). Here again, the eukaryotes' responses show strong similarities with aspects of the As response in prokaryotes (see above, section 8.3.1). This observation is consistent with the wide distribution in the living world of similar mechanisms of arsenic detoxification based on arsenite transporters like ArsB and Acr3 (Ghosh *et al.*, 1999).

In recent works on hyper-tolerance, a comparative transcriptomic approach has been used on the protist *Euglena mutabilis*, a photosynthesizing bio-indicator of acid mine drainage. To explain its hyper-tolerance to the arsenical compounds often found in its natural ecological niche, its response to As was compared with that of *E. gracilis*, a close relative that is not found in As-contaminated environments. Preliminary results revealed the overexpression in *E. mutabilis* – as compared with *E. gracilis* – of genes involved in As efflux, protein turnover and DNA repair. (This As-stimulated induction of As efflux and oxidative stress resistance is also well documented in prokaryotes: see sections 8.3.1 and 8.3.2).

Finally, proteomic analyses have been performed to investigate the role of arbuscular mycorrhizae, which are symbionts of the As hyperaccumulator fern *Pteris vittata*. The presence of fungi reduces As stress, resulting in a down-regulation of enzymes involved in oxidative stress defense, which suggests a plant-protective role for those mycorrhizae (Bona *et al.*, 2011). Moreover, this fern's hypertolerance may also be associated with the accumulation of large amounts of ascorbic acid and glutathione, two antioxidant compounds which are probably involved in As-induced oxidative stress defense mechanisms (Singh *et al.*, 2006). Although further investigations are needed, these observations suggest that one of the key mechanisms involved in As hypertolerance in eukaryotes may be linked to the resistance to oxidative stress.

8.3.4   *Environmental genomic approaches to the functioning of communities in situ*

Recently, the metaproteomic approach has emerged as a powerful tool for characterizing the protein content of whole microbial communities, with the aim of describing their general functioning and their responses to perturbations. The approach has two main results:

1) Identification of the main conserved proteins expressed by the community, and thus the taxa active in each component of the community's metabolism. For example, GroEL is a stress

protein widespread enough to be used as a taxonomic marker; the identification of orthologs has made it possible to identify the strains active in the As-rich Carnoulès mine (Bruneel *et al.*, 2011; Halter *et al.*, 2011b).

2) Identification of metabolic functions expressed *in situ* can clarify the functioning of the community as a whole. Such a strategy has been successfully used for example to study the As-rich Carnoulès creek. The creek contained significant levels of proteins such as rusticyanin and arsenite oxidase; metagenome sequencing revealed that these were encoded by genes in *At. ferrooxidans* and *Thiomonas* sp., respectively (see above – section 8.2.2); these strains are thus strongly implicated in the biomineralization of iron and As, respectively. In addition, this approach revealed that most proteins, i.e. 70% of all the identified proteins, were expressed by uncultured ecotypes of "*Candidatus* Fodinabacter communificans" (Bertin *et al.*, 2011). Although not directly involved in As attenuation, these bacterial strains may have a crucial role in the overall functioning of the Carnoulès ecosystem. Indeed, they have been shown to contain and express functions involved in the synthesis, or degradation of organic compounds, including cofactors, purines, and amino acids (Bertin *et al.*, 2011). Many proteins synthesized by these organisms may thus participate in the metabolism of organic matter, which may be of importance in a partly oligotrophic ecosystem such as AMD where this organism is found.

Even though metaproteomics may reveal only the few metabolically dominant organisms in any microbial community, it still represents a considerable step towards clarifying community metabolism. Eventually it may allow us to determine the relative importance of As-specific

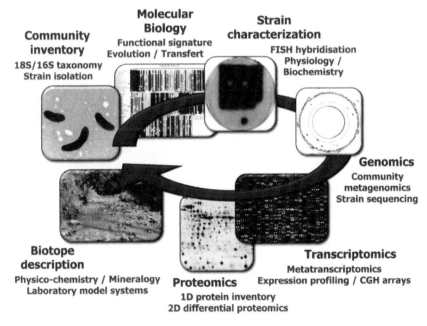

Figure 8.3.   The comprehensive analysis of a microbial ecosystem relies on a combination of complementary approaches. In a first step, the microorganisms are identified using molecular taxonomy and strain isolation. Strains of interest can then be studied by classical methods from physiology, biochemistry, molecular biology, and microscopy. It is at this point that the methods discussed in this chapter come into their own. Genomics, transcriptomics and proteomics can yield important insights into the functioning of cultured microorganisms in laboratory conditions, but they can now also be used to study complex microbial communities *in situ*. Data from community-level functional genomics – i.e. metatranscriptomics and metaproteomics – can then be used to construct a model of the ecosystem under study; ultimately, the aim would be to decipher the role of microorganisms in the biogeochemical cycles of elements, including As.

genes, stress-resistance, and generalized metabolism of organic compounds in As-contaminated ecosystems.

8.4   CONCLUSIONS

Recent developments in genomics have led to the publication of a large number of sequences from both cultured microorganisms and microbial communities. These data have revealed a widespread response to As, which includes the induction of detoxification proteins encoded by the *ars/acr* resistance genes. In contrast, other mechanisms induced by As have been identified in fewer microorganisms flourishing in As-rich ecological niches; these processes include proteins encoded by the *aio* and *arrA* genes, involved in As(III) oxidation and As(V) respiration, respectively. The genomic data have also revealed that genes required in As resistance and metabolism are often found in genomic islands, so they may have been spread among taxonomically unrelated prokaryotes though horizontal gene transfer. The nature of the genes neighboring the As-related genes (or even present in the same GEI) also suggests that the response to As includes other functions such as oxidative stress resistance, DNA repair or biofilm formation. It is to be hoped that the methods that are routinely applied to model organisms will be more widely used to characterize environmental isolates, or microbial communities in their environment. Associated with more classical methods of microbiology, genetics, and biochemistry (Fig. 8.3), these approaches will greatly improve our knowledge of the structure, the functioning, and the evolution of microbial genomes and of microbial communities colonizing As-rich ecosystems.

ACKNOWLEDGEMENTS

The authors thank Genoscope and the National Research Agency (ANR) for their financial support.

REFERENCES

Arsène-Ploetze, F., Koechler, S., Marchal, M., Coppée, J.-Y., Chandler, M., Bonnefoy, V., Brochier-Armanet, C., Barakat, M., Barbe, V., Battaglia-Brunet, F., Bruneel, O., Bryan, C.G., Cleiss-Arnold, J., Cruveiller, S., Erhardt, M., Heinrich-Salmeron, A., Hommais, F., Joulian, C., Krin, E., Lieutaud, A., Lièvremont, D., Michel, C., Muller, D., Ortet, P., Proux, C., Siguier, P., Roche, D., Rouy, Z., Salvignol, G., Slyemi, D., Talla, E., Weiss, S., Weissenbach, J., Médigue, C. & Bertin, P.N.: Structure, function, and evolution of the *Thiomonas* spp. genome. *PLoS Genet.* 6:2 (2010), e1000859.
Arsène-Ploetze, F., Carapito C., Plewniak, F. & Bertin, P.N.: Proteomics as a tool for the characterization of microbial isolates and complex communities. Proteomic applications in Biology. In J. Heazlewood, C.J. Petzold (eds): *Proteomics/Book 3*. InTech, Croatia, 2011, pp. 99–114.
Baar, C., Eppinger, M., Raddatz, G., Simon, J., Lanz, C., Klimmek, O., Nandakumar, R., Gross, R., Rosinus, A., Keller, H., Jagtap, P., Linke, B., Meyer, F., Lederer, H. & Schuster, S.C.: Complete genome sequence and analysis of *Wolinella succinogenes*. *PNAS* 100:20 (2003), pp. 11690–11695.
Bertin, P.N., Heinrich-Salmeron, A., Pelletier, E., Goulhen-Chollet, F., Arsène-Ploetze, F., Gallien, S., Lauga, B., Casiot, C., Calteau, A., Vallenet, D., Bonnefoy, V., Bruneel, O., Chane-Woon-Ming, B., Cleiss-Arnold, J., Duran, R., Elbaz-Poulichet, F., Fonknechten, N., Giloteaux, L., Halter, D., Koechler, S., Marchal, M., Mornico, D., Schaeffer, C., Smith, A.A.T., Van Dorsselaer, A., Weissenbach, J., Médigue, C. & Le Paslier, D.: Metabolic diversity among main microorganisms inside an arsenic-rich ecosystem revealed by meta- and proteo-genomics. *ISME J.* 5 (2011), pp. 1735–1747.
Bertin, P.N., Médigue, C. & Normand, P: Advances in environmental genomics: towards an integrated view of micro-organisms and ecosystems. *Microbiology* 154:2 (2008), pp. 347–359.
Bona, E., Marsano, F., Massa, N., Cattaneo, C., Cesaro, P., Argese, E., di Toppi, L.S., Cavaletto, M. & Berta, G.: Proteomic analysis as a tool for investigating arsenic stress in *Pteris vittata* roots colonized or not by arbuscular mycorrhizal symbiosis. *J. Proteomics.* In press (2011).
Borch, T., Kretzschmar, R., Kappler, A., Cappellen, P.V., Ginder-Vogel, M., Voegelin, A. & Campbell, K.: Biogeochemical redox processes and their impact on contaminant dynamics. *Environ. Sci. Technol.* 44:1 (2010), pp. 15–23.

Bruneel, O., Volant, A., Gallien, S., Chaumande, B., Casiot, C., Carapito, C., Bardil, A., Morin, G., Brown, G.E., Jr., Personné, C.J., Le Paslier, D., Schaeffer, C., Van Dorsselaer, A., Bertin, P.N., Elbaz-Poulichet, F. & Arsène-Ploetze, F.: Characterization of the active bacterial community involved in natural attenuation processes in arsenic-rich creek sediments. *Microb. Ecol.* 61:4 (2011), pp. 793–810.

Bryan, C.G., Marchal, M., Battaglia-Brunet, F., Kugler, V., Lemaitre-Guillier, C., Lièvremont, D., Bertin, P.N. & Arsène-Ploetze, F.: Carbon and arsenic metabolism in *Thiomonas* strains: differences revealed diverse adaptation processes. *BMC Microbiol.* 9 (2009), p. 127.

Carapito, C., Muller, D., Turlin, E., Koechler, S., Danchin, A., Van Dorsselaer, A., Leize-Wagner, E., Bertin, P.N. & Lett, M.-C.: Identification of genes and proteins involved in the pleiotropic response to arsenic stress in *Caenibacter arsenoxydans*, a metalloresistant beta-proteobacterium with an unsequenced genome. *Biochimie* 88:6 (2006), pp. 595–606.

Chauhan, N.S., Ranjan, R., Purohit, H.J., Kalia, V.C. & Sharma, R.: Identification of genes conferring arsenic resistance to *Escherichia coli* from an effluent treatment plant sludge metagenomic library. *FEMS Microbiol. Ecol.* 67:1 (2009), pp. 130–139.

Cleiss-Arnold, J., Koechler, S., Proux, C., Fardeau, M.-L., Dillies, M.-A., Coppee, J.-Y., Arsene-Ploetze, F. & Bertin, P.N.: Temporal transcriptomic response during arsenic stress in *Herminiimonas arsenicoxydans*. *BMC Genomics* 11:1 (2010), pp. 709.

Comte, S., Guibaud, G. & Baudu, M.: Biosorption properties of extracellular polymeric substances (EPS) towards Cd, Cu and Pb for different pH values. *J. Hazard. Mater.* 151:1 (2008), pp. 185–193.

Coppée, J.-Y: Do DNA microarrays have their future behind them? *Microbes Infect.* 10:9 (2008), pp. 1067–1071.

Cozen, A.E., Weirauch, M.T., Pollard, K.S., Bernick, D.L., Stuart, J.M. & Lowe, T.M.: Transcriptional map of respiratory versatility in the hyperthermophilic crenarchaeon *Pyrobaculum aerophilum*. *J. Bacteriol.* 191:3 (2009), pp. 782–794.

Denef, V.J., Kalnejais, L.H., Mueller, R.S., Wilmes, P., Baker, B.J., Thomas, B.C., VerBerkmoes, N.C., Hettich, R.L. & Banfield, J.F.: Proteogenomic basis for ecological divergence of closely related bacteria in natural acidophilic microbial communities. *Proc. Natl. Acad. Sci. U.S.A.* 107:6 (2010a), pp. 2383–2390.

Denef, V.J., Mueller, R.S. & Banfield, J.F.: AMD biofilms: using model communities to study microbial evolution and ecological complexity in nature. *ISME J.* 4:5 (2010b), pp. 599–610.

Dufresne, A., Ostrowski, M., Scanlan, D.J., Garczarek, L., Mazard, S., Palenik, B.P., Paulsen, I.T., de Marsac, N.T., Wincker, P., Dossat, C., Ferriera, S., Johnson, J., Post, A.F., Hess, W.R. & Partensky, F.: Unraveling the genomic mosaic of a ubiquitous genus of marine cyanobacteria. *Genome Biol.* 9:5 (2008), pp. R90.

Duval, S., Ducluzeau, A.L., Nitschke, W. & Schoepp-Cothenet, B.: Enzyme phylogenies as markers for the oxidation state of the environment: the case of respiratory arsenate reductase and related enzymes. *BMC Evol. Biol.* 8 (2008), p. 206.

Ettema, T.J.G., de Vos, W.M. & van der Oost, J.: Discovering novel biology by in silico archaeology. *Nat. Rev. Microbiol.* 3:11 (2005), pp. 859–869.

Fisher, E., Dawson, A.M., Polshyna, G., Lisak, J., Crable, B., Perera, E., Ranganathan, M., Thangavelu, M., Basu, P. & Stolz, J.F.: Transformation of inorganic and organic arsenic by *Alkaliphilus oremlandii* sp. nov. strain OhILAs. *Ann. N.Y. Acad. Sci.* 1125:1 (2008), pp. 230–241.

Foerstner, K.U., von Mering, C. & Bork, P.: Comparative analysis of environmental sequences: potential and challenges. *Philos. Trans. R. Soc. Lond.*, B, *Biol. Sci* 361:1467 (2006), pp. 519–523.

Gadd, G.M.: Metals, minerals and microbes: geomicrobiology and bioremediation. *Microbiology* 156:Pt 3 (2010), pp. 609–643.

Ghosh, M., Shen, J. & Rosen, B.P.: Pathways of arsenite detoxification in *Saccharomyces cerevisiae*. *PNAS* 96:9 (1999), pp. 5001–5006.

Guibaud, G., van Hullebusch, E., Bordas, F., d'Abzac, P. & Joussein, E.: Sorption of Cd(II) and Pb(II) by exopolymeric substances (EPS) extracted from activated sludges and pure bacterial strains: Modeling of the metal/ligand ratio effect and role of the mineral fraction. *Bioresour. Technol.* 100:12 (2009), pp. 2959–2968.

Halter, D., Casiot, C., Heipieper, H.J., Plewniak, F., Marchal, M., Simon, S., Arsène-Ploetze, F. & Bertin, P.N.: Surface properties and intracellular speciation revealed an original adaptive mechanism to arsenic in the acid mine drainage bio-indicator *Euglena mutabilis*. *Appl. Microbiol. Biotechnol.* In press (2011a).

Halter, D., Cordi, A., Gribaldo, S., Gallien, S., Goulhen-Chollet, F., Heinrich-Salmeron, A., Carapito, C., Pagnout, C., Montaut, D., Seby, F., Van Dorsselaer, A., Schaeffer, C., Bertin, P.N., Bauda, P. & Arsène-Ploetze, F.: Taxonomic and functional prokaryote diversity in mildly arsenic-contaminated sediments. *Res. Microbiol.* In press (2011b).

Heinrich-Salmeron, A., Cordi, A., Brochier-Armanet, C., Halter, D., Pagnout, C., Abbaszadeh-Fard, E., Montaut, D., Seby, F., Bertin, P.N., Bauda, P. & Arsène-Ploetze, F.: Unsuspected diversity of arsenite-oxidizing bacteria revealed by a widespread distribution of the *aoxB* gene in prokaryotes. *Appl. Environ. Microbiol.* 77:13 (2011), pp. 4685–4692.

Henne, A., Brüggemann, H., Raasch, C., Wiezer, A., Hartsch, T., Liesegang, H., Johann, A., Lienard, T., Gohl, O., Martinez-Arias, R., Jacobi, C., Starkuviene, V., Schlenczeck, S., Dencker, S., Huber, R., Klenk, H-P., Kramer, W., Merkl, R., Gottschalk, G. & Fritz, H-J.: The genome sequence of the extreme thermophile *Thermus thermophilus*. *Nat. Biotechnol.* 22:5 (2004), pp. 547–553.

Hoeft, S.E., Kulp, T.R., Stolz, J.F., Hollibaugh, J.T. & Oremland, R.S.: Dissimilatory Arsenate reduction with sulfide as electron donor: Experiments with Mono Lake water and isolation of strain MLMS-1, a chemoautotrophic arsenate respirer. *Appl. Environ. Microbiol.* 70:5 (2004), pp. 2741–2747.

Hoffman, L.R., D'Argenio, D.A., MacCoss, M.J., Zhang, Z., Jones, R.A. & Miller, S.I.: Aminoglycoside antibiotics induce bacterial biofilm formation. *Nature* 436:7054 (2005), pp. 1171–1175.

Holmes, D.E., O'Neil, R.A., Chavan, M.A., N'Guessan, L.A., Vrionis, H.A., Perpetua, L.A., Larrahondo, M.J., DiDonato, R., Liu, A. & Lovley, D.R.: Transcriptome of *Geobacter uraniireducens* growing in uranium-contaminated subsurface sediments. *ISME J.* 3:2 (2009), pp. 216–230.

Huber, R., Sacher, M., Vollmann, A., Huber, H. & Rose, D.: Respiration of arsenate and selenate by hyperthermophilic archaea. *Syst. Appl. Microbiol.* 23:3 (2000), pp. 305–314.

Inskeep, W.P., Macur, R.E., Hamamura, N., Warelow, T.P., Ward, S.A. & Santini, J.M.: Detection, diversity and expression of aerobic bacterial arsenite oxidase genes. *Environ. Microbiol.* 9:4 (2007), pp. 934–943.

Janssen, P.J., Van Houdt, R., Moors, H., Monsieurs, P., Morin, N., Michaux, A., Benotmane, M.A., Leys, N., Vallaeys, T., Lapidus, A., Monchy, S., Médigue, C., Taghavi, S., McCorkle, S., Dunn, J., van der Lelie, D. & Mergeay, M.: The complete genome sequence of *Cupriavidus metallidurans* strain CH34, a master survivalist in harsh and anthropogenic environments. *PLoS ONE* 5:5 (2010), pp. e10433.

Jin, Y.H., Dunlap, P.E., McBride, S.J., Al-Refai, H., Bushel, P.R. & Freedman, J.H.: Global transcriptome and deletome profiles of yeast exposed to transition metals. *PLoS Genet* 4:4 (2008), pp. e1000053.

Jo, W.J., Loguinov, A., Wintz, H., Chang, M., Smith, A.H., Kalman, D., Zhang, L., Smith, M.T. & Vulpe, C.D.: Comparative functional genomic analysis identifies distinct and overlapping sets of genes required for resistance to monomethylarsonous acid (MMAIII) and arsenite (AsIII) in yeast. *Toxicol. Sci.* 111:2 (2009), pp. 424–436.

Juhas, M., van der Meer, J.R., Gaillard, M., Harding, R.M., Hood, D.W. & Crook, D.W.: Genomic islands: tools of bacterial horizontal gene transfer and evolution. *FEMS Microbiol. Rev.* 33:2 (2009), pp. 376–393.

Koechler, S., Cleiss-Arnold, J., Proux, C., Sismeiro, O., Dillies, M-A., Goulhen-Chollet, F., Hommais, F., Lièvremont, D., Arsène-Ploetze, F., Coppée, J-Y. & Bertin, P.N.: Multiple controls affect arsenite oxidase gene expression in *Herminiimonas arsenicoxydans*. *BMC Microbiol.* 10 (2010), pp. 53.

Krafft, T. & Macy, J.M.: Purification and characterization of the respiratory arsenate reductase of *Chrysiogenes arsenatis*. *Eur. J. Biochem.* 255:3 (1998), pp. 647–653.

Kulp, T.R., Hoeft, S.E., Asao, M., Madigan, M.T., Hollibaugh, J.T., Fisher, J.C., Stolz, J.F., Culbertson, C.W., Miller, L.G. & Oremland, R.S.: Arsenic(III) fuels anoxygenic photosynthesis in hot spring biofilms from Mono Lake, California. *Science* 321:5891 (2008), pp. 967–970.

Li, B., Lin, J., Mi, S. & Lin, J.: Arsenic resistance operon structure in *Leptospirillum ferriphilum* and proteomic response to arsenic stress. *Bioresour. Technol.* 101:24 (2010), pp. 9811–9814.

Lin, Y.-C., Hsu, J.-Y., Shu, J.-H., Chi, Y., Chiang, S.-C. & Lee, S.T.: Two distinct arsenite-resistant variants of *Leishmania amazonensis* take different routes to achieve resistance as revealed by comparative transcriptomics. *Mol. Biochem. Parasitol.* 162:1 (2008), pp. 16–31.

Macy, J.M., Nunan, K., Hagen, K.D., Dixon, D.R., Harbour, P.J., Cahill, M. & Sly, L.I.: *Chrysiogenes arsenatis* gen. nov., sp. nov., a new arsenate-respiring bacterium isolated from gold mine wastewater. *Int. J. Syst. Bacteriol.* 46:4 (1996), pp.1153–1157.

Majtán, J., Majtánová, L., Xu, M. & Majtán, V.: In vitro effect of subinhibitory concentrations of antibiotics on biofilm formation by clinical strains of *Salmonella enterica* serovar *Typhimurium* isolated in Slovakia. *J. Appl. Microbiol.* 104:5 (2008), pp. 1294–1301.

Marchal, M., Briandet, R., Koechler, S., Kammerer, B. & Bertin, P.N.: Effect of arsenite on swimming motility delays surface colonization in *Herminiimonas arsenicoxydans*. *Microbiology* 156:Pt 8 (2010), pp. 2336–2342.

Mergeay, M., Monchy, S., Vallaeys, T., Auquier, V., Benotmane, A., Bertin, P.N., Taghavi, S., Dunn, J., van der Lelie, D. & Wattiez, R.: *Ralstonia metallidurans*, a bacterium specifically adapted to toxic metals: towards a catalogue of metal-responsive genes. *FEMS Microbiol. Rev.* 27:2–3 (2003), pp. 385–410.

Milde, K., Sand, W., Wolff, W. & Bock, E.: *Thiobacilli* of the corroded concrete walls of the Hamburg sewer system. *J. Gen. Microbiol.* 129:5 (1983), pp. 1327–1333.

Moore, C.M., Gaballa, A., Hui, M., Ye, R.W. & Helmann, J.D.: Genetic and physiological responses of *Bacillus subtilis* to metal ion stress. *Mol Microbiol.* 57:1 (2005), pp. 27–40.

Mukhopadhyay, R., Rosen, B.P., Phung, L.T. & Silver, S.: Microbial arsenic: from geocycles to genes and enzymes. *FEMS Microbiol. Rev.* 26:3 (2002), pp. 311–325.

Muller, D., Médigue, C., Koechler, S., Barbe, V., Barakat, M., Talla, E., Bonnefoy, V., Krin, E., Arsène-Ploetze, F., Carapito, C., Chandler, M., Cournoyer, B., Cruveiller, S., Dossat, C., Duval, S., Heymann, M., Leize, E., Lieutaud, A., Lièvremont, D., Makita, Y., Mangenot, S., Nitschke, W., Ortet, P., Perdrial, N., Schoepp, B., Siguier, P., Simeonova, D.D., Rouy, Z., Segurens, B., Turlin, E., Vallenet, D., Van Dorsselaer, A., Weiss, S., Weissenbach, J., Lett, M-C., Danchin, A. & Bertin, P.N.: A tale of two oxidation states: bacterial colonization of arsenic-rich environments. *PLoS Genet.* 3:4 (2007), pp. e53.

Ordax, M., Marco-Noales, E., López, M.M. & Biosca, E.G.: Exopolysaccharides favor the survival of *Erwinia amylovora* under copper stress through different strategies. *Res. Microbiol.* 161:7 (2010), pp. 549–555.

Oremland, R.S., Blum, J.S., Culbertson, C.W., Visscher, P.T., Miller, L.G., Dowdle, P. & Strohmaier, F.E.: Isolation, growth, and metabolism of an obligately anaerobic, selenate-respiring bacterium, strain SES-3. *Appl. Environ. Microbiol.* 60:8 (1994), pp. 3011–3019.

Oremland, R.S., Hoeft, S.E., Santini, J.M., Bano, N., Hollibaugh, R.A. & Hollibaugh, J.T.: Anaerobic oxidation of arsenite in Mono Lake water and by a facultative, arsenite-oxidizing chemoautotroph, strain MLHE-1. *Appl. Environ. Microbiol.* 68:10 (2002), pp. 4795–4802.

Páez-Espino, D., Tamames, J., Lorenzo, V. & Cánovas, D: Microbial responses to environmental arsenic. *Biometals* 22:1 (2009), pp. 117–130.

Patel, P.C., Goulhen, F., Boothman, C., Gault, A.G., Charnock, J.M., Kalia, K. & Lloyd, J.R.: Arsenate detoxification in a Pseudomonad hypertolerant to arsenic. *Arch. Microbiol.* 187:3 (2006), pp. 171–183.

Qin, J., Lehr, C.R., Yuan, C., Le, X.C., McDermott, T.R. & Rosen, B.P.: Biotransformation of arsenic by a Yellowstone thermoacidophilic eukaryotic alga. *PNAS* 106:13 (2009), pp. 5213–5217.

Qin, Q-L., Li, Y., Zhang, Y-J., Zhou, Z-M., Zhang, W-X., Chen, X-L., Zhang, X-Y., Zhou, B-C., Wang, L. & Zhang, Y-Z.: Comparative genomics reveals a deep-sea sediment-adapted life style of *Pseudoalteromonas* sp. SM9913. *ISME J.* 5:2 (2011), pp. 274–284.

Quéméneur, M., Heinrich-Salmeron, A., Muller, D., Lièvremont, D., Jauzein, M., Bertin, P.N., Garrido, F. & Joulian, C.: Diversity surveys and evolutionary relationships of aoxB genes in aerobic arsenite-oxidizing bacteria. *Appl. Environ. Microbiol.* 74:14 (2008), pp. 4567–4573.

Rubakhin, S.S., Romanova, E.V., Nemes, P. & Sweedler, J.V.: Profiling metabolites and peptides in single cells. *Nat. Meth.* 8:4s (2011), pp. S20–S29.

Saltikov, C.W. & Newman, D.K.: Genetic identification of a respiratory arsenate reductase. *PNAS* 100:19 (2003), pp. 10983–10988.

Saltikov, C.W., Cifuentes, A., Venkateswaran, K. & Newman D.K.: The *ars* detoxification system is advantageous but not required for arsenate respiration by the genetically tractable *Shewanella* species strain ANA-3. *Appl. Environ. Microbiol.* 69:5 (2003), pp. 2800–2809.

Santini, J.M., Sly, L.I., Schnagl, R.D. & Macy, J.M.: A new chemolithoautotrophic arsenite-oxidizing bacterium isolated from a gold mine: phylogenetic, physiological, and preliminary biochemical studies. *Appl. Environ. Microbiol.*, 66:1 (2000), pp. 92–97.

Santini, J.M. & vanden Hoven, R.N.: Molybdenum-containing arsenite oxidase of the chemolithoautotrophic arsenite oxidizer NT-26. *J. Bacteriol.* 186:6 (2004), pp. 1614–1619.

Santini, J.M., Kappler, U., Ward, S.A., Honeychurch, M.J., vanden Hoven, R.N. & Bernhardt, P.V.: The NT-26 cytochrome c_{552} and its role in arsenite oxidation. *Biochim. Biophys. Acta* 1767 (2007), pp. 189–196.

Shigeta, M., Tanaka, G., Komatsuzawa, H., Sugai, M., Suginaka, H. & Usui, T.: Permeation of antimicrobial agents through *Pseudomonas aeruginosa* biofilms: A simple method. *Chemotherapy* 43:5 (1997), pp. 340–345.

Silver, S. & Phung, L.T. A bacterial view of the periodic table: genes and proteins for toxic inorganic ions. *J. Ind. Microbiol. Biotechnol.* 32: 11 (2005), pp. 587–605

Simmons, S.L., Dibartolo, G., Denef, V.J., Goltsman, D.S.A., Thelen, M.P. & Banfield, J.F.: Population genomic analysis of strain variation in *Leptospirillum* group II bacteria involved in acid mine drainage formation. *PLoS Biol.* 6:7 (2008), pp. e177.

Singh, N., Ma, L.Q., Srivastava, M. & Rathinasabapathi, B.: Metabolic adaptations to arsenic-induced oxidative stress in *Pteris vittata* L and *Pteris ensiformis* L. *Plant Science* 170 (2006), pp. 274–282.

Switzer Blum, J., Burns Bindi, A., Buzzelli, J., Stolz, J.F. & Oremland, R.S.: *Bacillus arsenicoselenatis*, sp. nov., and *Bacillus selenitireducens*, sp. nov.: two haloalkaliphiles from Mono Lake, California that respire oxyanions of selenium and arsenic. *Arch. Microbiol.* 171:1 (1998), pp. 19–30.

Thorsen, M., Lagniel, G., Kristiansson, E., Junot, C., Nerman, O., Labarre, J. & Tamás, M.J.: Quantitative transcriptome, proteome, and sulfur metabolite profiling of the *Saccharomyces cerevisiae* response to arsenite. *Physiol. Genomics* 30:1 (2007), pp. 35–43.

Thorsen, M., Perrone, G., Kristiansson, E., Traini, M., Ye, T., Dawes, I., Nerman, O. & Tamas, M.: Genetic basis of arsenite and cadmium tolerance in *Saccharomyces cerevisiae*. *BMC Genomics* 10:1 (2009), pp. 105.

Tyson, G.W., Chapman, J., Hugenholtz, P., Allen, E.E., Ram, R.J., Richardson, P.M., Solovyev, V.V., Rubin, E.M., Rokhsar, D.S. & Banfield, J.F.: Community structure and metabolism through reconstruction of microbial genomes from the environment. *Nature* 428:6978 (2004), pp. 37–43.

vanden Hoven, R.N. & Santini, J.M.: Arsenite oxidation by the heterotroph *Hydrogenophaga* sp. str. NT-14: the arsenite oxidase and its physiological electron acceptor. *Biochim. Biophys. Acta* 1656 (2004), pp. 148–155.

Verstraeten, N., Braeken, K., Debkumari, B., Fauvart, M., Fransaer, J., Vermant, J. & Michiels, J.: Living on a surface: swarming and biofilm formation. *Trends Microbiol.* 16:10 (2008), pp. 496–506.

Wang, Z., Gerstein, M. & Snyder, M.: RNA-Seq: a revolutionary tool for transcriptomics. *Nat. Rev. Genet.* 10:1 (2009), pp. 57–63.

Weckwerth, W.: Metabolomics: an integral technique in systems biology. *Bioanalysis* 2:4 (2010), pp. 829–836.

Wei, X., Fang, L., Cai, P., Huang, Q., Chen, H., Liang, W. & Rong, X.: Influence of extracellular polymeric substances (EPS) on Cd adsorption by bacteria. *Environ. Pollut.* 159:5 (2011), pp. 1369–1374.

Weiss, S., Carapito, C., Cleiss, J., Koechler, S., Turlin, E., Coppee, J-Y., Heymann, M., Kugler, V., Stauffert, M., Cruveiller, S., Médigue, C., Van Dorsselaer, A., Bertin, P.N. & Arsène-Ploetze, F.: Enhanced structural and functional genome elucidation of the arsenite-oxidizing strain *Herminiimonas arsenicoxydans* by proteomics data. *Biochimie* 91:2 (2009), pp. 192–203.

Wilkins, M.J., Verberkmoes, N.C., Williams, K.H., Callister, S.J., Mouser, P.J., Elifantz, H., N'guessan, A.L., Thomas, B.C., Nicora, C.D., Shah, M.B., Abraham, P., Lipton, M.S., Lovley, D.R., Hettich, R.L., Long, P.E. & Banfield, J.F.: Proteogenomic monitoring of *Geobacter* physiology during stimulated uranium bioremediation. *Appl. Environ. Microbiol.* 75:20 (2009), pp. 6591–6599.

Wilmes, P., Bowen, B.P., Thomas, B.C., Mueller, R.S., Denef, V.J., Verberkmoes, N.C., Hettich, R.L., Northen, T.R. & Banfield, J.F.: Metabolome-proteome differentiation coupled to microbial divergence. *MBio* 1:5 (2010), pp. e00246-10.

Yuan, C., Lu, X., Qin, J., Rosen, B.P. & Le, X.C.: Volatile arsenic species released from *Escherichia coli* expressing the AsIII S-adenosylmethionine methyltransferase Gene. *Environ. Sci. Technol.* 42:9 (2008), pp. 3201–3206.

Zargar, K., Hoeft, S., Oremland, R. & Saltikov, C.W.: Identification of a novel arsenite oxidase gene, *arxA*, in the haloalkaliphilic, arsenite-oxidizing bacterium *Alkalilimnicola ehrlichii* Strain MLHE-1. *J. Bacteriol.* 192:14 (2010), pp. 3755–3762.

Zhang, Y-B., Monchy, S., Greenberg, B., Mergeay, M., Gang, O., Taghavi, S. & Lelie, D.: ArsR arsenic-resistance regulatory protein from *Cupriavidus metallidurans* CH34. *Antonie van Leeuwenhoek* 96:2 (2009), pp. 161–170.

Zhang, Y., Ma, Y-F., Qi, S-W., Meng, B., Chaudhry, M.T., Liu, S-Q. & Liu, S-J.: Responses to arsenate stress by *Comamonas* sp. strain CNB-1 at genetic and proteomic levels. *Microbiology* 153:11 (2007), pp. 3713–3721.

CHAPTER 9

Arsenite oxidation – regulation of gene expression

Marta Wojnowska & Snezana Djordjevic

9.1 INTRODUCTION

In contrast to the significant progress made in studies of arsenite [As(III)] metabolism in various organisms, and the structural and mechanistic characterisation of As(III) oxidase (Aio), we have only limited understanding of the regulation of Aio activity. The first report on this process came in 2005 when McDermott and his co-workers demonstrated complex regulation of Aio in *Agrobacterium tumefaciens*. Regulation in this organism involves As(III) exposure, quorum sensing and a two-component signal transduction system (Kashyap *et al.*, 2006). The transposon mutagenesis method used in that study has since been applied to other species, along with a variety of experimental approaches including transcriptomics, proteomics, and molecular and biochemical investigations, to reveal genes that are involved in the control of Aio. The picture now emerging is that the regulation of Aio is complex and that, apart from some common features such as a canonical two-component signal transduction system, different organisms have different genetic bases for the control of Aio activity, perhaps reflecting phylogeny and their different environments. In this chapter we will present the current knowledge and understanding of the multitude of mechanisms involved in control of As(III) oxidase gene expression. We will develop hypotheses on the unexplored regulatory mechanisms and suggest avenues for future investigation.

9.2 MULTIPLE MODES OF ARSENITE OXIDASE REGULATION

In many As(III)-oxidizing organisms, chemical conversion of As(III) to As(V) is upregulated by exposure to As(III). However, the As(III) ion itself does not directly affect the expression of As(III) oxidase genes; several molecular mechanisms have been implicated in the bacterial response to As(III) exposure. Table 9.1 summarizes the findings from studies of organisms in which at least one mode of regulation has been identified through either sequence analysis or functional investigations. Most commonly a putative two-component signal transduction system, encoded by *aioS* and *aioR* – together named *aioSR* (or *aroSR/aoxSR*), is associated with As(III) oxidase transcriptional control. However, in a few instances there is a clear correlation between expression levels of As(III) oxidase genes (*aioAB*) and the population's growth phase, leading to the suggestion that quorum sensing might also be involved in regulation.

   The most extensive analysis of the regulatory mechanism has been carried out on the heterotroph *Herminiimonas arsenicoxydans*. In addition to *aioSR* (*aoxSR*), As(III) oxidase activity is dependent on the expression of *rpoN*, coding for the alternative sigma factor (σ54) of RNA polymerase, and *dnaJ*, coding for the heat shock protein Hsp40, an Hsp70 co-chaperone (Koechler *et al.*, 2010; Weiss *et al.*, 2007). Temporal transcriptomics analysis showed that induction in *aioSR* and *aioABCD* operons takes place after prolonged As(III) exposure (6–8 hours) and is a specific As(III) response, while the induction of expression of heat-shock proteins is more immediate and is a part of a general As(III) stress response mechanism (Cleiss-Arnold *et al.*, 2010). At present, it is difficult to argue definitively that these various regulatory components are generally present across the range of microbial species, owing to the lack of comprehensive experimental analysis.

Table 9.1. Summary of all mechanisms and elements shown experimentally to regulate arsenite oxidase expression.

| Organism | Function of arsenite oxidation | Two-component system AioSR | | | | Other regulatory elements | Reference |
		Organisation of *aioSR* relative to *aio*	Expression	Role in *aio* expression	Quorum sensing		
Achromobacter sp. str. SY8	Detoxification	Separate (opposite direction) (*aioSRX*)	As(III)-induced	?	?		Cai *et al.*, 2009
Agrobacterium tumefaciens	Detoxification	Together as part of *aio* operon	As(III)-induced	Essential	Present	AioX	Kashyap *et al.*, 2006 Liu *et al.*, in press
Herminiimonas arsenicoxydans	Detoxification	Separate (opposite direction)	As(III)-induced	Essential	Likely	DnaJ, RpoN	Koechler *et al.*, 2010 Muller *et al.*, 2007 Cleiss-Arnold *et al.*, 2010
Ochrobactrum tritici str. SCII24	Detoxification	Separate	?	Not essential	Likely		Branco *et al.*, 2009
Rhizobium sp. str. NT-26	Chemolithoautotrophic growth	Separate	Constitutive	Essential – also for As(III) tolerance	?		Sardiwal *et al.*, 2010

9.3   AioSR AND THEIR INVOLVEMENT IN Aio REGULATION

9.3.1   *Diversity in gene organisation*

Two open reading frames, named *aioS* and *aioR*, lie upstream of the As(III) oxidase-coding region of many As(III)-oxidizing organisms (Kashyap *et al.*, 2006; Muller *et al.*, 2007). On the basis of the sequence homology of the proteins encoded by these genes, they were annotated as a putative two-component signal transduction system. Annotation and function of *aioS* and *aioR*, and their respective proteins, were confirmed at both gene and protein level (Sardiwal *et al.*, 2010). Although it is clear that in many cases these proteins are essential components for the regulation of As(III) oxidase activity, their respective gene organisation is not conserved across all of the organisms in which they have been identified. In some species *aioSR* genes are oriented in the same direction (NT-26, *A. tumefaciens*, *O. tritici*) as *aio* genes (Sardiwal *et al.*, 2010; Kashyap *et al.*, 2006; Branco *et al.*, 2009) while in others they are present in the opposite direction (*Alcaligenes faecalis*, *Achromobacter sp. str. SY8*, *H. arsenicoxydans*) (Muller *et al.*, 2008; Cai *et al.*, 2009). Similarly, there are differences at the transcriptional level such that in *A. tumefaciens aioSR* are cotranscribed with the *aio* cluster, comprising the same operon, while in NT-26 they constitute separate operons. Furthermore the presence of *aioSR* has not been confirmed in all bacteria where As(III) oxidase activity has been reported, although in some cases this may be due to a lack of full genome-sequence data.

9.3.2   *Two-component systems (TCS)*

Two-component signal transduction processes, common to many bacterial and archaeal species, involve two conserved protein components – a transmitter and a receiver (Stock *et al.*, 2000; Gao and Stock, 2009). These can be associated with various sensor/input and effector/output domains, thus allowing diverse signals to evoke a range of appropriate responses.

Most TCS proteins detect extracellular or membrane-integral stimuli, e.g. small molecules, nutrients, light and turgor (Mascher *et al.*, 2006). The evolution of TC proteins was most likely driven by the need to couple signal detection on one side of the membrane (periplasmic space) with an adaptive response inside the cell (cytoplasm). Hence a typical TCS comprises a membrane-bound transmitter protein with a sensor domain, and a cytoplasmic receiver protein associated with an effector domain (Stock *et al.*, 2000; Gao and Stock, 2009).

TCS use a phosphotransfer reaction as a means of transmitting a signal (Ninfa and Magasanik 1986; Stock *et al.*, 2000; Gao and Stock, 2009). Stimulus detection is linked to autophosphory-lation of the transmitter (a histidine kinase), with the phosphoryl group subsequently transmitted onto the receiver (response regulator). The general mechanism of TC signalling is depicted in Figure 9.1. The periplasmic sensor region (S) of a histidine kinase detects the stimulus (black

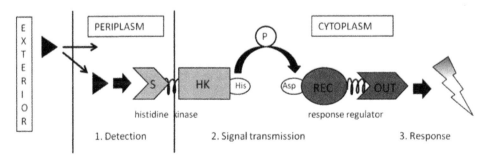

Figure 9.1.   Schematic diagram of a prototypical two-component stimulus transduction system. Sensory region of a histidine kinase is labeled S; kinase core is labeled HK. In a response regulator, the receiver domain is labeled as REC while the output domain is labeled OUT. A putative stimulus is depicted by the black triangle while the cellular response to that stimulus is shown by a lightning symbol.

triangle), which affects autokinase activity of the cytoplasmic kinase core (HK). Depending on the system, stimulus detection can induce or inhibit autophosphorylation. When the core autophosphorylates upon a conserved histidine residue (His), the receiver domain (REC) of its cognate response regulator recognises the phospho-His signal and catalyses the transfer of the phosphoryl group onto a conserved aspartate residue (Asp). The output domain (OUT) of the response regulator evokes an adaptive response (lightning). As in the case of kinase core, reception of phosphoryl group may either activate or repress the activity of the output domain.

High fidelity of signal transmission relies on specific interactions between the signalling proteins. Kinase core and receiver domains contain amino-acid regions ensuring specific interactions, which eliminates the possibility of non-specific cross-talk (Laub and Goulian, 2007). However, there are numerous examples of more complex signalling pathways, which require additional proteins, enzymatic activities and/or signalling steps (Laub and Goulian, 2007; Stock *et al.*, 2000; Gao and Stock, 2009). Branched pathways (one-to-many and many-to-one) have also been observed.

9.3.2.1 *Histidine kinase*

A typical histidine kinase (HK) is composed of a catalytic core located in the cytoplasm and a stimulus-sensing domain extending into the periplasmic space (Gao and Stock, 2009). The two domains are linked by at least one transmembrane helix.

Sensing domains vary in size and structural topology depending on the stimulus they detect (Mascher *et al.*, 2006). Multi-pass membrane-integral domains with no regions extending into the periplasm are likely to detect signals such as turgor changes. Domains with a substantial portion reaching into periplasmic space can bind either small molecule ligands (e.g. nutrients) or other proteins (e.g. transporters or periplasmic binding proteins). Stimulus can therefore be detected both directly and indirectly *via* an intermediate protein.

Stimulus detection affects the conformation of the cytoplasmic, C-terminal portion of the protein, either stimulating or inhibiting kinase activity (Gao and Stock, 2009). The kinase core is often connected to the transmembrane region *via* an extended helical region known as the HAMP domain, which plays an important role in transmitting conformational changes from periplasm into cytoplasm (Ferris *et al.*, 2011).

The kinase core is composed of two distinct structural domains (Marina *et al.*, 2005): 1) The site of phosphorylation, a histidine residue, is located within the Dimerisation and Histidine phosphotransfer domain (DHp), which comprises two antiparallel helices; this domain is responsible for homodimerisation of histidine kinase molecules. 2) The catalytic domain (CA) contains several highly conserved amino-acid motifs, which are essential for ATP binding and hydrolysis. When the kinase core is activated by conformational perturbations resulting from stimulus detection, the CA domain hydrolyses ATP and the phosphoryl group is transferred onto the specific histidine residue in the DHp domain. Autophosphorylation may take place either in *trans-*, between two units of the homodimer, or in *cis-*reaction mechanisms involving only one polypeptide unit (Marina *et al.*, 2005; Stock *et al.*, 2000; Gao and Stock, 2009).

It has been shown that many HKs also possess phosphatase activity: they can dephosphorylate their cognate response regulators when signalling needs to be silenced (Huynh *et al.*, 2010; Stock *et al.*, 2000; Gao and Stock, 2009).

9.3.2.2 *Response regulator*

A response regulator (RR) typically has an N-terminal receiver domain which contains several conserved acidic residues required for catalysis of the phosphotransfer reaction (Stock *et al.*, 1989; Stock *et al.*, 2000; Gao and Stock, 2009). The phosphoryl group is transferred from a cognate histidine kinase onto the conserved aspartate residue of the RR, causing conformational changes that result in activation (or inhibition, depending on the system studied) of the C-terminal effector domain. Most effector domains associated with RRs bind to DNA, thus inducing or suppressing gene expression. Other types of output domains are involved in protein/protein interaction or perform enzymatic activities.

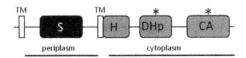

Figure 9.2. AioS domain organisation (not to scale). TM – transmembrane region, S – sensor domain, H – HAMP domain, DHp – Dimerisation and Histidine phosphotransfer domain, CA – catalytic domain. The asterisks (*) mark the domains that form the kinase core.

Figure 9.3. AioR domain organisation (not to scale). REC – receiver domain, AAA+ – ATPases Associated with a variety of cellular Activities, Db – DNA-binding domain.

9.3.2.3 *Histidine kinase* AioS *and arsenite sensing*

AioS (Fig. 9.2) is a classical membrane-bound HK, with a periplasmic sensor region anchored to the membrane by two transmembrane helices (Sardiwal *et al.*, 2010). One of these helices leads *via* a helical HAMP domain into the C-terminal cytoplasmic core, whose autophosphorylation activity has been confirmed in *Rhizobium* NT-26 (Sardiwal *et al.*, 2010).

Cysteine side chains are commonly involved in interaction with As(III) *via* their sulfhydryl groups, but there are no cysteine residues in the periplasmic region of AioS (Sardiwal *et al.*, 2010). However, several other amino-acid side chains have also been implicated in As(III) binding (Ellis *et al.*, 2001), so the possibility that the periplasmic domain of AioS interacts with, and responds to, As(III) directly was considerred despite the absence of thiol groups in AioS. The sensor/input region is not significantly homologous to any domains of known structure and function although its fold is distantly related to the topology of a PAS (Per/Arnt/Sim) domain (Sardiwal *et al.*, 2010). PAS domains are versatile sensors that bind a variety of small molecule ligands (Mascher *et al.*, 2006). In some systems PAS-like domains have also been shown to interact with periplasmic binding proteins, e.g. the quorum-sensing HK LuxQ (Neiditch *et al.*, 2006; Gao and Stock, 2009). Cai *et al.* (2009) postulated that a putative periplasmic binding protein, AioX (also known as AoxX), could bind As(III) in the periplasm and then interact with the sensor region of AioS. Transcription of *aioX*, which is frequently found alongside *aioS* and *aioR* genes in other organisms, is induced by As(III) in *Achromobacter*, and is cotranscribed with the TCS-encoding genes. Recently, Liu *et al.* demonstrated that in *A. tumefaciens* AioX does indeed bind As(III) directly, and that *aioX* mutants are defective in *aioBA* induction and As(III) oxidation (Liu *et al.*, in press). Therefore it is reasonable to propose that direct binding of As(III) to AioX leads to induction of As(III) oxidase through interaction of AioX with the periplasmic PAS-like domain of AioS, although there is not yet any biochemical evidence that these two proteins interact.

9.3.2.4 *Response regulator AioR and aio operon expression*

It has been shown that protein AioR (Fig. 9.3) is the cognate response regulator of AioS in *Rhizobium* NT-26: the N-terminal receiver domain of AioR receives the phosphoryl group from the kinase core domain of AioS (Sardiwal *et al.*, 2010). In the C-terminal region of AioR there is a helix-turn-helix DNA-binding motif. The central part of AioR comprises an ATPase domain belonging to the AAA+ superfamily (ATPases Associated with various cellular Activities). These versatile ATPases – which have been identified in several RR proteins – are involved in transcription activation regulated by σ54 transcription initiation factor (Gao and Stock, 2009). The specific σ54-dependent promoters lie upstream of AioR-controlled transcripts (Koechler *et al.*, 2010). Furthermore, AioR contains a conserved amino-acid motif required for binding to the σ54 factor of RNA polymerase (Sardiwal *et al.*, 2010; Koechler *et al.*, 2010).

In *Herminiimonas arsenicoxydans* a putative σ54 promoter has been identified upstream of the Aio operon, and transposon mutagenesis has shown that RpoN (also known as σ54) is necessary for *aio* expression (Koechler *et al.*, 2010). Notably, all organisms that possess AioSR-encoding genes (see Table 9.1) also contain the σ54 promoter sequence upstream of the As(III) oxidase genes (Koechler *et al.*, 2010). These findings strongly suggest that the mechanism of AioR-mediated induction of *aio* transcription involves an interaction of AAA+ ATPase domain of AioR with RpoN, that recognizes the σ54 promoter.

9.3.2.5   *Experimental evidence implicating AioSR in regulation of aio expression*
RT-PCR experiments indicated that *aioS* and *aioR* are constitutively expressed in *Rhizobium NT-26* (Sardiwal *et al.*, 2010), even in the absence of As(III), and that they do not belong to the same operon as the As(III) oxidase genes. Expression of the *aio* operon was detected only in the presence of As(III). Targeted gene disruption proved that both TCS proteins are necessary for As(III) oxidase to be expressed as no *aio* transcript was detected in either mutant (*aioS* or *aioR*) by RT-PCR. The mutants were unable to grow chemolithoautotrophically with 5 mM As(III), but their growth under heterotrophic conditions (yeast extract alone) was unaffected by As. Interestingly, under heterotrophic conditions with 5 mM As(III) the growth rate of *aioS* mutant was significantly reduced. This implies that AioS, in addition to controlling *aio* expression, may be involved in the regulation of As(III) tolerance, for instance by cross-regulating additional RR(s) in a one-to-many signalling mode.

In *Agrobacterium tumefaciens* str. 5A (Kashyap *et al.*, 2006) transposon-mediated *aioR* disruption eliminated the As(III) oxidation phenotype, suggesting a downstream effect on the *aio*-encoding genes. Both RT-PCR and mutant complementation experiments indicate that in *A. tumefaciens*, unlike its close relative NT-26, *aioS* and *aioR* are contranscribed with the *aio* genes, and their expression is As(III)-dependent. A non-oxidizing *aioR* mutant strain could not be complemented *in trans* with an *aioSR*-containing construct; the entire gene cluster *aioSRBA-cytC* was necessary to restore As(III) oxidation. In addition, as indicated by quantitative RT-PCR, As(III)-exposed *aioR* mutants showed significantly reduced expression of the upstream gene in the operon, *aioS*, compared with the levels detected in wild-type As(III)-exposed cells. This suggests an As(III)-dependent autoregulatory role for AioR in *aio* operon expression. Interestingly, quorum sensing has also been shown to affect *aio* expression in this bacterium, and it appears that the AioSR system may not be essential, at least in late-log growth phase.

The putative promoter sequence upstream of the Aio-encoding genes in *Ochrobactrum tritici str. SCII24* (Branco *et al.*, 2009) is homologous to the promoter found in NT-26; using RT-PCR it was shown that *aioS* and *aioR*, which lie upstream of that promoter, were transcribed separately from the *aio* gene cluster. The As(III) oxidation phenotype was generated in a non-oxidizing *O. tritici* str. 5bv11 that had been supplemented with a construct containing *aioBA-cytC* genes preceded by the putative promoter. The existence of native AioS and AioR in this strain was ruled out by Southern analysis. The fact that the genes encoding structural Aio proteins and the cytochrome were sufficient to turn 5bv11 into an As(III) oxidizer implies that, as in *A. tumefaciens*, in *O. tritici* the AioSR system is not indispensible and a quorum-sensing mechanism may also be involved in regulating *aio* expression. However, arsenate (AsV) production in the 5bv11 strain was observed in later growth phases than in the SCII24 strain, which suggests that AioSR proteins may play a role in earlier stages of bacterial growth. Interestingly, in contrast to NT-26 where the absence of AioS affected both *aio* expression and As(III) tolerance, the growth of *O. tritici* appears unaffected by As(III) (1 mM), regardless of the presence of AioSR and/or Aio. This suggests that tolerance to As(III) in *O. tritici* is regulated by another system, and not by AioS as postulated for NT-26.

In *Achromobacter sp. SY8* (Cai *et al.*, 2009) RT-PCR experiments revealed that *aioSR* are cotranscribed with *aioX*, a gene encoding a putative periplasmic protein, which was postulated to bind directly to As(III) and, potentially, interact with the periplasmic domain of AioS. The genes coding for As(III) oxidase lie in the opposite orientation and were shown to form a separate transcript.

The most extensive experimental evidence of *aio* regulation comes from studies on *Herminiimonas arsenicoxydans* (Koechler *et al.*, 2010; Cleiss-Arnold, *et al.*, 2010; Muller, *et al.*, 2007). Microarray analysis has indicated that in As(III)-exposed cells in the exponential-growth phase, several different genes are up-regulated, including *aioS* and *aioR* (Koechler *et al.*, 2010). In addition, a methyl-accepting chemotaxis protein gene also appears to be expressed in the presence of As(III), suggesting a role in coupling As(III) detection with motility changes. A previous study showed that transposon disruption of *aioS* and *aioR* resulted in a complete loss of As(III) oxidation (Muller *et al.*, 2007). Further transposon mutagenesis, performed in a more recent study, revealed that aside from *aioS* and *aioR* gene products, *rpoN*, encoding an alternative sigma factor (σ54) of RNA polymerase, was also necessary for As(III) oxidase synthesis. As the *aio* operon is preceded by a σ54 promoter in *H. arsenicoxydans*, as well as in other known AioSR-possessing As(III) oxidizers, it is likely that the AAA+ ATPase domain of AioR interacts with RpoN thereby activating transcription of the *aio* operon. Another gene identified in the microarray experiments was *dnaJ*, encoding a heat-shock protein, Hsp40, which might facilitate folding of AioR. None of the *aioS*, *aioR*, *dnaJ* and *rpoN* mutants was capable of synthesising Aio enzyme, as was confirmed by Western immunoblotting analysis. Quantitative RT-PCR was used to show that *aio* expression levels increased significantly on exposure to As(III), but this increase was not observed in any of the mutants.

In temporal transcriptomics analysis, numerous genes were regulated by As(III) exposure in either early or late stages of the response to the metalloid (Cleiss-Arnold *et al.*, 2010). Western immunoblotting detection indicated that Aio was not synthesised within 15 minutes of the addition of As(III), when many genes involved in the immediate stress-response were upregulated. In contrast, As(III) oxidase, AioS and AioR can be detected in the later stages of cell growth, after prolonged exposure (6–8 hours) to As(III).

9.4   QUORUM SENSING

The term 'quorum sensing' describes the phenomenon that enables the bacterial cell to sense cell density in its surroundings. Bacteria release a range of small cell-to-cell signalling molecules that are referred to as 'pheromones' or 'autoinducers'. Once the threshold concentration of an autoinducer is reached, a coordinated change in behaviour is initiated by a specific two-component signal-transduction system (for a recent review see Frederix and Downie, 2011).

In *Agrobacterium tumefaciens* str. 5A (Kashyap *et al.*, 2006) As(III) oxidase is expressed in a late-log phase cell culture even in the absence of As(III) in the growth medium. Furthermore, when filter-sterilised culture broth or ethyl acetate extracts from cell cultures in late log phase were added to the growth medium of cells in early log phase, RT-PCR clearly showed induction of As(III) oxidase expression. Ethyl acetate extraction is a commonly used procedure for isolation of N-acylhomoserine lactones (AHLs) from *A. tumefaciens* cultures. AHLs are well characterised autoinducers, and they may play a role in quorum-sensing As(III)-independent regulation of As(III) oxidase expression in this bacterium.

Quorum-sensing control of As(III)-oxidase expression is also implicated in *Ochrobactrum tritici* str. SCII24 (Branco *et al.*, 2009). While As(III) is oxidized by wild-type *O. tritici* str. SCII24, strain 5bv11 cannot oxidize As(III). The introduction of a plasmid carrying the structural As(III) oxidase and cytochrome *c* genes (P*aoxABC*) to 5bv11 enabled it to oxidize As(III). However, in this strain As(V) production was detected only in late log phase, a characteristic feature of quorum-sensing-linked control. Further confirmation that As(III)-oxidase expression in *O. tritici* is also regulated by quorum-sensing mechanisms comes from the observation that the presence of the AioSR system was not necessary for As(III)-oxidase synthesis. It should be stressed that the importance of As(III) oxidation in this heterotrophic alphaproteobacterium is not completely clear. While As(III) oxidation does not contribute to energy metabolism, and seems to be primarily a detoxification process, an insertion mutation in the As(III)-oxidase gene does not affect cell growth in the presence of As(III) (Branco *et al.*, 2008).

In *Herminiimonas arsenicoxydans*, although AioSR was not induced until 6–8 hours after initial exposure to As(III), at that point cells were still in the log phase of growth and no induction of quorum-sensing proteins was detected (Cleiss-Arnold *et al.*, 2010).

Whether the quorum-sensing signalling circuit converges with the AioSR TCS or forms an entirely independent signalling pathway remains to be determined. It is tempting to suggest that in *A. tumefaciens* the two signalling systems would have to be linked as *aioSR* and *aioAB* appear to be co-transcribed. However, Kashyap *et al.* (2006) demonstrated co-transcription in As(III)-exposed late-log phase cells, when a quorum-sensing mechanism might already be involved in regulation. In addition, it would be interesting to see if the As(III)-oxidase activity of *aoxR* mutant MSUAt1 is restored in the late log phase even though it is not detected in the first 15h of cell growth. An alternative scenario would be that AioS responds to both As(III) and an autoinducer such as AHL that would bind to AioS *via* their respective periplasmic binding proteins; however this cannot be the case in *O. tritici*, as in this organism As(III)-oxidase activity has been detected even in the mutant strains lacking AioSR.

9.5   HEAT-SHOCK PROTEIN DnaJ

One of the three Tn5 transposon-insertion mutations in *Herminiimonas arsenicoxydans* (which were found to lack As(III)-oxidase activity) disrupted the *dnaJ* gene (Koechler *et al.*, 2010). DnaJ is a heat-shock protein, Hsp40, which functions as a co-chaperone in an Hsp70 complex comprising DnaK-DnaJ-GrpE. The Hsp70 complex is required for folding of specific substrate proteins but has also been implicated in mRNA stability (Yoon *et al.*, 2008). Koechler *et al.* (2010) showed that the loss of the ability to produce As(V) is associated with the absence of *aioAB* transcription, suggesting that DnaJ/Hsp70 is somehow linked to transcriptional regulation rather than protein folding of AioAB. Koechler *et al.*, proposed several mechanisms for DnaJ involvement: DnaJ might assist in the folding of AioR, without which there would be no induction of the *aioAB* genes; alternatively, the RpoN sigma factor, which was shown to be needed for initiation of *aioAB* transcription, might be a DnaJ substrate. They also suggested that DnaJ might affect the stability of *aioAB* mRNA – some heat-shock proteins play a significant role in mRNA stabilisation and protection from nucleases. Interestingly, Koechler *et al.* also propose that DnaJ may be involved in post-translational regulation of AioAB, given that the AioA sequence of *H. arsenicoxydans* contains a Tat (Twin-arginine translocation) signal. Proteins targeted to the periplasm through the Tat-protein export pathway depend on Hsp70 activity for secretion (Tullman-Ercek *et al.*, 2007). While the involvement of DnaJ in AioAB folding cannot be excluded, DnaJ is clearly involved in regulation of As(III)-oxidase synthesis in *H. arsenicoxydans*.

The need for DnaJ for the folding of proteins that are exported to the periplasm through the Tat pathway provides another avenue for DnaJ involvement in As(III)-oxidase regulation. The periplasmic binding protein AioX, which is present in many As(III) oxidizers including *H. arsenicoxydans*, also contains the Tat sequence so its export may be DnaJ-dependent. If DnaJ supports the folding or export of AioX which is involved in direct As(III) sensing, as suggested by Cai *et al.* (2009) and demonstrated by Liu *et al.* (2011), then the AioSR-regulated expression of As(III) oxidase will be abrogated in the absence of DnaJ. This hypothesis needs to be tested experimentally. Notably, however, in *Rhizobium* NT-26 a putative periplasmic binding protein homologous to AioX, lacks the N-terminal Tat-signal peptide, suggesting that at least in this organism DnaJ may not be required for the regulation of As(III) sensing (Djordjevic, Santini and Wojnowska, unpublished observation). Furthermore, the absence of an obvious signal-peptide sequence raises the questions of whether and how AioX in NT-26 and its close relatives is exported to the periplasm.

9.6   CONCLUSIONS

Figure 9.4 represents a composite, species-independent view of the molecular components involved in regulation. As(III) enters the periplasm and is detected directly by AioX and indirectly by AioS, which initiates two-component signal transduction. Activated AioR induces *aio*

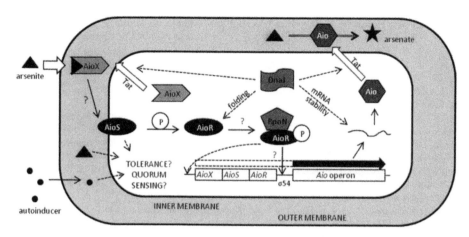

Figure 9.4.   Schematic diagram representing proteins and mechanisms implicated in the control of arsenite oxidase expression. P – phosphoryl group, Tat – twin-arginine translocation export pathway; question marks indicate unkown mechanisms, dashed arrows signify putative pathways.

operon expression with a σ54 promoter, most likely with the assistance of an alternative σ factor, RpoN. In some organisms the transcript may also include *aioSR*-mRNA. DnaJ may be essential for 1) maintaining *aioAB* mRNA stability, 2) AioR folding, or 3) translocation of As(III) oxidase and/or AioX into the periplasm. Quorum sensing – which triggers *aio* gene expression – and tolerance to As(III) may be regulated by AioS, As(III) itself or an unknown autoinducer.

The molecular basis of regulation of As(III)-oxidase gene expression involves a multitude of processes. Regulatory components may vary between organisms or depend on the phase of cell growth and the metabolic state. At the moment, very few organisms have been thoroughly investigated. More comprehensive analysis, applying a range of techniques to diverse organisms, is needed to expand our knowledge and understanding of the regulation of As(III)-oxidase expression. Future experiments should not only expand the information on the various genes and protein components of regulatory systems, but also explore how the multiple levels of control integrate their modes of action to produce the specific response.

ACKNOWLEDGEMENTS

Marta Wojnowska is funded by the Wellcome Trust PhD Interdisciplinary Programme in Structural, Computational and Chemical Biology at the Institute of Structural and Molecular Biology; London, UK.

REFERENCES

Branco, R., Francisco, R., Chung, A.P. & Morais, P.V.: Identification of an *aox* system that requires cytochrome c in the highly arsenic-resistant bacterium *Ochrobactrum tritici* SCII24. *Appl. and Env. Microbiol.* 75:15 (2009), pp. 5141–5147.

Cai, L., Rensing, C., Li, X. & Wang, G.: Novel gene clusters involved in arsenite oxidation and resistance in two arsenite oxidizers: *Achromobacter* sp. SY8 and *Pseudomonas* sp. TS44. *Appl. Microbiol. Biotechnol.* 83 (2009), pp. 715–725.

Cleiss-Arnold, J., Koechler, S., Proux, C., Fardeau, M.-L., Dillies, M.-A., Coppee, J.-Y., Arsène-Ploetze, F. & Bertin, P.N.: Temporal transcriptomic response during arsenic stress in *Herminiimonas arsenicoxydans*. *BMC Genomics* 11:709 (2010)

Ellis, P. J., Conrads, T., Hille, R. & Kuhn, P.: Crystal structure of the 100 kDa arsenite oxidase from *Alcaligenes faecalis* in two crystal forms at 1.64 Å and 2.03 Å. *Structure* 9 (2001), pp. 125–132.

Ferris, H.U., Dunin-Horkawicz, S., Mondéjar, L.G., Hulko, M., Hantke, K., Martin, J., Schultz, J.E., Zeth, K., Lupas, A.N. & Coles, M. The mechanisms of HAMP-mediated signaling in transmembrane receptors. *Structure*. 19:3 (2011), pp. 378–385.

Frederix, M. & Downie, A.J.: Quorum sensing regulating the regulators. *Adv. Microb. Physiol.* 58 (2011), pp. 23–80.

Gao, R. & Stock, A.M.: Biological insights from structures of two-component proteins. *Annu. Rev. Microbiol.* 63 (2009), pp. 133–154.

Huynh, T.N., Noriega, C.E. & Stewart, V.: Conserved mechanism for sensor phosphatase control of two-component signalling revealed in the nitrate sensor NarX. *PNAS* 107 (2010), pp. 21140–21145.

Kashyap, D.R., Botero, L.M., Franck, W.L., Hassett, D.J. & McDermott, T.R.: Complex regulation of arsenite oxidation in *Agrobacterium tumefaciens*. *J. Bacteriol.* 188:3 (2006), pp. 1081–1088.

Koechler, S., Cleiss-Arnold, J., Proux, C., Sismeiro, O., Dillies, M.-A., Goulhen-Chollet, F., Hommais, F., Lièvremont, D., Arsène-Ploetze, F., Coppée, J.-Y. & Bertin, P.: Multiple controls affect arsenite oxidase gene expression in *Herminiimonas arsenicoxydans*. *BMC Microbiol.* 10:53 (2010)

Laub, M.T. & Goulian, M.: Specificity in two-component signal transduction pathways. *Annu. Rev. Genet.* 41 (2007), pp. 121–145.

Liu, G., Liu, M., Kim, E.-H., Matty, W., Bothner, B., Lei, B., Rensing, C., Wang, G. & McDermott, T.R. *Environ. Microbiol.* Dec 19. doi: 10.1111/j.1462-2920.2011.02672.x. [Epub ahead of print]

Marina, A., Waldburger, C.D. & Hendrickson, W.A.: Structure of the entire cytoplasmic portion of a sensor histidine-kinase domain. *EMBO J.* 24 (2005), pp. 4247–4259.

Mascher, T., Helmann, J.D. & Unden, G.: Stimulus perception in bacterial signal-transducing histidine kinases. *Microbiol. Mol. Biol. Reviews* 70:4 (2006), pp. 910–938.

Muller, D., Médigue, C., Koechler, S., Barbe, C. *et al.*: A tale of two oxidation states: bacterial colonization of arsenic-rich environments. *PLoS Genetics* 3:4 (2008), pp. 518–530.

Neiditch, M.B., Federle, M.J., Pompeani, A.J., Kelly, R.C., Swem, D.L., Jeffrey, P.D., Bassler, B.L. & Hughson, F.M.: Ligand-induced asymmetry in histidine sensor kinase complex regulates quorum sensing. *Cell* 126 (2006), pp. 1095–1108.

Ninfa, A.J. & Magasanik, B.: Covalent modification of the glnG product, NRI, by the glnL product, NRII, regulates the transcription of the glnALG operon in *Escherichia coli*. *PNAS* 83 (1986), pp. 5909–5913.

Tullman-Ercek, D., DeLisa, M.P., Kawarasaki, Y., Iranpour, P., Ribnicky, B., Palmer, T. & Georgiou G.: Export pathway selectivity of *Escherichia coli* twin arginine translocation signal peptides. *J. Biol. Chem.* 282:11 (2007), pp. 8309–8316.

Sardiwal, S., Santini, J.M., Osborne, T.H. & Djordjevic, S.: Characterization of a two-component signal transduction system that controls arsenite oxidation in the chemolithoautotroph NT-26. *FEMS Microbiol. Lett.* 313 (2010), pp. 20–28.

Stock, A.M., Mottonen, J.M., Stock, J.B. & Schutt, C.E.: Three-dimensional structure of CheY, the response regulator of bacterial chemotaxis. *Nature* 337 (1989), pp. 745–749.

Stock, A.M., Robinson, V.L. & Goudreau, P.N.: Two-component signal transduction. *Annu. Rev. Biochem.* 69 (2000), pp. 183–215.

Weiss, S., Carapito, C., Cleiss, J., Sandrine Koechler, S. *et al.*: Enhanced structural and functional genome elucidation of the arseniteoxidizing strain *Herminiimonas arsenicoxydans* by proteomics data. *Biochimie* 91 (2009), pp. 192–203.

Yoon, H, Hong, J. & Ryu, S.: Effects of chaperones on mRNA stability and gene expression in *Escherichia coli. J. Microbiol. Biotechnol.* 18 (2008), pp. 228–233.

CHAPTER 10

Evolution of arsenite oxidation

Robert van Lis, Wolfgang Nitschke, Simon Duval & Barbara Schoepp-Cothenet

10.1 INTRODUCTION

Despite its low crustal abundance arsenic (As) is widely distributed in nature. As is commonly found in an insoluble form associated with rocks and minerals but also occurs as soluble inorganic oxyanions, arsenite (H_3AsO_3, more common in anoxic environments) and arsenate ($HAsO_4^{2-}/H_2AsO_4^-$, occurring primarily in aerobic environments) (see Chapter 1 for details). Arsenate [As(V)], and arsenite [As(III)] are both toxic to complex life, As(III) considered 100 times more toxic than As(V). As(V), a phosphate analog, interferes with normal phosphorylation processes by replacing phosphate, whereas As(III) binds to sulfhydryl groups of cysteine residues in proteins, thereby inactivating them (Smedley and Kinniburgh, 2002). Owing to the toxicity of As compounds, organisms, including prokaryotes, have evolved detoxifying energy-consuming systems to limit levels of these oxyanions in the cytoplasm (see Mukhopadhyay et al., 2002). However, prokaryotes have also been demonstrated to use As(V) and even As(III) as electron acceptor and donor, respectively, for their energy-conserving systems (see Stolz et al., 2006; Zannoni et al., 2009). The first report of bacterial As(III) oxidation dates back to 1918 (Green, 1918), and during the last decade, an increasing number of phylogenetically diverse As(III)-oxidizing bacteria have been isolated from different environments. Although As(III) oxidation may appear as a relatively exotic bioenergetic mechanism owing to the still comparatively small number of species known to perform this reaction, this paucity may reflect a sampling bias rather than a true scarcity of corresponding representatives. An important role for As(III) oxidation as a bioenergetic mechanism in the early Archaean era (more than 3 billion years ago) was already suggested a decade ago (Mukhopadhyay et al., 2002). Arsenic compounds are present in substantial amounts in hydrothermal environments which are frequently considered to resemble vestiges of primordial biochemistry. In the archaean environment these compounds will certainly have occurred predominantly in their As(III) state, owing to the absence of O_2.

The two enzymes currently known to carry out bioenergetic As(III) oxidation are arsenite oxidase (Aio; see also Chapter 7 for detailed description) and Arx, a variant of arsenate reductase (Arr). The fact that Arr, an enzyme initially shown to be a terminal reductase of energy-conserving electron-transfer chains, also mediates physiological oxidation of As(III) in certain species (as Arx) poses an evolutionary riddle which we will address in this contribution.

AioAB is a heterodimeric enzyme (Ellis et al., 2001) facing the periplasm. AioA belongs to the Complex Iron Sulfur Molybdoenzymes (CISM) superfamily (see Rothery et al., 2008) and AioB is a member of the Rieske protein superfamily (Lebrun et al., 2006). As for the scarcely characterized ArrAB enzymes, amino acid sequences and metal analyses of ArrA show that this protein is also a member of the CISM superfamily (Krafft and Macy, 1998; Saltikov and Newman, 2003; Afkar et al., 2003; Stolz et al., 2006; Malasarn et al., 2008). In several cases, the arr gene cluster contains a third gene, arrC. The corresponding gene product, ArrC, predicted to be devoid of any cofactor, probably serves as a membrane anchor and as the site of quinone redox reactions (see Stolz et al., 2006; Duval et al., 2008).

Clues to the ancient past of As(III) oxidation can be gathered from biology and geochemistry. While all groups working on this subject agree on the paleogeochemistry of As, the literature contains two differing points of view as to the nature of the enzymes oxidizing As(III) in the Archaean. Aio and Arx have both been suggested to be the ancestral As(III)-oxidizing enzyme.

Aio has been proposed on the basis of phylogenetic data (Lebrun *et al.*, 2003) and Arx based on biochemical and microbiological results (Richey *et al.*, 2009; Zargar *et al.*, 2010). Molecular phylogeny is in general the most appropriate way of inferring the evolutionary history of an enzymatic system. However, the reliability of phylogenetic tree reconstruction depends crucially on the quality of the dataset, i.e. the analysed sequences and their multiple alignments. Most sequences used in such an approach do not correspond to well-characterised enzymes and are thus only putative homologs, suggested by sequence similarity to a well-studied enzyme. In the case of enzyme superfamilies especially, it is essential to distinguish actual homologs from paralogs. This task is often achieved by taking into consideration the genomic context of a given gene and specific sequence idiosyncrasies related to specific functions (see Duval *et al.*, 2008). The reliability of multiple sequence alignments can in turn be improved by using available 3D structures and corresponding structural alignments. However, even in the best of all possible sequence worlds, phylogenetic trees reflect probabilities and thus need to be confronted with independently derived information to assess their chances of corresponding to reality. Consequently, in what follows we will briefly present the relevant structural and genetic data needed to interpret the analyses of the genomics and phylogenies of both enzymatic systems. For more details readers are invited to consult the Chapters dedicated to these aspects. The external parameters which we have in mind will be presented at the end of this contribution and pertain mainly to functional and thermodynamic constraints. Taking all these considerations into account we will try to establish an evolutionary timeline for the evolution of enzymatic oxidation of As(III). We propose a scenario which distinguishes processes that occurred already in the forebears of all living cells, now frequently called the Last Universal Common Ancestor (LUCA), i.e. prior to the divergence of the domains Archaea/Bacteria more than 3.4 billion years ago, from those arising after this divergence but still in the Archaean and finally from evolutionary innovations introduced or modified when oxygen levels started to rise – i.e. at the end of the Archaean era.

10.2   MOLECULAR DESCRIPTION OF ARSENIC BIOENERGETIC ENZYMES

10.2.1   *Arsenite oxidase*

Aio (formerly Aox, Aro or Aso; see Lett *et al.*, 2011) was first purified and structurally characterized from *Alcaligenes faecalis* (Anderson *et al.*, 1992; Ellis *et al.*, 2001). Its catalytic subunit AioA (formerly AoxB, AroA or AsoA) (about 825 residues) carries a molybdopterin cofactor together with a [3Fe-4S] center; this classifies the enzyme as a member of the CISM superfamily, though it does diverge from the rest of the members in that no amino acid coordinates to the molybdenum. The second sequence motif specific to AioA reflects the fact that it binds a [3Fe-4S] center instead of the more usual [4Fe-4S] centers of other members of the CISM superfamily. This [3Fe-4S] cluster is coordinated by the Cys21-X_2-Cys24-X_3-Cys28 motif (Fig. 10.1a). In all cases, the AioB (AoxA, AroB or AsoB) subunit contains the Twin-arginine translocation (Tat) signal sequence RRXFL, responsible for the translocation of the enzyme to the periplasm (Fig. 10.1b). The detailed structure of Aio is described in Chapter 7 but overall structure and location of Aio in the cell are illustrated in Figure 10.2 since they are necessary to the understanding of integration of the enzyme in the cellular metabolism. AioB (around 170 residues), is a member of the Rieske protein superfamily by virtue of its [2Fe-2S] center and protein fold (see also Lebrun *et al.*, 2003; Duval *et al.*, 2010). It contains a specific sequence motif CXHX$_n$CX$_2$H (Fig. 10.1c), – shared with the Rieske proteins from *bc*-complexes (Ellis *et al.*, 2001; Lebrun *et al.*, 2003; Lebrun *et al.*, 2006; Duval *et al.*, 2010) – which provides the two Cys and two His residues responsible for cluster binding. Aio has been purified and/or characterized from several bacteria (Lebrun *et al.*, 2003; Santini and vanden Hoven 2004; vanden Hoven and Santini, 2004; Silver and Phung, 2005; Duquesne *et al.*, 2008; Prasad *et al.*, 2009; Lieutaud *et al.*, 2010).

Several *aio* operons have been sequenced, revealing heterogeneity in their organization (see Chapters 5 and 8). Although the order *aioBA* is conserved, the positions of other elements are not

```
a   AioA_AerPe    MHVYPR-MGRVPLPPKGAQHYTTMCQFCNVGCGYDVYVWP
    AioA_PyrCa    -MSFRF-TGRVPLPPPDAERYTTMCQFCNVGCGYDVYVWP
    AioA_ChlPh    MSLFDR-KDTLPIPPKDAEKYTTVCQYCSAGCGYNVYVWP
    AioA_TheTh    MALIPR-RDRLPIPPKNAKVYNQVCQYCTVGCGYKVYVWP
    AioA_ChlAu    MPVYRP-ADRMVLPPVDAEKYQTVCHYCIVGCGYHVYKWP
    AioA_NT-26    -MAFKRHIDRLPIIPADAKKHNVTCHFCIVGCGYHAYTWP
    AioA_AlcFa    ----GCPNDRITLPPANAQRTNMTCHFCIVGCGYHVYKWP
    AioA_S22      ---MSAPKDRITLPPKDAARTNMTCHFCIVGCGYHVYKWP
    AioA_HerAr    ---MSKNRDRVALPPVNAQKTNHTCHFCIVGCGYHVYKWD
    AioA_ThiSp    ---MSQFKDRVALPPADAQKTNLTCHFCIVGCGYHAYTWD
```

```
b  AioB_AerPe    ---------------ITRREFLAAAGAG      c   AioB_AerPe    GPDKDIVAFSMLCTHMGGFLIFD-GNTKTLICPLHFSQFD
   AioB_PyrCa    MSEGEKKEGVKRPDSTRRTVVAAAGAL           AioB_PyrCa    GPDGDVVAFVNVCTHMGGPLIYVPDTN-CAVCQLHYTQFD
   AioB_ChlPh    ---------------ISRRKFIRTSAAS          AioB_ChlPh    GPKKNVVAFSTLCTHMGCPATYN-NG--RLVCKCHYSMFD
   AioB_TheTh    ---------------MTRRRFVQLTAAA          AioB_TheTh    GRERDIVAFSALCTHMGCPVQYE-EG--RFICRCHYSMFD
   AioB_ChlAu    ---------------LDRRSFLKLSSIS          AioB_ChlAu    GPDGDIVAYSGLCTHMGCPMTYD-PATKIFVCPCHYSHFD
   AioB_NT-26    ---------------IGRRQFLRGGALA          AioB_NT-26    GPDGDIVGFSTICPHKGFPLSYS-ADNKTFNCPGHFSVFD
   AioB_AlcFa    --------MSDTINLTRRGFLKVSGSG           AioB_AlcFa    GPDDDIVAYSVLCTHMGCPTSYD-SSSKTFSCPCHFTEFD
   AioB_S22      --------MSDPQIFTRRGFLKLSGTG           AioB_S22      GPDHDIVAFSTMCTHMGCPTVFDNKTK-TFKCPCHFSEFD
   AioB_HerAr    ---------------TSRRNFLKIAGSS          AioB_HerAr    GPNNDIVAHSILCTHMGCPVSYD-ASAKTFKCPCHFSVFD
   AioB_ThiSp    ----------MTEKVSRRIFLKVAGTS           AioB_ThiSp    GPDGDIVAYSNLCTHMGCPLMYDPATQR-FKCPCHYSMFD
```

Figure 10.1.   Conserved sequence motifs of AioA and AioB. (a) the [3Fe-4S] cluster binding motif in AioA, (b) the Tat sequence and (c) the [2Fe-2S] cluster binding motif in AioB. Selected sequences are from *Aeropyrum pernix* (AerPe), *Pyrobaculum calidifundis* (PyrCa), *Chlorobium phaeobacteroides* (ChlPh), *Thermus thermophilus* (TheTh), *Chloroflexus aurantiacus* (ChlAu), *Rhizobium* NT-26 (NT-26), *Alcaligenes faecalis* (AlcFa), *Ralstonia* S22 (S22), *Herminiimonas arsenicoxydans* (HerAr) and *Thiomonas* 3As (ThiSp).

(Silver and Phung 2005; Stolz *et al.*, 2006). A gene coding for a *c*-type cytochrome occurs in most, but not all, of the known proteobacterial *aio* clusters (Kashyap *et al.*, 2006; Duquesne *et al.*, 2008; Cai *et al.*, 2009; Lieutaud *et al.*, 2010; Schoepp-Cothenet unpublished data), and does not occur in the clusters of species from any other phylogenetic lineages. In several proteobacterial strains, the operon contains *aioS* and *aioR* genes, coding for a two-component system, AioS (sensor kinase)/ AioR (regulator), which regulates the expression of the *aio* genes (Kashyap *et al.*, 2006; Branco *et al.*, 2009; Cai *et al.*, 2009; Koechler *et al.*, 2010; Sardiwal *et al.*, 2010). However, these genes are not present in other phyla. *aioBA* are therefore the only constant elements of the *aio* cluster that can be used as markers for a genomic analysis covering all species.

10.2.2   Arsenate reductase

Anaerobic microbes capable of respiring As(V) were first discovered more than a decade ago (Ahmann *et al.*, 1994; Laverman *et al.*, 1995). The enzyme performing this respiration was shown to be arsenate reductase (Arr). So far, Arrs from three organisms have been characterized: *Chrysiogenes arsenatis* (Krafft and Macy, 1998), *Bacillus selenitireducens* (Afkar *et al.*, 2003) and *Shewanella* sp. ANA-3 (Malasarn *et al.*, 2008). Recently, an enzyme closely related to Arr in sequence has been isolated from *Alkalilimnicola ehrlichii* MLHE-1 and partially characterized (Richey *et al.*, 2009; Zargar *et al.*, 2010); despite its sequence similarity it differs from Arr in that it oxidizes As(III) rather than reducing As(V); consequently it has been named Arx. No structure is yet available but sequence and metal analyses have established that ArrA/ArxA (around 850 residues) also belongs to the CISM family. Indeed, these enzymes exhibit the conserved motifs of a molybdopterin binding (Fig. 10.3c) and a [4Fe-4S] binding domain (Fig. 10.3b). In addition, ArrA/ArxA contains a Tat signal peptide (Fig. 10.2a) which allows for its translocation to the periplasm (Krafft and Macy 1998; Afkar *et al.*, 2003; Malasarn *et al.*, 2008). From sequence analysis, ArrB/ArxB (around 230 residues) is predicted to contain four [4Fe-4S] clusters, or one [3Fe-4S] plus three [4Fe-4S] clusters (Saltikov and Newman 2003; Richey *et al.*, 2009).

All purified Arr/Arx contain only two subunits, whereas genomic analyses have revealed several strains in which the *arr/arx* cluster includes a third gene named *arrC/arxC*, coding for

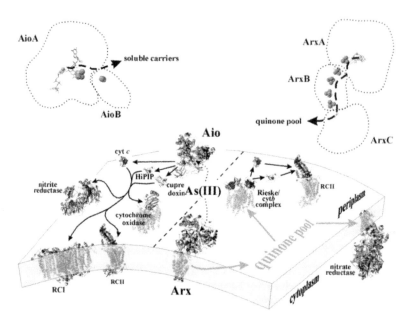

Figure 10.2.   Integration of Aio and Arx into energy converting metabolisms. Top: Cofactor arrangement within Aio and Arx and catalytic electron transfer pathways (broken arrows). Outlines of individual subunits are indicated by dotted lines. Bottom: The scheme features possible integrations of Aio (left half) and Arx (right half) into diverse bioenergetic chains. We show only those chains for which likely candidate organisms have been isolated or are suggested by genomic data. The structure of Arx has been modeled based on the 3D structure of the Psr enzyme from *Thermus thermophilus*. Electron transfers are denoted by continuous arrows (in light shade if occurring within the membrane and mediated by lipophilic carriers) and the delivery of the ultimate electron donating substrate As(III) is shown by dotted arrows. High Potential Iron Sulphur Protein (HiPIP), cytochrome *c* (cytc) or cupredoxin are possible electron acceptors for Aio. Cytochrome *c* oxidase, nitrite reductase and photosynthetic reaction centres (type I or II; RCI or RCII) may serve as terminal oxidases in Aio-based chains. Only quinones can shuttle electrons from Arx to the Rieske/cyt*b* complex which in turn again reduces the small carriers (HiPIP, cyt*c* or cupredoxins as mentioned above). RCII or nitrate reductase are possible terminal oxidases in Arx-based chains.

a structural protein (Malasarn *et al.*, 2004; Stolz *et al.*, 2006; Richey *et al.*, 2009). This additional subunit (around 290 residues and apparently devoid of any cofactor) shows a close homology to PsrC, a membranous quinone-oxidizing protein that is a constituent of the polysulfide reductase (Psr) complex (Dietrich and Klimmek 2002). By analogy, it has therefore been proposed that *arrC/arxC* might anchor ArrAB /ArxAB to the cytoplasmic membrane and oxidise or reduce quinones (Duval *et al.*, 2008, see Fig. 10.3). Whenever the *arr* cluster contains the *arrC* gene it shows a CAB order, whereas the three known *arx* clusters all show an ABC order. Some additional heterogeneity in *arr/arx* cluster organization has emerged from genomic analyses (Stolz *et al.*, 2006) since genes encoding regulatory components are sometimes but not always part of the cluster. Either *arrCAB* or *arxABC* or *arrAB* are therefore the organizations to be expected in genomic analyses.

10.3   FUNCTION OF THE ENZYMES

10.3.1   *Function of Aio*

Only scant data have so far been reported on the enzymology of Aio and its precise role in bioenergetic chains (see Chapter 7). We have, however, tried to summarise its possible roles, in

Figure 10.3. Conserved sequence motifs of Arr/ArxA. (a) the Tat sequence, (b) the [4Fe-4S]-binding domain, (c) the molybdenum-coordinating motif. Selected sequences are from *Wolinella succinogenes* (Wolsu), *Desulfitobacterium hafniense* (Desha), *Alkaliphilus oremlandii* (Closp), *Alkaliphilus metallidigenes* (Alkme), *Chrysiogenes arsenatis* (Chrar), *Shewanella* sp. ANA-3 (Shean), delta proteobacterium MLMS-1 (MLMS), *Halarsenatibacter silvermanii* (Halsi), *Alkalilimnicola ehrlichii* (Alkeh), and *Halorhodospira halophila* (Halha).

Figure 10.2. Bacteria using Aio to oxidize As(III) can be divided into two groups: (i) chemolithoautotrophs (aerobes or anaerobes, using As(III) as the electron donor and CO_2/HCO_3^- as the sole carbon source) and (ii) heterotrophs, growing in the presence of organic matter [for recent reviews see Silver and Phung (2005) and Stolz *et al.* (2006)].

Since As(III) is a two-electron-donating substrate, the catalytic turnover is assumed to start with oxidation of As(III) at the molybdenum center (which can accept up to two electrons). Several lines of evidence suggest that As(III) oxidation ultimately results in the reduction of a soluble cytochrome in most Proteobacteria: the co-purification of cytochromes with the enzyme (Anderson *et al.*, 1992; vanden Hoven and Santini, 2004), the presence of cytochrome-encoding genes in *aio* gene clusters (see above); and the fact that As(III) oxidation in *Ochrobactrum tritici* requires the cytochrome encoded in the *aio* operon (Branco *et al.*, 2009). The electron acceptor of Aio in other organisms is unknown. As azurin, which is reduced by Aio (Anderson *et al.*, 1992, Lieutaud *et al.*, 2010), auracyanin, another cupredoxin-type copper protein, may be the electron acceptor in *Chloroflexus aurantiacus* (see below). Since High Potential Iron Sulfur (HiPIP) has been shown to be the soluble electron acceptor in several organisms replacing cytochrome, it may equally be the electron acceptor in some As(III)-oxidizing bacteria.

Most As(III) oxidizers characterized are mesophilic aerobes, in which the final electron acceptor of the metabolic pathway is O_2, probably *via* a cytochrome c/HiPIP/cupredoxin oxidase (see vanden Hoven and Santini 2004). There are also several strains that contain *aio* genes and oxidize As(III) under various anaerobic conditions (Rhine *et al.*, 2007; Chang *et al.*, 2010; Sun *et al.*, 2009; Sun *et al.*, 2010a and b). Under these conditions, Aio may be part of an anaerobic respiratory pathway. For example, two proteobacteria are known to use Aio to chemoautotrophically oxidize As(III) with nitrate (NO_3^-) as electron acceptor (Rhine *et al.*, 2007; Sun *et al.*, 2009) suggesting a link between Aio and the denitrification pathway. More recently, several strains have been shown to use Aio in association with the denitrification pathway (Sun *et al.*, 2010a). The connection of Aio with periplasmic soluble electron carriers suggest the soluble periplasmic nitrite reductase as most probable denitrification enzyme linked to As(III) oxidation. A link between Aio activity and chlorate reduction has also been shown (Sun *et al.*, 2010b). Finally, it has been suggested that Aio is linked to photosynthesis in certain organisms. Genomic surveys have revealed three photosynthetic bacteria with *aio* genes (Lebrun *et al.*, 2003; Duval *et al.*, 2008). One of them, *Chloroflexus aurantiacus*, has been shown to express the enzyme during photosynthetic growth (Lebrun *et al.*, 2003; Duval *et al.*, 2010). No data are yet available to establish Aio-controlled As(III) oxidation that reduces photooxidized reaction centres; but light has been shown to induce oxidation of both auracyanine and Aio-Rieske (Zannoni and Ingledew 1985; McManys *et al.*, 1992), suggesting that these components are involved in the photosynthetic pathway.

10.3.2 *Function of Arr/Arx*

Historically, the enzyme Arr was first shown to play the role of an anaerobic terminal oxidase in prokaryotic energy conversion, and thus to be situated at the opposite end of bioenergetic electron transfer chains with respect to its "sister-enzyme" Aio. In most As(V)-respirers, various carbon sources (e.g. lactate, succinate, pyruvate) are the initial electron donors. As discussed above, the presence of the membrane anchor protein ArrC and its strong relatedness to the PsrC subunit of Psr suggest that in species where ArrAB is complemented by ArrC, the electrons used to reduce arsenate are derived from oxidation of a quinol binding at the ArrC subunit. In *Shewanella* species, which have no C-subunit, CymA, another membrane-attached protein interacting with quinone (Murphy *et al.*, 2007) is known to play the role of the quinol oxidizer. Arr is therefore part of a typical anaerobic respiratory chain, with organic donors feeding electrons into the liposoluble quinone pool from which Arr draws the electrons required for As(V) reduction. Structural homology between Arr and Psr, inferred from the similarity of all three of their constitutive subunits, suggests homologous function, even for intra-enzyme electron transfer, i.e. that electrons are transferred *via* the four Fe-S centres in the ArrB subunit to the [4Fe-4S] and molybdenum centres of the catalytic A subunit and then on to As(V).

An Arr-type enzyme has also been suggested to act as arsenite oxidase in the anaerobic respiration of either nitrate (Hoeft *et al.*, 2007) or selenate (Fisher and Hollibaugh, 2008), or in a photosynthetic process (Kulp *et al.*, 2008). If Arr could reverse its function, as in the case of succinate dehydrogenase/fumarate reductase (Maklashina and Cecchini, 1999), the function of the reversed enzyme, called Arx, would be to reduce quinones by oxidizing As(III). Since NO_3^- reductase (Nar) oxidizes quinols to reduce NO_3^- (for a review see Gonzales *et al.*, 2006), the link between Arx and Nar appears straightforward, though still to be established. Although their functioning is suggested by growth experiments, the detailed molecular make-up of photosynthetic and selenate respiration pathways implying Arx also still have to be determined. PHS-1, a γ-Proteobacterium of the family Ectothiorhodospiracea, contains Arx which has been shown to oxidise As(III) during photosynthesis. This strain nevertheless uses H_2S as electron donor to support photosynthetic growth (Kulp *et al.*, 2008). Our own results on *Halorhodospira halophila* (results to be published elsewhere) furthermore suggest that H_2S is indispensable for phototrophic growth on As(III). The electrons derived from the oxidation of As(III) by Arx therefore appear incapable of reducing CO_2 for incorporation into biomass by a reverse electron flow; the corresponding bioenergetic chain remains unclear. Questions also remain as to how strain ML-SRAO (Fisher and Hollibaugh, 2008) couples As(III) oxidation to selenate reduction. We tentatively propose a pathway similar to that established in *Thauera selenatis*, in which electrons originating from the Arx enzyme are channeled towards selenate reduction *via* the liposoluble quinone pool, the Rieske/*cytb* complex and a soluble cytochrome, and eventually into selenate reductase (Lowe *et al.*, 2010).

10.4   PHYLOGENETIC ANALYSIS OF Aio AND Arr

10.4.1 *The phylogenetic approach; caveats and solutions*

10.4.1.1 *Data mining and sequence analyses*
When only a few representatives of an enzyme family have been characterized, phylogenetic analysis is difficult. As discussed in the paragraph above, this is the case for Aio and even more

for Arr and Arx. Even so, a number of phylogenetic analyses have been published on partial Aio and Arr/Arx sequences amplified from environmental samples. These sequences will probably represent enzymes performing the respective redox conversions of As, but they cannot be directly compared with one another since they have been obtained with different primers; nor can they be aligned with paralogous sequences (see below: 10.4.1.2) since they always correspond to less than 150 residues of the full-length sequence. It is tempting, therefore, to retrieve sequences for analyses through BLAST searches of genome databases. However, since it is premature to assign a physiological function to a given gene product based only on sequence homology, one has to be careful when performing simultaneous phylogenetic analysis on both Aio and Arr/Arx. First, some retrieved sequences are incorrectly annotated. Identifying enzymes based solely on annotation would produce distorted phylogenies and should therefore be avoided. As previously reported (Duval *et al.*, 2008), BLAST scores yield strong similarities between the catalytic A-subunits of Arr/Arx and of the enzyme Psr, as well as of tetrathionate reductase. Only in-depth examination of gene cluster organization and scrutinization of retrieved sequences for conserved motifs (see Duval *et al.*, 2008) can help to avoid misinterpreting gene clusters. Lack of such studies explains the high frequency of annotation errors. For example, Kulp *et al.* (2008) mistakenly identified a Psr sequence from *Pyrobaculum aerophilum* as Arr, and inferred a misleading tree topology [as discussed in Schoepp-Cothenet *et al.* (2009a)]. Kulp *et al.* based their identification solely on proximity in BLAST searches, whereas phylogenetic gene cluster organization and the lengths of the B and A subunits all show that the enzyme is part of a *psr* cluster rather than a member of an *arr* clade.

Similar problems arise while searching for *aio* sequences by genomic analyses. Genome surveys detected four gene clusters similar to *aioBA* (Duval *et al.*, 2008) where the gene coding for a Rieske protein is upstream of that for the molybdenum-subunit. These sequences differ significantly from those of true *aio* genes by several distinctive stretches of residues and by the total length of both *aioA* and *aioB*. The fact that none of their parent species has been shown to oxidize As(III) suggests that this clade represents a new subclass of molybdopterin-enzymes with a hitherto unknown function.

10.4.1.2 *Sequence alignment, tree reconstruction and interpretation*
Although automated multiple alignment algorithms and the resulting tree reconstructions have correctly revealed the clades of Aio, Arr/Arx and Psr, the phylogenetic relationships between these subfamilies has lacked robustness, so the positions of the roots has been unclear. A similar problem occurred in the case of the Rieske protein superfamily (Lebrun *et al.*, 2006), including AioB and PetA (from the Rieske/cyt*b* complexes). To improve the reliability of the multiple alignments, AioB/PetA sequences alignments were performed based on available 3D structures when possible, and the same approach was taken for the molybdoenzymes. X-ray crystal structures have been reported for members of three of the four enzyme families dealt with in this contribution, *i.e.* Nar (1Y4Z), Aio (Q7SIF4) and Psr (2VPZ). Unfortunately, no structure is yet available for Arr. Since the Mo-subunit is the only protein common to all these enzymes, our phylogenetic analysis has been performed on this subunit only. The catalytic Mo-subunits of the respective enzymes were structurally superimposed, and the sequence alignments generated.

In several articles on the evolution of As metabolism, the mere presence of As redox-converting activities in both Archaea and Bacteria has been taken as evidence for an ancient origin of this metabolism (Kulp *et al.*, 2008; Stolz *et al.*, 2010). Frequently, the presence of well-separated clusters of archaeal and bacterial sequences in phylogenetic trees (Kulp *et al.*, 2008), and some-times even just the existence of both archaeal and bacterial representatives, is taken as evidence for the occurrence of the respective enzyme in LUCA. However, early but post-divergence hori-zontal gene transfer between Archaea and Bacteria would yield equivalent phylogenies. To avoid this ambiguity, the respective trees must be rooted by paralogous sequences whenever possible. In our case, each of the subfamilies (Aio, Arr, Psr and Nar) constitutes a root for the others. The position of the root between Archaea and Bacteria is then used as a criterion for likely pre-divergence origins. Furthermore, the global congruence of the analyzed tree with current species

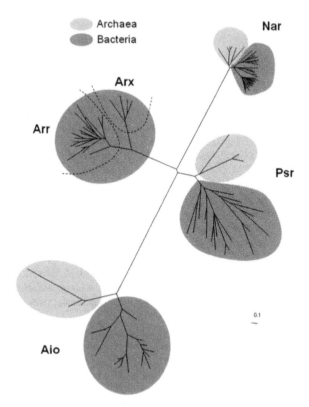

Figure 10.4.   Schematic plylogenetic (unrooted neighbor joining) tree of Psr, Nar, Aio and Arr/Arx. The clades corresponding to polysulfide reductase (Psr), arsenite oxidase (Aio) and nitrate reductase (Nar) resemble species trees and feature a prominent Archaea/Bacteria cleavage, and their roots lie between the archaeal and bacterial subtrees. The combined occurrence of these features suggests that these enzymes were present in LUCA. This is not the case for Arr/Arx.

trees should also be taken into account. There is, of course, no perfectly "vertical-inheritance" tree but in the case of rampant horizontal gene transfer in a given tree, we will remain cautious towards interpreting this tree as indicating pre-divergence origins, even if the root-criterion mentioned above is fulfilled.

10.4.2   *Phylogenetic distribution of Aio*

Phylogenetically, Aio is much more widespread than Arr, and the screening of new environments such as hydrothermal, saline and/or sulfide rich environments (Donahoe-Christiansen *et al.*, 2004; D'Imporio *et al.*, 2007; Connon *et al.*, 2008; Hamamura *et al.*, 2009; Handley *et al.*, 2010) and the use of newly designed primers (Hamamura *et al.*, 2009) reveal an ever wider occurrence. After the new sequences are included, the phylogeny of "genuine AioA" (excluding "Aio-like" sequences from Duval *et al.*, 2008), corroborates the conclusions arrived at in previous works suggesting that this enzyme was already present in LUCA (see Lebrun *et al.*, 2003; Duval *et al.*, 2008).

As shown in Figure 10.4, the Aio clade does indeed split into clear-cut and distinct bacterial and archaeal clusters, and the root in this composite Mo-enzyme tree lies between Archaea and Bacteria. Although specific lateral gene transfer events have been detected (see for example Lieutaud *et al.*, 2010), the tree shows a general congruence of the AioA tree with current species trees. An early origin of Aio is also implied by the position of its Rieske subunit (AioB) on a tree encompassing Aio and Rieske/cyt*b* complexes shown in Figure 10.5 (see also Lebrun *et al.*, 2006). Arr and Arx seem to be evolutionary variants of an ancient sulphur-reducing enzyme related to

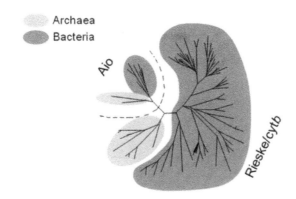

Figure 10.5. Schematic phylogenetic (unrooted neighbor joining) tree of Rieske proteins from Aio (AioB) and Rieske/cyt*b* complex (PetA). Both proteins show features (see legend of Fig. 10.4) suggesting that the corresponding enzymes were present in LUCA.

Psr, with a conserved structural build-up from the A, B and C subunits. Aio, by contrast, has a substantially different molecular architecture, with only the catalytic molybdopterin subunit in common with the Arr/Arx/Psr-type enzymes. These two distinct classes of enzymes provide good examples of the evolution of new enzyme complexes from universal redox protein building blocks (Baymann *et al.*, 2003).

Although none of the three Archaea strains known to have *aio* genes (*Pyrobaculum calidifondis*, *Sulfolobus tokodaii* and *Aeropyrum pernix*) has been characterized for its capacity to oxidise As(III), *Sulfolobus acidocaldarius* BC, a close relative of *S. tokodaii*, has been shown to do so (Sehlin and Lindström 1992). Unfortunately, no genome sequence is yet available for *S. acidocaldarius*.

10.4.3 *Phylogenetic distribution of Arr*

The phylogenetic tree reconstructed from sequences of the catalytic molybdopterin subunits is shown in Figure 10.4. Another version, differing in sequence sampling but agreeing in tree topology, has been published by Richey *et al.* (2009) and Zargar *et al.* (2010). The Arr-type sequences branch into two separate clades: "genuine Arr" and Arx. There are a large number of "genuine Arr" sequences; these come from Beta, Gamma, Epsilon and Deltaproteobacteria, Firmicutes, Chrysiogenetes and Deferribacteres, and thus cover a large part of the bacterial domain while so far being absent from Archaea (see Fig. 10.6). However, the topology of the Arr subtree differs significantly from that of the bacterial species tree. The Arx clade is typified by the enzyme encoded by the Mlg0216 gene from the γ-proteobacterium *Alkalilimnicola ehrlichii*. The product of this gene has been shown to oxidize As(III) instead of reducing As(V). Further members of this clade are the Mlg_2416 gene sequence, also from *A. ehrlichii*, a sequence from the Gammaproteobacterium *Halorhodospira halophila* and one from the Alphaproteobacterium *Magnetospirillum magnetotacticum* (Richey *et al.*, 2009; Zargar *et al.*, 2010). As detailed above, we did not include in the tree the Arx sequence identified in PHS-1 (Kulp *et al.*, 2008) since it contains only 150 residues. We recently characterized (van Lis *et al.* to be published) the As conversion capacities of *H. halophila* and observed that this strain not only converted As(III) to As(V) but also expressed the enzyme identified by genomic analysis by Richey *et al.* (2009) as Arx. It is therefore highly probable that this clade is homogenous for enzymes oxidising As(III).

No Archaeal *arr/arx* sequence was found in the genomic databases. Despite the fact that the Archaea *Pyrobaculum arsenaticum* and *P. aerophilum* have been shown to respire As(V) (Huber *et al.*, 2000), no amplification of *arrA* genes has been obtained (Malasarn *et al.*, 2004). The absence of any Archaeal ArrA/ArxA sequences strongly argues against the deep ancestry of the enzyme proposed by Oremland and co-workers (Hoeft *et al.*, 2010; Zargar *et al.*, 2010;

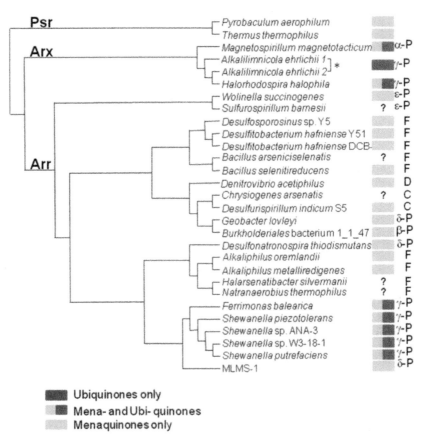

Figure 10.6.   Neighbor Joining-phylogram of the Arr/Arx clade rooted by Psr sequences showing distribution of quinone usage among Arr and Arx harboring species. Only the Arr clade contains species synthesizing MK alone; only the Arx clade contains a species synthesizing only UQ. Question marks denote unknown. *A. ehrlichii* possesses two *arx* gene clusters – named 1 and 2 in the tree – but only one has been shown to be involved in As(III) oxidation (Richey *et al.*, 2009). Phylogenetic affiliations: β-P, γ-P, δ-P, ε-P denote β-, γ-, δ-, and ε-Proteobacteria; F denotes Firmicutes; D denotes Defferribacteres and C denotes Chrysiogenetes.

Stolz *et al.*, 2010). Of course, the present failure to identify archaeal Arr/Arx sequences might be due to sampling bias (far fewer archaeal than bacterial genomes have so far been sequenced) rather than to true absence of Arr/Arx in Archaea. However, the internal topology of the Arr/Arx clade bears no resemblance to species trees (see Fig. 10.6), so Arr/Arx provides a textbook example of an enzyme probably distributed over a range of species by horizontal gene transfer. The currently available sequences would therefore suggest a late origin in a bacterium followed by dispersal into other bacterial phyla by lateral gene transfer. The composite phylogeny suggests that the original Arr evolved from a sulfur-compound-reducing enzyme (as indicated by its phylogenetic proximity to Psr) already present in LUCA (Duval *et al.*, 2008; Fig. 10.4).

10.5   TAKING BIOENERGETICS INTO ACCOUNT

10.5.1   *Lessons from quinone usage among Arr/Arx-harboring bacteria*

The structural similarity of Arr/Arx with the closely related Psr suggests that the former also uses electrons from quinol oxidation for As(V) reduction. The fact that in *Shewanella* the presence of CymA is indispensible for As(V) reduction (Murphy and Saltikov, 2007) further supports the notion that Arr/Arx activity is linked to a reaction with quinone. Several types of pool-quinones, such as ubi (UQ)-, plasto-, mena (MK)-, rhodo (RQ)-, caldariella- or sulfolobus-quinones have been identified so far that can be present either as sole quinone in a given species or coexisting in the same organism (Collins and Jones, 1981; Hiraishi and Hoshino, 1984; Schäfer *et al.*, 2001). While two distinct MK biosynthesis pathways are known, only one such pathway is known for UQ biosynthesis (for a recent and in-depth review see Nowicka and Kruk, 2010). We have performed a genomic survey searching for quinone biosynthesis pathway genes in order to establish the types of quinones used in the various species containing Arr/Arx genes (Fig. 10.6). In both the Arr and the Arx clades, several species synthesize both UQ and MK (Gammaproteobacteria such as *H. halophila*, *Shewanella* strains). However, species synthesizing only MK are observed only in the Arr clade (e.g. *Burkholderiales bacterium* 1_1_47, *Wolinella succinogenes*). And on the other hand, the only species that uses only the UQ biosynthesis pathway, i.e. *A. ehrlichii*, is a member of the Arx clade.

The As(III)/As(V) redox couple has an electrochemical midpoint potential of $+60\,\mathrm{mV}$ (with respect to the standard hydrogen electrode, SHE). To be energetically relevant, a quinone must be substantially more reducing than the As(III)/As(V) couple in the As(V) reduction pathway and substantially more oxidizing than the As(III)/As(V) couple in the As(III) oxidation pathway. Whereas MK- and RQ-quinones feature a similar redox midpoint potential of $-70\,\mathrm{mV}$, UQ-, plasto-, caldariella- and sulfolobus-quinones have more positive redox midpoint potentials at $+100\,\mathrm{mV}$. Only the first two quinones therefore appear energetically favourable for As(V) reduction. The latter four quinones, by contrast, seem energetically well-suited for As(III) oxidation. We thus hypothesize that all the Arr-harboring strains might oxidize MK pool-quinones *via* an As(V) reduction process whereas the Arx-harbouring strains might reduce UQ pool-quinones *via* an As(III) oxidation process (Fig. 10.7). This would be comparable to the observation that quinol-fumarate reductase in *E. coli* uses MK for reduction of fumarate and UQ for oxidation of succinate (Maklashina and Cecchini, 1999).

Analysis of quinone-usage among prokaryotes indicates that MK is evolutionarily older than the higher-redox-potential quinones (see for example Schütz *et al.*, 2000). UQ appeared in the Gammaproteobacteria (Schoepp-Cothenet *et al.*, 2009b) as a response to the global environmental oxidation. From an inspection of the distributions of UQ and MK on the Arr/Arx phylogenetic tree (Fig. 10.6) two hypotheses can be proposed:

1) After the accumulation of O_2, i.e. in the proterozoic era after the Bacteria/Archaea cleavage, a Psr-type enzyme (from S metabolism already present in the Archaean) evolved into Arr to catalyse the reduction of the newly accumulating As(V), functioning as a MK/As(V) oxido-reductase. This Arr was then laterally transferred into bacteria with UQ (also newly evolved in response to the accumulation of O_2). In this new context, Arr would require little modification to acquire an As(III)/UQ oxido-reductase function (Arx).
2) Alternatively, the enzyme may have arisen in the Gammaproteobacteria in the proterozoic era, with an initial function as As(III)/UQ oxido-reductase (Arx), and then spread by lateral gene transfer throughout the bacterial domain, where (in the presence of MK) it acquired the role of MK/As(V) oxido-reductase (Arr).

Whichever of these hypotheses is correct, the Arr/Arx would be a post-O_2 enzyme, not an ancestral one.

10.5.2 *Was NO the first deep electron sink for Aio?*

To be biologically useful, the electron transfer reaction from As(III) to its ultimate electron acceptor must span a large enough difference of electrochemical potential. Based on the phylogeny results discussed in the previous section, Arr/Arx does not appear a likely candidate for As(III) oxidation in the Archaean. Aio, on the other hand, has been suggested to have had this function (Lebrun *et al.*, 2003; Duval *et al.*, 2008). In extant As(III) oxidizers with Aio, however, the terminal electron sink is typically O_2, which is generally considered to have been absent from the primordial environment. The prevalence of an oxidizing environment results from oxygenic photosynthesis' pumping O_2 into the biosphere only since around 2.5 billion years ago (Nitschke *et al.*, 1996; Xiong *et al.*, 2000; Baymann *et al.*, 2001; Tomitani *et al.*, 2006). Genomic surveys (see section 10.4.3) suggest that Aio was initially integrated into an anaerobic metabolism. Bacterial photosynthesis was proposed by Oremland and co-workers (Kulp *et al.*, 2008) to have been the ancestral anaerobic metabolism, providing the electron sink (i.e. the photooxidized reaction center pigment) for the reducing electrons abstracted from As(III). Photosynthesis, however, seems not to be ancient, i.e. pre-LUCA (see for example Baymann *et al.*, 2001). The fact that until 2007 As(III) oxidizers were only known to use O_2 or anoxygenic photosynthetic reaction centres as terminal electron acceptors made early-Archaean As(III) oxidation seem implausible. That view was changed by the discovery of proteobacteria able to chemoautotrophically oxidize As(III) using Aio with NO_3^- as electron acceptor (Rhine *et al.*, 2007; Sun *et al.*, 2009).

Reinforcing the message of the molecular phylogeny of Aio, the likely pre-LUCA origin of the Rieske-*cytb* complex (Fig. 10.5) also suggests that a highly oxidizing bioenergetic electron sink already existed in the early Archaean, since electron acceptors of the Rieske/*cytb* complex typically feature redox potentials higher than +100 mV. On the basis of a reevaluation of the phylogeny of the heme copper oxidase (HCO) superfamily and recent insights into structures and functions of the major subgroups of HCOs, we proposed nitric oxide (NO) reductase as the ancestor of the HCO superfamily (Ducluzeau *et al.*, 2009). NO would therefore be the likely ancestral substrate to the HCO superfamily in the earliest Archaean. In contrast to the case of oxygen, a source for mass production of NO and its derivatives (NO_3^-, NO_2^-) did exist on the early Earth (Navarro-Gonzàlez *et al.*, 2001). The denitrification pathway (or at least part of it) may thus represent an anaerobic respiratory process present already in the Archaean.

We have recently performed phylogenetic analyses of the other enzymes in the denitrification pathway. The phylogenies of Nar (see Fig. 10.4) and NO_2^- reductase (van Lis *et al.*, 2011) suggest that both enzymes were present in LUCA. Together these results suggest that anaerobic respiration of nitrogen oxides formed part of LUCA's bioenergetic respiration. It seems therefore that As(III) oxidation took place in the Archaean *via* Aio activity, connected not to aerobic respiration but to a partial denitrification pathway (see Fig. 10.7).

10.5.3 *Molecular constraints on Aio- and Arx-mediated electron transfer*

Chemiosmotic energy-converting mechanisms in all life on this planet are variations on a strikingly conserved theme. Electrons derived from reduced substrates enter a chain of membrane-integral and/or -associated enzymes and are channelled on towards terminal electron-accepting substrates. Some of the free energy available during individual electron transfer steps is stored in a transmembrane proton motive gradient. This proton motive gradient is subsequently used by ATP synthase to produce ATP.

The Aio and the Arx enzymes, although performing similar substrate conversions, are integrated into their parent bioenergetic chains in markedly different ways. Within Aio, the reducing equivalents resulting from As(III) oxidation are transferred from the catalytic Mo-center towards a [3Fe-4S] and then a [2Fe-2S] Rieske cluster to eventually reduce small soluble periplasmic electron shuttle proteins, such as type I cytochromes, cupredoxin-type copper proteins and possibly also HiPIPs (see sections 10.2 and 10.3; illustrated in Fig. 10.2). From then on, electrons directly flow to the terminal oxidases which may be denitrification enzymes (Fig. 10.7, columns 1 and 5), photosynthetic reaction centres (Fig. 10.7, columns 3 and 9) or O_2-reductases (column 7).

As discussed above, all Arx-containing clusters detected so far contain a gene coding for a PsrC-like membrane-integral subunit. Quinones are thus reduced by this PsrC-related membrane-anchor protein. We can conclude that Arx exchanges the electrons arising from As(III) oxidation with the liposoluble quinone pool of their respective host species (see Fig. 10.2 and Fig. 10.7, columns 2, 4, 6, 8 and 10). This dichotomy – between Aio on the one hand and Arx on the other – involves basic differences in the way they are embedded in particular types of electron

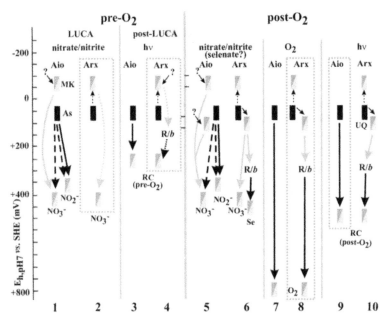

Figure 10.7.   Redox midpoint potential diagram for putative electron transfer chains performing the oxidation of arsenite mediated by Aio or Arx. Grey and black arrows stand for membrane-integral (i.e. quinone-mediated) and periplasmic electron transfer, respectively. The dashed arrows in the slots dealing with nitrate reduction denote pathways possible only in Archaea (see text). Dotted arrows mark electron transfer steps which are mechanistically conceivable but thermodynamically doubtful or impossible. The dotted gray boxes indicate cases for which no extant organism has been isolated so far. R/b, P$^+$/P, Se and hν stand for the Rieske/cyt*b* complex, the photooxidised reaction centre pigments, selenate and photons of visible light, respectively. The full black boxes denote the redox potential of the substrate arsenite whereas the mixed-coloured boxes denote the potentials of major redox compounds involved in the respective chains. The arrows terminated by a question mark stand for putative alternative or supplementary electron donors in columns 1, 4 and 5.

transfer chains. In terms of their function, Aio and Arx are strongly dissimilar in their interaction with bioenergetic chains. In Figures 10.2 and 10.7 we try to visualise possible schemes for the integration of Aio and Arx into As(III)-oxidising metabolisms. For example, having Aio and Arx work in denitrifying chains (Fig. 10.7, columns 1, 2, 5 and 6) predicts completely different pathways for the involved reducing equivalents in the two distinct systems as well as differences between Archaea and Bacteria. Aio in fact would not be able to transfer electrons to bacterial Nar, which faces the cytoplasm and takes its reductants from the quinone pool *via* a membrane-integral *b*-type cytochrome, but only to the periplasmic NO_2^- reductase (Fig. 10.2 and Fig. 10.7, columns 1 and 5). We would predict, therefore, that Aio-mediated growth on NO_3^- needs an additional, probably lower-potential, electron-donating substrate to reduce nitrate (indicated by the question mark arrows and the subsequent light gray arrows in columns 1 and 5); the subsequent step – reduction of NO_2^- to NO – can be driven by Aio turnover, so it is reasonable to expect to find growth on NO_2^- with As(III) as only electron source. In most Archaea, by contrast (see van Lis *et al.*, 2011), Nar is a periplasmic enzyme drawing its reducing equivalents from periplasmic electron carriers, so it is quite conceivable that As(III) can reduce NO_3^- *via* Aio (dashed arrows in Fig. 10.7, columns 1 and 5). The problem does not arise in Arx-run bacterial As(III) oxidation since the Arx enzyme injects electrons into the quinone pool (Fig. 10.7, columns 2 and 6). The archaeal case need not be considered since Arx seems to be restricted to Bacteria, as discussed above.

The reverse of this argument applies to the integration of Aio and Arx into photosynthetic systems (Fig. 10.7, columns 3, 4, 9 and 10). The photo-oxidized pigment is always re-reduced by periplasmic electron shuttle proteins. Aio is perfectly suited to perform this task (columns 3 and 9), whereas the reducing equivalents produced by Arx and injected into the quinone pool require an additional enzyme to mediate their transfer into the periplasm (see Fig. 10.2 and Fig. 10.7, columns 4 and 10). Typically, this role is played by the proton-pumping Rieske/cyt$b$ complexes.

None of these considerations apply to the involvement of Aio and Arx in aerobic respiration (Fig. 10.7, columns 7 and 8), since both quinol- and cytochrome-oxidising terminal O_2 reductases exist in most Bacteria and Archaea.

10.5.4 *Arsenite oxidation; no escape from thermodynamics*

Although all of the schemes shown in Figure 10.2 comply with the *molecular* requirements of electron transfer from As(III) to terminal acceptors, not all of them are thermodynamically feasible. Rieske/cyt$b$ complexes, for example, transfer electrons uphill, against the gradient of membrane potential, in order to translocate protons. In healthy active cells this membrane potential is about 150 mV, and enough redox energy must be available from the reductant-oxidant couple to drive a Rieske/cyt$b$ complex through at least one turnover. To obtain a viable bias of redox equilibria towards the forward reaction of this complex, i.e. ensuring that it can turn over continuously, roughly another 100 mV are required. Post-O_2 photosynthetic reaction centers provide photoinduced oxidants at about +400 to +500 mV and the redox energy contained in electron transfer from As(III) to these photooxidised pigments thus is more than sufficient, no matter whether Aio or Arx mediate this reaction (Fig. 10.7, columns 9 and 10). So far, no bacterium using Aio (column 9) under such conditions has been found. A column 10 chain is energetically more favourable for the organism since the operation of the Rieske/cyt$b$ complex enhances membrane potential generation. As discussed below, the PHS-1 case might correspond to such a chain. By contrast, for pre-O_2 photosynthetic reaction centers, the <200 mV difference in redox midpoint potential between the arsenic couple (+60 mV) and the photo-oxidised pigment (with average redox potentials of +250 mV; see Nitschke and Dracheva, 1995) is not sufficient to drive electrons through a chain involving a Rieske/cyt$b$ complex. Thus, while Aio-mediated As(III) oxidation in pre-O_2 (MK-based) photosynthesis, independent of the Rieske/cyt$b$ complex seems viable (Fig. 10.7, column 3), the corresponding mechanism performed by Arx, necessarily involving this enzyme (Fig. 10.7, column 4), is energetically challenging. Since the midpoint potentials of all cofactors of the Rieske/cyt$b$ complex strictly follow the redox upshift from menaquinone to ubiquinone (Schoepp-Cothenet *et al.*, 2009b), the position of the enzyme in Figure 10.7 is different in MK- and UQ-systems. Data on the roles of the two enzymes in photosynthetic systems conform well to this reasoning and that in 10.5.3:

1) The genomes of MK-only photosynthetic green sulphur and green filamentous bacteria contain Aio, but never Arx, that is, these species correspond to column 3 of Figure 10.7.
2) The Gammaproteobacterium *H. halophila*, a phototroph containing a light-driven electron transfer chain operating with MK (Schoepp-Cothenet *et al.*, 2009b), contains Arx, oxidizes As(III) during photosynthetic growth but requires a further electron donor, H_2S, to be able to grow (indicated by the question mark and the arrow pointing towards menaquinone MK in column 4 of Fig. 10.7).

It is noteworthy that another photosynthetic Gammaproteobacterium bacterium, PHS-1, has been reported to oxidize As(III) while growing photosynthetically and contains Arx (Kulp *et al.*, 2008). Neither quinone content nor the redox midpoint potential for the photo-oxidized pigment of this species has been determined so far. At present it is therefore unknown whether PHS-1 falls into the pre-O_2 (MK) (Fig. 10.7, column 4) or the post-O_2 (MK/UQ) category of Figure 10.7 (column 10). If the former, we would predict growth requirements similar to those of *H. halophila*. If the latter, then As(III) might indeed be a bioenergetic reducing substrate. As mentioned above, pre-O_2 systems in general contain only low-potential MK with redox potentials of -70 mV. For Arx to function in such a system, the electrons arising from As(III) oxidation would have to flow uphill against a $> 100\,mV$ redox barrier, to be transferred eventually to terminal acceptors. Although this is not thermodynamically impossible, such a redox-energy landscape would substantially slow the throughput of electrons (see Dietrich and Klimmek, 2002) to an extent that probably rules it out in practice. We therefore postulate from first bioenergetic principles that Arx cannot work in a pre-O_2 MK-based electron transfer chain. Both the phylogenetic distribution of Arx and the experimental quinone analysis of corresponding species so far support this scenario (Fig. 10.6).

Apart from the energetic constraints on individual redox reactions involving Aio and Arx, autotrophy relying on As(III) oxidation faces a more general thermodynamic problem. Living cells not only need ATP, produced almost universally *via* chemiosmotic electron transfer chains, but also require reducing equivalents in the form of NADH for a plethora of essential metabolic reactions, but the As(V)/As(III) couple is not nearly reducing enough to directly reduce NAD^+ to NADH ($E_m = -320\,mV$). In this respect, As(III) resembles Fe(II), another prominent chemolithotrophic electron donor. Electron transfer based on Fe(II) has been shown to resort to reverse electron transfer uphill from Fe(II) to NAD^+, made energetically possible by the overwhelming downhill electron transfer from Fe(II) to strong oxidants such as O_2 (Bird *et al.*, 2011). Electrons from weakly reducing donors, such as Fe(II) or As(III), might enter electron transfer chains either at the level of the quinone pool (i.e. upstream of the Rieske/*cytb* complex) or into the pool of soluble periplasmic electron carriers (i.e. downstream of the Rieske/*cytb* complex). Fe(II) oxidation uses the latter option, probably because the midpoint potential of the Fe(II)/Fe(III) couple (substantially more positive than the As(III)/As(V) couple) is insufficient to provide high reduction levels of the quinone pool required to drive electrons uphill towards NAD^+. The As(III)/As(V) couple, slightly more reducing than UQ quinones, may be just potent enough for the quinone pool to be used as entry point for reverse electron transfer. Furthermore, anoxygenic photosynthetic chains might in theory be able to generate sufficient chemiosmotic potential for Arx-mediated reverse electron flow to occur. The above-mentioned phototroph PHS-1 may carry out this reaction, but we know too little about it to decide.

By contrast, it appears extremely unlikely to us that As(III) could induce high reduction levels of the low-potential MK pool. Thermodynamics would then preclude autotrophic growth *via* Arx with As(III) as sole electron source, at least in species using MK as pool quinones such those using column 4-type chains. Although the phylogenetic picture may admittedly still change as more and more genomes are sequenced, the lack of autotrophic As(III) oxidation mediated by Arx fully supports this line of reasoning.

Aio, however, reduces soluble periplasmic electron carriers, so the resulting bioenergetic chain is analogous to the Fe(II) oxidation pathway (Fig. 10.7, column 7). Underscoring the predictive power of thermodynamic considerations, we cannot help noticing that the Aio-mediated, O_2-reducing As(III) oxidation of the proteobacterium NT-26 is the only pathway where chemoautotrophy using As(III) as sole electron donor has been unambiguously demonstrated (Santini *et al.*, 2000). So far, no example of Arx-mediated electron transfer towards oxygen (although thermodynamically favourable) has been reported (Fig. 10.7, column 8).

10.6 EVOLUTIONARY SCENARIO OF ARSENITE OXIDATION

A counterintuitive feature of the prokaryotic bioenergetic As metabolism is that As(V), while occurring mainly in aerobic environments, is exclusively converted *via* anaerobic metabolisms whereas As(III), more common in anoxic environments, is in most extant species oxidized *via* O_2-respiration. Proposing an evolutionary scenario for As(III) therefore is not straightforward. Two distinct scenarios have so far been proposed to explain the evolutionary history of this process. The scenario proposed by Oremland and co-workers postulates that Arr, originally functioning in reverse as observed with extant Arx, was responsible for anaerobic As(III) oxidation in the Archaean before acquiring the As(V)-reducing function seen in most currently known prokaryotes. In this scenario anoxygenic photosynthetic reaction centers would have provided the ultimate

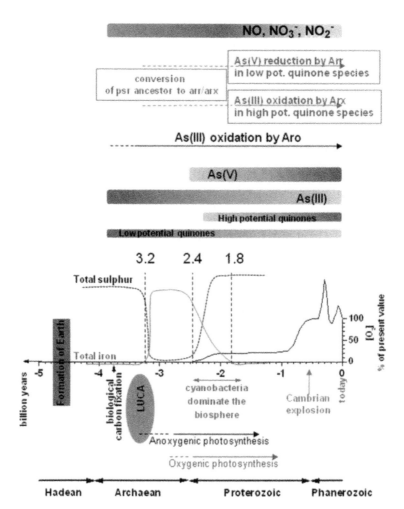

Figure 10.8.   Evolutionary scenario of As(III) oxidation by Aio and Arx. Phylogenetic analysis of Aio and Arr/Arx enzyme, together with phylogenetic analysis of sulphur metabolism and nitrogen metabolism enzymes (see section 10.4) suggest that Aio was already present in LUCA whereas Arx is a recent evolutionary innovation evolved from a Ttr or Psr enzyme. Functional analysis reveals 1) an implication of Aio in anaerobic respiration involving nitrogen oxides 2) a link between Arx activity and quinol reduction (see section 10.3). The evolution of Aio therefore appears to be closely linked to the evolutionary history of denitrification enzymes whereas the emergence of Arx appears to be dependent on the evolutionary history of quinones. Taking furthermore paleogeochemical data such as nitrogen oxides' and arsenic's abundances and redox states into account, we propose that Aio, connected to the denitrification pathway, has carried out the primordial biological oxidation of arsenite in the Archean in the presence of large amounts od nitrogen oxides. More recently, Arx, as a consequence to the appearance of high redox potential quinols, would contribute to the oxidation of arsenite performing an arsenite:quinone oxidoreduction reaction.

electron acceptor for As(III) oxidation. Aio would have appeared later in response to the global accumulation of O_2 (see for example Kulp *et al.*, 2008), O_2 providing the ultimate electron acceptor for As(III) oxidation by Aio. As we have pointed out above, the putative ancestral Arx-based bioenergetic chains resulting from this scenario face a number of mechanistic and thermodynamic difficulties (see sections 10.5.3 and 10.5.4). The data assembled in this chapter favor a very different scenario, summarized in Figure 10.8. Aio, integrated in the anaerobic respiration of nitrogen oxides, would have existed already in the anaerobic early Archaean as far back as LUCA. Only after the divergence of Bacteria and Archaea has Aio been recruited into more recent pathways such as photosynthesis, a bioenergetic process found only in Bacteria and of course their endosymbiotic descendants, the plastids. Aio activity would have been integrated into aerobic respiration only later – i.e. after the appearance of atmospheric oxygen defining the beginning of the proterozoic era. The global oxidation of the environment dated to about 2.5 billion years ago would then have led not only to the accumulation of the oxidized species As(V) but also to the emergence of high-potential quinones. Subsequently, an enzyme initially dedicated to reduction of sulfur compounds and related to the extant Psr enzyme evolved into the As(V)-reducing Arr enzyme. At roughly the same time, Arx originated by reversing the catalytic reaction of an Arr/Psr-related enzyme to oxidize As(III), resulting in reduction of a UQ pool. As far as we can see, this scenario does not suffer from any of the aforementioned mechanistic and thermodynamic inconsistencies.

REFERENCES

Afkar, E., Lisak, J., Saltikov, C., Basu, P., Oremland, R.S. & Stolz, J.F.: The respiratory arsenate reductase from *Bacillus selenitireducens* strain MLS10. *FEMS Microbiol. Lett.* 226:1 (2003), pp. 107–12.

Ahmann, D., Roberts, A.L., Krumholz, L.R. & Morel, F.M.: Microbe grows by reducing arsenic. *Nature* 371 (1994), p. 750.

Anderson, G.L., Williams, J. & Hille, R.: The purification and characterization of arsenite oxidase from *Alcaligenes faecalis*, a molybdenum-containing hydroxylase. *J. Biol. Chem.* 267 (1992), pp. 23674–23682.

Baymann, F., Brugna, M., Mühlenhoff, U. & Nitschke, W.: Daddy, where did (PS) I come from? *Biochim. Biophys. Acta* 1507 (2001), pp. 291–310.

Baymann, F., Lebrun, E., Brugna, M., Schoepp-Cothenet, B., Giudici-Orticoni, M.T. & Nitschke, W.: The redox protein construction kit: pre-last universal common ancestor evolution of energy-conserving enzymes. *Phil. Trans. R. Soc. Lond. B* 358 (2003), pp. 267–274.

Bird, L.J., Bonnefoy, V., Newman, D.K.: Bioenergetic challenges of microbial iron metabolisms. *Trends Microbiol.* 19(7) (2011), pp. 330–40.

Branco, R., Francisco, R., Chung, A.P. & Morais, P.V.: Identification of an *aox* system that requires cytochrome *c* in the highly arsenic-resistant bacterium *Ochrobactrum tritici SCII24*. *Appl. Environ. Microbiol.* 75:15 (2009), pp. 5141–5147.

Cai, L., Rensing, C., Li, X. & Wang, G.: Novel gene clusters involved in arsenite oxidation and resistance in two arsenite oxidizers: *Achromobacter* sp. SY8 and *Pseudomonas* sp. TS44. *Appl. Microbiol. Biotechnol.* 83:4 (2009), pp. 715–725.

Chang, J.S., Yoon, I.H., Lee, J.H., Kim, K.R., An, J. & Kim, K.W.: Arsenic detoxification potential of *aox* genes in arsenite-oxidizing bacteria isolated from natural and constructed wetlands in the Republic of Korea. *Environ. Geochem. Health.* 32:2 (2010), pp. 95–105.

Collins, M.D. & Jones, D.: Distribution of isoprenoid quinone structural types in bacteria and their taxonomic implications. *Microbiol. Rev.* 45 (1981), pp. 316–354.

Connon, S.A., Koski, A.K., Neal, A.L., Wood, S.A. & Magnuson, T.S.: Ecophysiology and geochemistry of microbial arsenic oxidation within a high arsenic, circumneutral hot spring system of the Alvord Desert. *FEMS Microbiol. Ecol.* 64 (2008), pp. 117–128.

Dietrich, W. & Klimmek, O.: The function of methyl-menaquinone-6 and polysulfide reductase membrane anchor (PsrC) in polysulfide respiration of *Wolinella succinogenes. Eur. J. Biochem.* 269 (2002), pp. 1086–1095.

D'Imperio, S., Lehr, C., Breary, M. & McDermott, T.R.: Autecology of an arsenite chemolithotroph: sulfide constraints on function and distribution in a geothermal spring. *Appl. Environ. Microbiol.* 73:21 (2007), pp. 7067–7074.

Donahoe-Christiansen, J., D'Imperio, S., Jackson, C.R., Inskeep, W.P. & McDermott, T.R.: Arsenite-oxidizing *Hydrogenobaculum* strain isolated from an acid-sulfate-chloride geothermal spring in Yellowstone National Park. *Appl. Environ. Microbiol.* 70:3 (2004), pp. 1865–1868.

Ducluzeau, A.-L., van Lis, R., Duval, S., Schoepp-Cothenet, B., Russell, M.J. & Nitschke, W.: Was NO the first deep electron sink? *Trends in Biochem. Sci.* 34 (2009), pp. 9–15.

Duquesne, K., Lieutaud, A., Ratouchniak, J., Muller, D., Lett, M.-C. & Bonnefoy, V.: Arsenite oxidation by a chemoautotrophic moderately acidophilic *Thiomonas* sp.: from the strain isolation to the gene study. *Environ. Microbiol.* 10:1 (2008), pp. 228–237.

Duval, S., Ducluzeau, A.-L., Nitschke, W. & Schoepp-Cothenet, B.: Enzyme phylogenies as markers for the oxidation state of the environment: the case of respiratory arsenate reductase and related enzymes. *BMC Evol. Biol.* 8 (2008), pp. 206.

Duval, S., Santini, J.M., Nitschke, W., Hille, R. & Schoepp-Cothenet, B.: The small subunit AroB of arsenite oxidase: lessons on the [2Fe-2S]-Rieske protein superfamily. *J. Biol. Chem.* 285:27 (2010), pp. 20442–20451.

Ellis, P.J., Conrads, T., Hille, R. & Kuhn, P.: Crystal structure of the 100 kDa arsenite oxidase from *Alcaligenes faecalis* in two crystal forms at 1.64 Å and 2.03 Å. *Structure* 9 (2001), pp. 125–132.

Fisher, J.C. & Hollibaugh J.T.: Selenate-dependent anaerobic arsenite oxidation by a bacterium from Mono lake, California. *Appl. Environn. Micriobiol.* 74 (2008), pp. 2588–2594.

Gonzales, P.J., Correia, C., Moura, I., Brondino, C.D. & Moura, J.J.G.: Bacterial nitrate reductases: Molecular and biological aspects of nitrate reduction. *J. Inorg. Biochem.* 100 (2006), pp. 1015–1023.

Green, H.H.: Isolation and description of a bacterium causing oxidation of arsenite to arsenate in cattle-dipping baths. *S. Afr. J. Sci.* 14 (1918), pp. 465–467.

Hamamura, N., Macur, R.E., Korf, S., Ackerman, G., Taylor, W.P., Kozubal, M., Reysenbach, A.-L. & Inskeep, W.P.: Linking microbial oxidation of arsenic with detection and phylogenetic analysis of arsenite oxidase genes in diverse geothermal environments. *Environ. Microbiol.* 11:2 (2009), pp. 421–431.

Handley, K.M., Boothman, C., Mills, R.A., Pancost, R.A. & Lloyd, J.R.: Functional diversity of bacteria in a ferruginous hydrothermal sediment. *ISME J.* 4 (2010), pp. 1193–1205.

Hiraishi, A. & Hoshino, Y.: Distribution of rhodoquinone in *Rhodospirillaceae* and its taxonomic implication. *J. Gen. Appl. Microbiol.* 30 (1984), pp. 435–448.

Hoeft, S.E., Blum, J.S., Stolz, J.F., Tabita, F.R., Witte, B., King, G.M,. Santini, J.M. & Oremland, R.S.: *Alkalilimnicola ehrlichii* sp. nov., a novel, arsenite-oxidizing haloalkaliphilic gammaproteobacterium capable of chemoautotrophic or heterotrophic growth with nitrate or oxygen as the electron acceptor. *Int. J. Sys. Evol. Microbiol.* 57(2007), pp. 504–512.

Hoeft, S.E., Kulp, T.R., Han, S., Lanoil, B., Oremland, R.S.: Coupled arsenotrophy in a hot spring photosynthetic biofilm at Mono Lake, California. *Appl. Environ. Microbiol.* 76:14 (2010), pp. 4633–4639.

Huber, R., Sacher, M., Vollmann, A., Huber, H. & Rose, D.: Respiration of arsenate and selenate by hyperthermophilic Archaea. *System. Appl. Microbiol.* 23 (2000), pp. 305–314.

Kashyap, D.R., Botero, L.M., Franck, W.L., Hasset, D.J. & McDermott, T.R.: Complex regulation of arsenite oxidation in *Agrobacterium tumefaciens. J. Bacteriol.* 188:3 (2006), pp. 1081–1088.

Krafft, T. & Macy, J.M.: Purification and characterization of the respiratory arsenate reductase of *Chrysiogenes arsenatis. Eur. J. Biochem.* 255 (1998), pp. 647–653.

Koechler, S., Cleiss-Arnold, J., Proux, C., Sismeiro, O., Dillies, M.-A., Goulhen-Chollet, F., Hommais, F., Lièvremont, D., Arsène-Ploetze, F., Coppée, J.-Y. & Bertin, P.N.: Multiple controls affect arsenite oxidase gene expression in *Herminiimonas arsenicoxydans.BMC Microbiol.* 10 (2010), p. 53.

Kulp, T.R., Hoeft, S.E., Asoa, M., Madigan, M.T., Hollibaugh, J.T., Fisher, J.C., Stolz , J.F., Culbertson, C.W., Miller, L.G. & Oremland, R.S.: Arsenic (III) fuels anoxygenic photosynthesis in hot spring biofilms from Mono Lake, California. *Science* 321 (2008), pp. 967–970.

Laverman, A.M., Blum, J.S., Schaefer, J.K., Phillips, E., Lovley, D.R. & Oremland, R.S.: Growth of strain SES-3 with arsenate and other diverse electron acceptors. *Appl. Environ. Microbiol.* 61 (1995), pp. 3556–3561.

Lett, M.-C., Lievremont, D. Muller, D., Silver, S. & Santini, J.M. : Unified nomenclature for genes involved in prokaryotic aerobic arsenite oxidation. *J. Bacteriol.* 194 (2012), pp. 207–208.

Lebrun, E., Brugna, M., Baymann, F., Muller, D., Lièvremont, P., Lett, M.-C. & Nitschke, W.: Arsenite oxidase, an ancient bioenergetic enzyme. *Mol. Biol. Evol.* 20 (2003), pp. 686–693.

Lebrun, E., Santini, J.M., Brugna, M., Ducluzeau, A.-L., Ouchane, S., Schoepp-Cothenet, B., Baymann, F. & Nitschke, W.: The Rieske protein: a case study on the pitfalls of multiple sequence alignments and phylogenetic reconstruction. *Mol. Biol. Evol.* 23:6 (2006), pp. 1180–1191.

Lieutaud, A., van Lis, R., Duval, S., Capowiez, L., Muller, D., Lebrun, R., Lignon, S., Fardeau, M.L., Lett, M.C., Nitschke, W. & Schoepp-Cothenet, B.: Arsenite oxidase from *Ralstonia* sp. S22 characterization of the enzyme and its interaction with soluble cytochromes. *J. Biol. Chem.* 285:27 (2010), pp. 20433–20441.

Lowe, E.C., Bydder, S., Hartshorne, R.S., Tape, H.L.U., Dridge, E.J., Debieux, C.M., Paszkiewicz, K., Singleton, I., Lewis, R. J., Santini, J.M., Richardson, D.J. & Butler, C.: Quinol-cytochrome *c* oxidoreductase and cytochrome c_4 mediate electron transfer during selenate respiration in *Thauera selenatis. J. Biol. Chem.* 285:24 (2010), pp. 18433–18442.

McManus, J.D., Brune, D.C., Han, J., Sanders-Loehr, J., Meyer, T.E., Cusanovich, M.A., Tollin, G. & Blankenship, R.E.: Isolation, characterization, and amino acid sequences of auracyanins, blue copper proteins from the green photosynthetic bacterium *Chloroflexus aurantiacus*. *J. Biol. Chem.* 267:10 (1992), pp. 6531–6540.

Maklashina, E. & Cecchini, G.: Comparison of catalytic activity and inhibitors of quinone reactions of succinate dehydrogenase (Succinate-ubiquinone oxidoreductase) and fumarate reductase (Menaquinol-fumarate oxidoreductase) from *Escherichia coli*. *Arch. Biochem. Biophys.* 369 (1999), pp. 223–232.

Malasarn, D., Keefe, J.R. & Newman, D.K.: Characterization of the arsenate respiratory reductase from *Shewanella* sp. strain ANA-3. *J. Bacteriol.* 190 (2008), pp. 135–142.

Mukhopadhyay, R., Rosen, B.P., Phung, L.T. & Silver, S.: Microbial arsenic: from geocycles to genes and enzymes. *FEMS Microbiol. Rev.* 26 (2002), pp. 311–325.

Muller, D., Lievremont, D., Simeonova, D.D., Hubert, J.C. & Lett., M.C.: Arsenite oxidase *aox* genes from a metal-resistant beta-proteobacterium. *J. Bacteriol.* 185 (2003), pp. 135–141.

Murphy, J.M. & Saltikov, C.W.: The *cymA* gene, encoding a tetraheme *c*-type cytochrome, is required for arsenate respiration in *Shewanella* species. *J. Bacteriol.* 189 (2007), pp. 2283–2290.

Navarro-González, R., McKay, C.P & Nna Mvondo, D.: A possible nitrogen crisis for Archaean life due to reduced nitrogen fixation by lightning. *Nature* 412 (2001), pp. 61–64.

Nitschke, W., Mattioli, T. & Rutherford, A.W.: The FeS-type photosystems and the evolution of photosynthetic reaction centres. In: H. Baltscheffsky (ed.): *Origin and evolution of biological energy conservation*. VCH Publ. Inc, New York, 2001, pp. 177–203.

Nitschke, W. & Dracheva, S.M.: Reaction center associated cytochromes. In: R.E. Blankenship, M.T. Madigan, C.E. Bauer (eds): *Anoxygenic photosynthetic bacteria*. Kluwer Academic Publishers, Dordrecht, 1995, pp. 775–805.

Nowicka, B. & Kruk, J.: Occurrence, biosynthesis and function of isoprenoid quinones. *Biochim. Biophys. Acta* 1797 (2010), pp. 1587–1605.

Prasad, K.S., Subramanian, V. & Paul, J.: Purification and characterization of arsenite oxidase from *Arthrobacter* sp. *Biometals* 22:5 (2009), pp. 711–721.

Richey, C., Chovanec, P., Hoeft, S.E., Oremland, R.S., Basu, P. & Stolz, J.F.: Respiratory arsenate reductase as a bidirectional enzyme. *Biochem. Biophys. Res. Commun.* 382:2 (2009), pp. 298–302.

Rhine, E.D., Phels, C.D. & Young, L.Y.: Anaerobic arsenite oxidation by novel denitrifying isolates. *Environ. Microbiol.* 8:5 (2006), pp. 899–908.

Rhine, E.D., Ni Chadhain, S.M., Zylstra, G.L. & Young, L.Y.: The arsenite oxidase genes (*aroAB*) in novel chemoautotrophic arsenite oxidizers. *Biochem. Biophys. Res. Com.* 354 (2007), pp. 662–667.

Rothery, R.A., Workun, G.J. & Weiner, J.H.: The prokaryotic complex iron-sulfur molybdoenzyme family. *Biochim. Biophys. Acta* 1778:9 (2008), pp. 1897–1929.

Saltikov, C.W. & Newman, D.K.: Genetic identification of a respiratory arsenate reductase. *Proc. Nat. Acad. Sci. USA* 100 (2003), pp. 10983–10988.

Santini, J.M., Sly, L.I., Schnagl, R.D. & Macy J.M.: A New Chemolithoautotrophic arsenite-oxidizing bacterium isolated from a gold mine: phylogenetic, physiological, and preliminary biochemical studies. *Appl. Environ. Microbiol.* 66:1 (2000), pp. 92–97.

Santini, J.M. & vanden Hoven, R.N.: Molybdenum-containing arsenite oxidase of the chemolithoautotropic arsenite oxidizer NT-26. *J. Bacteriol.* 186:6 (2004), pp. 1614–1619.

Santini, J.M., Kappler, U., Ward, S.A., Honeychurch, M.J., vanden Hoven, R. & Bernhardt, P.V.: The NT-26 cytochrome c_{552} and its role in arsenite oxidation. *Biochim. Biophys. Acta* 1767 (2007), pp. 189–196.

Sardiwal, S., Santini, J.M., Osborne, T.H. & Djordjevic, S.: Characterization of a two-component signal transduction system that controls arsenite oxidation in the chemolithoautotroph NT-26. *FEMS lett.* 313 (2010), pp. 20–28.

Sehlin, H.M. & Lindström E.B.: Oxidation and reduction of arsenic by *Sulfolobus acidocaldarius* strain BC. *FEMS Microbiol. Lett.* 93 (1992), pp. 87–92.

Schäfer, G., Moll, R. & Schmidt, C.L.: Respiratory enzymes from *Sulfolobus acidocaldarius*. *Methods Enzymol.* 331(2001), pp. 369–410.

Schoepp-Cothenet, B., Duval, S., Santini, J.M. & Nitschke, W.: Comment on "Arsenic (III) fuels anoxygenic photosynthesis in hot spring biofilms from Mono Lake, California". *Science* 323 (2009a), p. 583c.

Schoepp-Cothenet, B., Lieutaud, C., Baymann, F., Verméglio, A., Friedrich, T., Kramer, D.M. & Nitschke, W.: Menaquinone as pool quinone in a purple bacterium. *Proc. Natl. Acad. Sci. USA* 106:21 (2009b), pp. 8549–8554.

Schütz, M., Brugna, M., Lebrun, E., Baymann, F., Huber, R., Stetter, K.-O., Hauska, G., Toci, R., Lemesle-Meunier, D., Tron, P., Schmidt, C. & Nitschke, W.: Early evolution of cytochrome *bc*-complexes. *J. Mol. Biol.* 300 (2000), pp. 663–676.

Silver, S. & Phung, L.T.: A bacterial view of the periodic table: genes and proteins for toxic inorganic ions. *Appl. Environ. Microbiol.* 71 (2005), pp. 599–608.

Smedley, P.L. & Kinniburgh, D.G.: A review of the source, behavior and distribution of arsenic in natural waters. *Appl. Geochem.* 17 (2002), pp. 517–568.

Stolz, J.F., Basu, P., Santini, J.M. & Oremland, R.S.: Arsenic and selenium in microbial metabolism. *Annu. Rev. Microbiol.* 60 (2006), pp. 107–130.

Stolz, J.F., Basu, P. & Oremland, R.S.: Microbial arsenic metabolism: new twists on an old Poison. *Microbe* 5 (2010), pp. 53–59.

Sun, W., Sierra, R. & Field, J.A.: Anoxic oxidation of arsenite linked to denitrification in sludges and sediments. *Water Res.* 42:17 (2008), pp. 4569–4577.

Sun, W., Sierra-Alvarez, R., Fernandez, N., Sanz, J.L., Amils, R., Legatzki, A., Maier, R.M. & Field, J.A.: Molecular characterization and *in situ* quantification of anoxic arsenite-oxidizing denitrifying enrichment cultures. *FEMS Microbiol. Ecol.* 68 (2009), pp. 72–85.

Sun, W., Sierra-Alvarez, R., Hsu, I., Rowlette, P. & Field, J.A.: Anoxic oxidation of arsenite linked to chemolithotrophic denitrification in continuous bioreactors. *Biotechnol. Bioeng.* 105:5 (2010a), pp. 909–917.

Sun, W., Sierra-Alvarez, R., Milner, L. & Field, J.A.: Anaerobic oxidation of arsenite linked to chlorate reduction. *Appl. Environ. Microbiol.* (2010b), pp. 6804–6811.

Tomitani, A., Knoll, A.H., Cavanaugh, C.M. & Ohno, T.: The evolutionary diversification of cyanobacteria: molecular-phylogenetic and paleontological perspectives. *Proc. Natl. Acad. Sci. USA* 103 (2006), pp. 5442–5447.

vanden Hoven, R.N. & Santini, J.M.: Arsenite oxidation by the heterotroph *Hydrogenophaga* sp. str. NT-14: the arsenite oxidase and its physiological electron acceptor. *Biochim. Biophys. Acta* 1656 (2004), pp. 148–155.

van Lis, R., Ducluzeau, A.-L., Nitschke, W. & Schoepp-Cothenet, B.: The nitrogen cycle in the Archaean; an intricate interplay of enzymatic and abiotic reactions. In: J.W.B. Moir (ed.): *Nitrogen cycling in bacteria: molecular analysis*. Caister Academic Press, Portland, USA, 2011, chapter 1.

Xiong, J., Fischer, W.M., Inoue, K., Nakahara, M. & Bauer, C.E.: Molecular evidence for the early evolution of photosynthesis. *Science* 289 (2000), pp. 1724–1730.

Zannoni, D., Schoepp-Cothenet, B. & Hosler, B.: Respiration and respiratory complexes. In: N. Hunter, F. Daldal & M. Thurnauer (eds): *The purple phototrophic bacteria*. Springer, Dordrecht, The Nederlands, 2008, pp. 537–561.

Zannini, D. & Ingledew, W.J.: A thermodynamic analysis of the plasma membrane electron transport component in photoheterotrophically grown cells of *Chloroflexus aurantiacus*. *FEBS Lett.* 193:1, pp. 93–98.

Zargar, K., Hoeft, S., Oremland, R. & Saltikov, C.W.: Identification of a novel arsenite oxidase gene, *arxA*, in the haloalkaliphilic, arsenite-oxidizing bacterium *Alkalilimnicola ehrlichii* strain MLHE-1. *J. Bacteriol.* 92:14 (2010), pp. 3755–3762.

CHAPTER 11

Remediation using arsenite-oxidizing bacteria

François Delavat, Marie-Claire Lett & Didier Lièvremont

11.1 INTRODUCTION

Arsenic (As) is present in water worldwide. Populations in developing countries in various parts of the world are particularly at risk, with chronic exposure to this element resulting in a range of cancers and other diseases. The World Health Organisation recommended Maximum Contaminant Level (MCL) for arsenic in water is $10\,\mu g\,L^{-1}$, but drinking water supplies in many countries contain much higher levels than this, so there is an urgent need for efficient methods of removing As.

Arsenic is present in the environment in both organic and inorganic forms in four oxidation states $(-3, 0, +3, +5)$. Inorganic species are the most toxic, especially arsenite, As(III). This reduced trivalent form dominates under anoxic conditions, e.g. in groundwater, while the oxidized form arsenate [As(V)] is generally dominant in oxygenated or surface waters over the pH range typically encountered in water treatment (Jiang, 2001).

As(III) is much more mobile than As(V), which is readily adsorbed to solid materials such as Fe oxides and hydroxides, clays, or alumina. Thus, oxidation of As(III) to As(V) results in As(V) immobilization in environmental compartments e.g. soils. In water, methods like coagulation/coprecipitation, sorption/ion-exchange, precipitation, or filtration are more effective at removing As(V) than As(III). To remove As(III) from contaminated environments it must first be oxidized. This oxidation can be performed abiotically using oxygen, but the reaction rate is very low. In oxygenated water, conversion of As(III) to As(V) is thermodynamically favoured, but adequate conversion may take days, weeks or months, depending on the specific conditions (Jiang, 2001). Usually, in remediation processes, this As(III) oxidation is carried out by the addition of strong oxidants such as ozone, hydrogen peroxide or chlorine (Kim and Nriagu, 2000), but because of undesirable byproducts, the use of chemical reagents is not recommended, especially in drinking water.

A large number of autotrophic and heterotrophic As(III)-oxidizing bacteria have been isolated from various environments in the last two decades (Cullen and Reimer, 1989; Weeger et al., 1999; Santini et al., 2000; Battaglia-Brunet et al., 2002; Oremland et al., 2002; Rhine et al., 2006; Stolz et al., 2006; Inskeep et al., 2007; Sun et al., 2010c) and a first psychrotolerant As(III) oxidizer has been recently isolated by Osborne et al. (2010). Bacterial As(III) oxidation may offer a safe alternative to chemical methods of oxidation, but to date, most As(III) biological oxidation studies have been conducted using small-scale or pilot-scale bioreactors; the literature contains no data on full- or industrial-scale applications. Perhaps owing to the stigma associated with bacterial contamination, microbial treatments of drinking water are not yet widely accepted, although this may change as they become more practical (Brown, 2007; Lytle et al., 2007). Moreover, the cost-effectiveness and ease of implementation of biological treatment processes depend on numerous factors, e.g. the choice of reactor design and operational conditions. Bespoke systems, tailored for one set of conditions, are subject to strict constraints. Nevertheless, these active systems are promising and will certainly work in the near future, at least in combination with passive systems designed to enhance naturally occurring physical, chemical and biological processes.

11.2   ARSENITE OXIDATION-BASED REMEDIATION BIOPROCESSES

Arsenic treatment processes need to be diverse since they must target waters, soils or sediments, with different geochemical conditions and As concentrations. Usually, two major steps are involved (Fig. 11.1): oxidation of As(III); and the subsequent removal of the As(V) produced.

In section 11.2.1 we review active systems for exploiting the biological oxidation of As(III). Passive remediation, which needs no energy input, is discussed in section 11.2.2. This natural attenuation (NA) of As may represent a cost-effective alternative if appropriate conditions can be combined.

11.2.1   *Active remediation systems*

The implementation of active systems requires an understanding of bacterial physiology and metabolism. In these systems, all of which require continuous energy input, arsenic speciation is controlled by microbial processes, so a first step is to choose a suitable pure culture of a previously isolated arsenite-oxidizing bacterial strain, or a naturally occurring arsenite-oxidizing bacterial consortium. The design of the system must then depend on whether As(III) oxidation occurs under oxic or anoxic conditions. To date, the use of pure cultures has been investigated mainly at the bench scale, while bacterial consortia have been studied at the pilot scale.

11.2.1.1   *Anoxic As(III) oxidation-based processes*
In anoxic environments, microorganisms play important roles in the mobilization of adsorbed As (Oremland *et al.*, 2005), in particular, through microbial reduction of As(V) and Fe(III) (Smedley and Kinniburgh, 2002). As(III) is the predominant species in these reducing environments; it is less strongly adsorbed than As(V) on minerals such as Fe, Mn or Al (hydr)oxides (Bhumbla and Keefer, 1994; Manning and Goldberg, 1997; Lin and Wu, 2001; Goldberg, 2002). Therefore, oxidation of As(III) to As(V) could contribute to an improved immobilization of As and thus help to mitigate As contamination in groundwater. This anoxic oxidation may use one of three electron acceptors: nitrate, chlorate and selenate.

Several authors have shown that under anaerobic conditions, nitrate (NO_3^-) acts as an electron acceptor in the oxidation of As(III) to As(V) by denitrifying bacteria, in lakes (Oremland *et al.*, 2002; Hoeft *et al.*, 2007) or soil (Rhine *et al.*, 2006). In these studies, bacteria were isolated from sites containing high concentrations of As (225–375 mg L^{-1}). Sun *et al.* (2008) inoculated sludges and sediments, never exposed to As contamination, in batch bioessays supplied with a basal mineral medium, As(III) (37.5–375 mg L^{-1}) and nitrate (620 mg L^{-1}). They reported the anoxic biological oxidation of As(III) (37.5–375 mg L^{-1}) by denitrifying microorganisms.

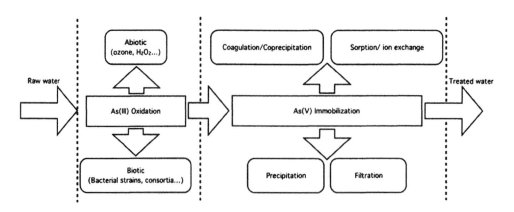

Figure 11.1.   Synopsis of the mechanisms involved in As treatment processes.

The same authors (Sun *et al.*, 2009a) used enrichment cultures from diverse anaerobic environmental samples. They isolated strains of *Azoarcus*, known for their ability to oxidize aromatic substrates under anaerobic conditions (Kasai *et al.*, 2006; Reinhold-Hurek and Hurek, 2006; Szaleniec *et al.*, 2007), and strains of *Diaphorobacter.* These strains were able to perform nitrate-dependent As(III) oxidation. These experiments demonstrated that in the water-saturated subsurface, NO_3^- could be used to oxidize As(III) to As(V), leading to As immobilization. Sun *et al.* (2009b) then examined the controlled reoxidation of As(III) and Fe(II) in anoxic environments. They injected nitrate ($150 \, mg \, L^{-1} \, NO_3^-$ supplied as KNO_3) into continuous flow columns packed with sand and inoculated with volatile suspended solids from anaerobic sludge, simulating a natural anaerobic groundwater-sediment system, with co-occuring As(III) ($500 \, \mu g \, L^{-1}$) and Fe(II). Microbial nitrate-dependent oxidation of Fe(II) and As(III) enhanced the adsorption of As on the newly biologically formed solid-phase Fe(III) (hydr)oxides. The efficiency of the system allowed the effluents not to exceed the MCL of $10 \, \mu g \, L^{-1}$ of As.

Sun *et al.* (2010b) then showed the anoxic biological oxidation of As(III) linked to chemolithotrophic denitrification in continuous bioreactors – a process that remained stable throughout a period of operation of over three years. The bioreactors were fed with a chemolithotrophic denitrifying granular sludge and a basal mineral medium containing As(III) (up to $610 \, mg \, L^{-1}$) nitrate (155 and $397 \, mg \, L^{-1}$) and bicarbonate. The process had a high conversion efficiency of As(III) to As(V) (92%) even with high loadings of As(III) (265–384 mg As $L_{reactor}^{-1} \, d^{-1}$). The microbial population tolerated As(III)-inflow concentrations up to $390 \, mg \, L^{-1}$, but was greatly inhibited with As concentrations of $570–610 \, mg \, L^{-1}$. The molar stoichiometric ratio of As(V) formed to nitrate removed indicated complete denitrification of NO_3^- to N_2 gas in the reactor with an As(III)-inflow concentration of $265 \, mg \, L^{-1}$ and a nitrate-inflow concentration of $397 \, mg \, L^{-1}$.

Using the same granular sludge, the same authors carried out experiments in upflow columns, packed with a non-Fe metal (hydr)oxide, i.e. activated alumina (Sun *et al.*, 2010a). The bioreactors were fed with As(III) ($265 \, mg \, L^{-1}$) as electron donor, NO_3^- ($403 \, mg \, L^{-1}$) as electron acceptor and $NaHCO_3$ ($672 \, mg \, L^{-1}$) as major carbon source. As(III) was oxidized and the As(V) produced was immobilized on the activated alumina by adsorption.

All of these experiments showed that in reducing or anoxic environments, where As(III) is the dominant species, the biologically nitrate-dependent oxidation of As(III) can greatly enhance the immobilization of As. The strong sorption of As on metal (hydr)oxides generally observed in natural environments such as soils or sediments (Lin and Wu, 2001; Dixit and Hering, 2003) is always counterbalanced by the opposite process, i.e. desorption. In the case of Al (hydr)oxides, As desorption importance decreased as sorption ageing time increased (Lin and Puls, 2000). Thus, injecting NO_3^- into the subsurface (e.g. in anaerobic groundwaters) as a means to immobilize As on Al (hydr)oxides or biogenic Fe (hydr)oxides represents a potential bioremediation strategy.

Having demonstrated that nitrate boosted biological As(III) oxidation (see above), Sun *et al.* (2011) developed an As(III) oxidation process, using a bench-scale upflow anaerobic sludge bed reactor, with chlorate (ClO_3^-) as an alternative electron acceptor. The bioreactor inoculum was an anaerobically digested sewage sludge obtained from a municipal wastewater treatment plant, fed with a basal mineral medium supplied with As(III) as the sole energy source, ClO_3^- as the sole electron acceptor, and $NaHCO_3$ as an additional carbon source. As(III) loadings varied according to operation periods (from 33 to $144 \, mg \, L_{reactor}^{-1} \, d^{-1}$). Over a 550 days of operation, 98% of As(III) was converted to As(V). This As(III) oxidation was demonstrated to be linked to the complete reduction of ClO_3^- to Cl^- and H_2O. An autotrophic enrichment culture sampled on day 530 was established from the bioreactor biofilm and a 16S rRNA gene clone library was constructed. The Betaproteobacteria accounted for the majority of all the clones (genera *Dechloromonas*, *Acidovorax* and *Alicycliphilus*). Gammaproteobacteria (*Stenotrophomonas*) and Alphaproteobacteria (*Rhodobacter*) were also represented. By exploring the potential use of ClO_3^- as terminal electron acceptor by anaerobic bacteria for As(III) oxidation, the same authors had showed that sludge samples from different sources, enrichment cultures and pure autotrophic microbial cultures (*Dechloromonas* sp. strain ECC1-pb1, and *Azospira* sp. strain ECC1-pb2) oxidized As(III) to As(V) while reducing chlorate under anaerobic conditions (Sun *et al.*, 2010c); the biological oxidation of As(III) to As(V) in absence of exogenous O_2 was dependent on the presence of chlorate.

These works show that the addition of the highly water soluble chlorate may open interesting perspectives to remediate As contamination e.g. in groundwater. However, we have to keep in mind that chlorate and chlorite induced oxidative damage to red blood cells in humans (WHO 2005).

Finally, a third electron acceptor, selenate (SeO_4^{2-}), can be used as an electron acceptor for As(III) oxidation. To date, its use has been demonstrated only in a Mono Lake water sample (Fisher and Hollibaugh, 2008): under anaerobic conditions, As(V) was produced when nitrate or selenate were supplied. In the experimental conditions used, selenate ($286\,mg\,L^{-1}$) supported the same As(III) oxidation rate as nitrate ($310\,mg\,L^{-1}$). Fisher and Hollibaugh (2008) went on to isolate a pure culture (strain ML-SRAO belonging to the *Bacillus* genera) which was shown, in a defined-salts medium (amended with $715\,mg\,L^{-1}$ selenate, $375\,mg\,L^{-1}$ As(III) and $450\,mg\,L^{-1}$ lactate), to be responsible for the anaerobic As(III) oxidation. No As(III) oxidation occurred with nitrate or oxygen, indicating that the process was selenite-dependent. Selenate has not received as much attention as chlorate or nitrate but the potential to use it to accelerate the As(III) oxidation and the subsequent As(V) immobilization exists.

These experiments on anaerobic As(III) oxidation, carried out at the bench-scale in small bioreactors, showed an unusual way to partially remediate arsenic. Experiments in oxic conditions are generally preferred as they are less demanding. Air sparging techniques are tried and tested techniques for *in-situ* organic contaminants bioremediation principally, even if supplying oxygen is often technically difficult. Thus, why not envisage bioremediation systems taking advantages of anaerobic As(III) oxidation?

11.2.1.2   Oxic As(III) oxidation-based processes

In oxic environments, As(III) can be converted abiotically to As(V), in a slow reaction that may take days or weeks depending on the conditions. Numerous As(III) bacterial oxidizers have been isolated from these environments and have been used in the development of remediation processes. Two major approaches are taken: a pure culture-based approach, mainly at the bench-scale, and a consortium-based approach at bench- or pilot-scale. This biological As(III)-oxidation must be combined with the removal of the As(V) produced, on an efficient adsorbent.

11.2.1.3   As(III) oxidation using bacterial pure cultures

A wide range of bacteria have been involved in As(III)-oxidation systems: *Herminiimonas arsenic-oxydans* (Weeger *et al.*, 1999; Lièvremont *et al.*, 2003; Simeonova *et al.*, 2005), *Haemophilus* sp., *Micrococcus* sp., *Bacillus* sp. (Ike *et al.*, 2008), *Microbacterium lacticum* (Mokashi and Paknikar, 2002), *Alcaligenes faecalis* (Suttigarn and Wang, 2005; Wang and Suttigarn, 2007; Wang *et al.*, 2009), *Pseudomonas putida* (Yang *et al.*, 2010), *Ralstonia eutropha* MTCC 2487 (Mondal *et al.*, 2008). All of these bacteria are heterotrophic. Only two chemoautotrophic bacteria have been used in pure culture systems: *Thiomonas arsenivorans* B6 (Dastidar and Wang, 2009, 2010; Wan *et al.*, 2010) and SDB1 bacteria which showed 100% identity with *Ensifer adhaerens* or *Sinorhizobium* sp. CAF63 (Lugtu *et al.*, 2010). These heterotrophic and chemoautotrophic bacteria are all reported to be efficient As(III) oxidizers.

All but one of these bacteria were isolated from As-containing environments. The exception was Yang *et al.*'s (2010) construction of a more efficient As(III)-oxidizing *Pseudomonas putida*, isolated from the wastewater of an electronic factory, by cloning the genes of the As(III) oxidase of *Thermus thermophilus* HB8. Its As(III) oxidative ability was good but the recombinant plasmid was unstable. In any case, it is questionable whether this method is really promising owing to strict regulations and the need for contained use of these bacteria.

The As(III) oxidation efficiency of strains in bioreactors depends on a number of parameters, such as the quality of influent, the cell-immobilizing material, and the initial As(III) concentration.

The second step, consisting of the As(V) chemical adsorption, requires a careful choice of the adsorbent.

Different technical options have been implemented: up-flow fixed bed reactors (Mondal *et al.*, 2008; Wan *et al.*, 2010), fluidized bed reactors (Wang *et al.*, 2009), stirred tank reactors (Lièvremont *et al.*, 2003; Simeonova *et al.*, 2005; Wang and Suttigarn, 2007; Dastidar and Wang, 2009, 2010; Lugtu *et al.*, 2010) and small glass columns (Mokashi and Paknikar, 2002). Batch operation mode (Lièvremont *et al.*, 2003; Simeonova *et al.*, 2005; Dastidar and Wang, 2009), in which reactors are fed once, is usually the preferred mode of operation when the systems are in a development phase. Continuous operation (Wang and Suttigarn, 2007; Wang *et al.*, 2009; Dastidar and Wang, 2010; Sun *et al.*, 2010b) increases the risks of contamination or cell washouts, as bioreactors are continuously fed.

All of these above studies were conducted in small bench-scale bioreactors, up to a few liters. They treated As(III)-contaminated influents in concentrations ranging from 0.5 (Mokashi and Paknikar, 2002) to 4,000 mg L^{-1} (Dastidar and Wang, 2010). Numerous experiments were carried out with a 10–100 mg L^{-1} As(III) concentration (Lièvremont *et al.*, 2003; Simeonova *et al.*, 2005; Wan *et al.*, 2010). These initial concentrations are often arbitrarily chosen either to illustrate the high oxidation capacity of the strain or to simulate *in-situ* environmental concentrations such as in Acid Mine Drainages (AMDs). However, they are all far higher than the MCL recommended for arsenic in water (10 µg L^{-1}).

Since surface or groundwaters were targeted in these systems, the influents to which As(III) was added were synthetic calcareous water (Wan *et al.*, 2010), simulated groundwater (Mokashi and Paknikar, 2002) or nutrient media (Lièvremont *et al.*, 2003; Simeonova *et al.*, 2005; Wang and Suttigarn, 2007) (Simulated water means that compositions were chosen according to the representative composition of the As-contaminated water of interest).

A carbon supplement was provided for heterotrophic bacteria, e.g. citrate (Wang and Suttigarn, 2007), lactate (Lièvremont *et al.*, 2003), sauerkraut brine (Simeonova *et al.*, 2005); while Mokashi and Paknikar (2002) investigated several carbon sources (acetate, lactate, citrate, methanol, sucrose, glucose). The work of Simeonova *et al.* (2005) showed that free or low-cost organic "wastes" could replace more commonly used carbon sources.

Operating conditions – pH and temperature – also varied among studies. They were set at the estimated optimum for each strain – near neutrality (Lièvremont *et al.*, 2003; Wang *et al.*, 2009), slightly acid (Mondal *et al.*, 2008; Dastidar and Wang 2009, 2010) – or to values that reflected the *in-situ* conditions. Some authors examined the effects of changing pH and temperature, e.g. Mokashi and Paknikar (2002), Simeonova *et al.* (2005) and Ike *et al.* (2008) chose respectively pH ranges of 5.5–10, 7–8, 7–10, and temperature ranges of 20–45°C, 4–25°C and 25–35°C. In the experimental conditions of their systems, Simeonova *et al.* (2005) demonstrated that temperature was the most important parameter while Mokashi and Paknikar (2002) reported an optimum pH of 7.5 and an optimum temperature of 30°C for As(III) oxidation. Ike *et al.* (2008) observed similar As(III) oxidation rates over the pH and temperature ranges studied.

In all reactors i.e. packed bed, fixed bed, fluidized bed, continuously operated or not, As(III) is oxidized by immobilized cells. Thus, carriers that can maintain high cell densities are of crucial importance. The supports used in these studies included sieved sand (Wan *et al.*, 2010), granular activated carbon (Mondal *et al.*, 2008), aluminosilicate zeolite (Lièvremont *et al.*, 2003), Ca-alginate beads (Simeonova *et al.*, 2005; Lugtu *et al.*, 2010), carragenan gel beads (Wang *et al.*, 2009), glass beads (Dastidar and Wang, 2010), and brick pieces (Mokashi and Paknikar, 2002). Once As(III) has been transformed into As(V) by the encapsulated or fixed cells, the As(V) must be adsorbed to prevent the release of As in the effluent. Various As(III) and As(V) adsorbents have been tested for their efficiency, e.g. biological materials, mineral oxides, activated carbons, polymer resins, by-products and industrial wastes (Mohan and Pittman, 2007). Minerals and mineral oxides are also widely used in these studies: e.g. zero-valent Fe (Mokashi and Paknikar, 2002), activated alumina (Ike *et al.*, 2008; Dastidar and Wang, 2010; Wan *et al.*, 2010) and carbonates such as kutnahorite (Lièvremont *et al.*, 2003).

In addition to immobilizing the cells it is important to maintain the As(III) efficiency of the strain used over long periods of operation time. But cell density is difficult to stabilize and, depending on the immobilization material, cells can leach out in reactor effluent. Moreover, reactors are fed with an As-contaminated influent whose condition may fluctuate.

The As(III) oxidation efficiency in these different systems varies considerably. *H. arsenicoxydans* was able to oxidize $30 \, mg \, L^{-1} \, h^{-1}$ of As(III) as free cells (Weeger *et al.*, 1999). Simeonova *et al.* (2005) observed an As(III) oxidation rate of $100 \, mg \, L^{-1} \, h^{-1}$ when *H. arsenicoxydans* was entrapped in Ca-alginate beads. Mokashi *et al.* (2002) reported $600 \, mg \, L^{-1} \, h^{-1}$ of As(III) oxidized with *M. lacticum* immobilized on brick pieces in the reactor operated in continuous mode. Depending on the residence time of the influent in the reactor, Wan *et al.* (2010) showed that $2.5 \, mg \, L^{-1} \, h^{-1}$ of As(III) loading was completely oxidized after 15 days of operation. In a continuous stirred-tank reactor, 99% of high inputs of As(III) ($2,000–4,000 \, mg \, L^{-1}$) was oxidized under optimal growth conditions for *Tm. arsenivorans* after 42 days of operation (Dastidar and Wang, 2010).

These systems require that the As(III)-oxidizing strain remains active whatever the composition of the influents. The challenge is then to control the bacterial population.

11.2.1.4 *As(III) oxidation using bacterial consortia*

Studies using bacterial consortia to remediate As(III) have taken two approaches. In the first, As(III) oxidizers were isolated from As-contaminated environments and then inoculated into bioreactors. The second approach relied on the natural formation of a biofilm containing As(III) oxidizers from the influents; this was then inoculated into the bioreactors. Thus, the bacterial community performing the oxidation could be specific to the As-contaminated materials. The use of biofilms instead of pure cultures has been reported to provide bacteria with an increased resistance to environmental stress (Stoodley *et al.*, 2002; Teitzel and Parsek, 2003), making them interesting for As bioremediation (Battaglia-Brunet *et al.*, 2002).

In mining environments, inorganic forms of As (i-As) are widespread, particularly in effluents (Acid Mine Drainages) resulting from the bioleaching of arsenic-bearing minerals. In these liquors, As concentrations are often very high ($>1 \, g \, L^{-1}$), with a significant proportion of As(III), while pH is low (<2). Most of the As(III) oxidizers from these environments are acid-tolerant, i.e. with an optimum growth pH near neutrality but able to survive at low pH, except for *Thiomonas* strains (Duquesne 2004; Battaglia-Brunet *et al.*, 2006b), which grow at an optimum of around pH 5.

CAsO1 is a bacterial consortium originally isolated from a disused gold mine situated in Saint-Yrieix, France (Dictor *et al.*, 2003). It is a mixed community of mesophilic bacteria containing the As(III)-oxidizing chemolithoautotroph *Tm. arsenivorans*, as well as a strain close to *Ralstonia pickettii* (identified by the 16S rDNA). The consortium (i.e. *Tm. arsenivorans*) was able to oxidize As(III) with CO_2 as carbon source, As(III) as electron donor and O_2 as terminal electron acceptor. *R. pickettii*, a ubiquitous bacterium found in water and soil, was able to survive in low nutrient conditions but was not autotrophic (Ryan *et al.*, 2007).

This CAsO1 consortium was shown to be resistant to As(III) concentrations of up to $13.5 \, g \, L^{-1}$ in As-rich liquors (Battaglia-Brunet *et al.*, 2011) and its ability to oxidize As(III) was conserved even in low As(III) concentrations, i.e. $50 \, \mu g \, L^{-1}$ (Challan Belval *et al.*, 2009), and between pH 3 and pH 8 (Dictor *et al.*, 2003). These low As(III) concentrations reflect those measured in various waters. Except for the As-rich mine drainage effluents (Battaglia-Brunet *et al.*, 2011), operating As(III) concentrations were in the order of micrograms. A system effective at such low As concentrations might be useful for treating not only industrial waters but also drinking water. The CAsO1 consortium was used in different bioreactor designs included bench-scale stirred bioreactors to treat As(III)-rich liquors (Battaglia-Brunet *et al.*, 2011), bench-scale fixed-bed bioreactors fed with synthetic effluents supplemented with As (Dictor *et al.*, 2003), and pilot-scale fixed-bed bioreactors fed with unprocessed As contaminated water (Michon, 2006; Michon *et al.*, 2010).

In the work of Michon (2006), bioreactors were supplied with unprocessed contaminated water containing $100\,\mu g\,L^{-1}$ As(III). Oxic conditions were maintained and a 90% As(III) oxidation rate was obtained within 3-h after a 5-day adaptation phase. The same author (Michon *et al.*, 2010) fed fixed-bed reactors with synthetic water spiked with As(III) ($25-100\,\mu g\,L^{-1}$). They demonstrated a crucial point, i.e. the persistence of the consortium in the pilot unit during operation, meaning that the efficiency of the system is conserved. This point is essential as As-contaminated influents contained bacteria, which could affect the life and activity of a biofilm.

Using fixed-bed reactors, Michel *et al.* (2007) compared the formation and activity of biofilms of CAsO1 with those of a pure *Tm. arsenivorans* culture, in order to optimize the process. When a support medium was added, e.g. polystyrene microplates or pozzolana (a volcanic basaltic ash material), the rate of As(III) oxidation by the pure culture declined relative to that of planktonic cells. The biofilm seemed to be a physical barrier limiting the diffusion of As(III) to the cells. In contrast, results obtained with CAsO1 during a six-weeks period showed that pozzolana enhanced bacterial colonization, and improved As(III) oxidation at low As(III) concentration ($<100\,\mu g\,L^{-1}$) (Challan Belval *et al.*, 2009). In all assays, pozzolana was a suitable material to immobilize CAsO1 or *Tm. arsenivorans* (Dictor *et al.*, 2003; Michon, 2006; Michel *et al.*, 2007; Challan Belval *et al.*, 2009; Michon *et al.*, 2010), as it did not interfere with As(V) adsorption and As(III) oxidation.

CAsO1 proved to be efficient at As(III) oxidation, even at low As concentrations; the consortium persisted for a long time in a continuously run bioreactor fed with As-contaminated influents.

The efficiency of this consortium was also tested by Battaglia-Brunet *et al.* (2011), using stirred reactors fed with As-rich liquors originating from a disused gold mine in Salsigne (France). An abiotic treatment with lime was followed by inoculation of CAsO1, which resulted in a decrease of the total As (t-As) concentration in the liquid phase. The As(V) produced by the biological As(III) oxidation was immediately adsorbed onto solids resulting from the lime treatment. The rate of As leaching from the solids was reduced relative to that in the abiotic treatment. Not only could the consortium be used to treat As-contaminated effluents, it could also reduce further As leaching from As-containing wastes. In Lopérec (France), Battaglia-Brunet *et al.* (2006a) sampled indigenous bacterial consortia along an As-containing mine drainage-water stream in which As(III) oxidation was observed. Laboratory experiments carried out in column reactors fed with synthetic mine water showed that the bacterial consortia oxidized As(III) and Fe(II). The As(III) removal rate reached $1900\,\mu g\,L^{-1}\,h^{-1}$. As(III)-oxidizing strains *Variovorax paradoxus* and *Leptothrix cholodnii* were isolated from the bioreactor and site sludges.

Since As-contaminated influents contain indigenous bacteria, the development of a natural biofilm in the bioreactors may represent an elegant solution for As remediation. One way of using this natural biofilm – fixed on a support medium – is to use Biologically active filters (BAFs).

A BAF process was developed to treat groundwater containing Fe and Mn and supplemented with As(III). The fixed-bed filtration unit consisted of two columns filled with a filtration medium i.e. polystyrene beads (Katsoyiannis *et al.*, 2002; Katsoyiannis *et al.*, 2004). The natural biofilm, coating the filtration medium, was observed to contain Fe-oxidizing and Mn-oxidizing bacteria. The mechanism of combining Fe or Mn oxidation with chemical As(III) oxidation, allowing As to be adsorbed on the insoluble oxides formed, is widely documented (Oscarson *et al.*, 1983; Edwards, 1994; Driehaus *et al.*, 1995; Manning *et al.*, 2002). Thus, ingeniously, Katsoyiannis *et al.* (2004) transformed the abiotic reaction into a biotic mechanism. The polystyrene beads contained entrapped bacteria i.e. *Gallionella ferruginea*, *Leptothrix ochracea* – common inhabitants of freshwater iron seeps and iron-rich wetlands. In this system, which required stringently controlled dissolved-oxygen conditions, As(III) oxidation ($100\,\mu g\,L^{-1}$) was almost complete in a few minutes, suggesting that bacteria played an important role in both the oxidation of As(III) and the generation of the Mn oxides that adsorbed As(III) and As(V).

A similar fixed-bed biological filtration unit was set up in the village of Ambacourt (France) (Casiot *et al.*, 2006). In this pilot scale study, the low-oxygen ($0.4-1.4\,mg\,O_2\,L^{-1}$) underground water was contaminated with Fe(II) and low As(III) concentrations ($10-40\,\mu g\,L^{-1}$). These two elements were removed efficiently (100% for Fe, 70–90% for As) from the solution. The high-Fe oxide natural biofilm that had developed in the fixed-bed filter was sampled for As(III) oxidation ability. The B2 bacterial strain was isolated and was able to oxidize As(III). Based on its 16S rDNA sequence, B2 belonged to a new genus showing only 95% sequence identity to *Leptothrix*. Katsoyiannis and Zouboulis (2004) suggested that *Leptothrix* strains, responsible for Fe oxidation, were also involved in As(III) oxidation.

Interestingly, Lytle *et al.* (2007) also reported, in an Ohio water treatment plant, oxidation of As(III) by naturally occurring bacteria that were concentrated in filters. Thus, natural bacterial consortia that develop in municipal and industrial wastewater treatment units, i.e. sludges, could be potential As(III)-oxidizing consortia. Andrianisa *et al.* (2006, 2008) decided to investigate the potential of activated sludges to catalyse the oxidation of As(III) in batch laboratory experiments. (In this process air or oxygen is forced into a sewage liquor to develop a biological floc.) They showed that As(III) was readily transformed to As(V) with no organic carbon source supplied and under aerobic conditions at pH 7 and 25°C. Various aerobic As(III)-oxidizing strains and one chemoautotrophic As(III)-oxidizing strain were isolated, though none was identified. Andrianisa *et al.* showed that As(III) was completely oxidized to As(V) when oxygen was supplied by an aerator in a full-scale oxidation ditch plant receiving As-contaminated water. The oxidized As(V) was then co-precipitated in a second step with Fe hydroxide. The system was efficient enough to reduce the residual As concentration in the supernatant from $200 \,\mu g \,L^{-1}$ to less than $5 \,\mu g \,L^{-1}$. Thus, this activated sludge-based treatment proved to be very efficient in oxidizing As(III) under the recommended MCL. Stasinakis and Thomaidis (2010) came to the same conclusion: Biological Wastewater Treatment Systems (BWTS) i.e. systems involving microorganisms, globally used to treat municipal and industrial wastewater, could be a reliable technology for the oxidation of As(III) to As(V) and its removal.

Studies of As(III)-oxidizing approaches based on pure cultures or consortia demonstrate that both are efficient. The research showed that biological As(III) oxidation systems can be performed on a wide range of As(III) concentrations. Work has been done principally on bench-scale systems and, to our knowledge, full-scale systems have not yet been implemented. Arsenic-contaminated influents can be quantitatively and qualitatively extremely variable, even for a given effluent over a short period of time, and this means that universal systems are unlikely to be feasible. Nevertheless, data collected from these experiments conducted under different operating conditions will help to develop treatment systems appropriate to local conditions.

11.2.2 *Passive remediation*

"Passive treatments" are processes with minimal human intervention and maintenance. They employ natural construction materials, promote the growth of natural vegetation and utilize naturally available energy sources e.g. microbial metabolic energy, photosynthesis and chemical energy (Pulles *et al.*, 2004). All such treatments rely on natural attenuation (NA): the combination of *in-situ* "physical, chemical, and biological processes that, under favorable conditions, act without human intervention to reduce the mass, toxicity, mobility, volume, or concentration of contaminants in soil or groundwater" (EPA, 1999). Here, we distinguish three types of NA in passive systems: "abiotic" remediation, phytoaccumulation and microbial remediation.

11.2.2.1 *"Abiotic" remediation*
Passive treatments, generally involving wetlands and oxidation ponds, are commonly used to treat mine drainage waters to remove high concentrations of metals, principally at low flow rates (less than $1000 \,m^3 \,d^{-1}$) and in acid conditions. They are attractive in the post-closure phase of mining, since they require only intermittent supervision, maintenance and monitoring of self-sustaining processes. They function without an external supply of power, but the treated water quality can be variable. The efficiency of pollutant removal depends on abiotic conditions, e.g. dissolved-oxygen content, pH, and the retention time of the water in the wetland. Mechanisms of removal of metals are diverse and include oxidation, precipitation, complexing, adsorption, ion exchange and uptake by plants (see next section). In aerobic wetlands, metal oxidation and hydrolysis are promoted, causing precipitation and physical retention of Fe, Al and Mn oxyhydroxides.

In the case of As, Chang *et al.* (2010) showed that bacteria isolated from wetlands could oxidize As(III) to As(V). Thus, it is reasonable to think that biological oxidation occurs *in situ* even if no direct evidence has been found. Several authors have reported t-As retention in such systems (Rahman *et al.*, 2011), i.e. the pollutant is held on adsorbent materials and not released; no information is available on the speciation of the element. In a disused gold mine in New Zealand, where no rehabilitation was undertaken after mine closure, Haffert and Craw (2010) identified As mineral phases downstream in Skippers Creek. They used these results to predict As mobility and to estimate the overall impact of the mine on the catchment water quality. Unfortunately, biological reactions were not taken into account. To date, there is little evidence to support the idea that abiotic remediation has played a significant role in remediation of any As-contaminated sites.

11.2.2.2   *Phytoaccumulation*

Phytoaccumulation of As offers a good alternative for As removal. The discovery of As accumulation by the chinese fern *Pteris vittata* (Ma *et al.*, 2001) has opened a new field for bioremediation. Various plants, including other ferns like *P. cretica*, *P. longifolia*, and *Pityrogramma calomelanos*, as well as aquatic macrophytes (Rahman and Hasegawa, 2011) have proven to be efficient at As uptake, suggesting that phytoaccumulation may be a useful component in the remediation of As-contaminated soil or water (Wang and Mulligan, 2006; Camacho, *et al.*, 2011).

The role of microorganisms in phytoaccumulation is poorly understood. A recent study (Mathews *et al.*, 2010), conducted in hydroponic conditions, showed that *P. vittata* can take up both As(III) and As(V). Regardless of which form was supplied, As(V) was dominant in the roots while As(III) was dominant in the rhizome and fronds. This study demonstrated that As(III) oxidation occurred in the growth medium, i.e. the rhizosphere, as well as in the roots of the plant: it must involve both microflora in the rhizosphere and As(III)-oxidizing enzymes directly associated with the plant.

Control of As uptake by these plants needs to be further studied. On the microbiological side, the rhizosphere bacterial community associated with these ferns certainly plays an important role in the uptake of As(III) and As(V). However, more investigations are needed to evaluate the relative importance of the As(III)-oxidizing and As(V)-reducing bacteria in the immobilization of As *via* phytoaccumulation.

11.2.2.3   *Microbial As(III) oxidation and natural attenuation*

As already mentioned, the two inorganic As species, As(III) and As(V), are the most abundant in water and soil, being released naturally from As-enriched minerals or from anthropogenic sources, including mining and smelting industries and agriculture. Depending on pH and redox conditions (Eh), oxidation of As(III) to As(V) reduces As bioavailability and thus participates in NA. As(V) can then be immobilized *via* sorption to solids such as organic matter (Grafe *et al.*, 2002; Redman *et al.*, 2002), clay minerals (Manning and Goldberg, 1996, 1997), and (hydr)oxides of Fe (Bowell, 1994; Driehaus *et al.*, 1998), Mn (Ouvrard *et al.*, 2002; Katsoyiannis *et al.*, 2004) or Al (Anderson *et al.*, 1976; Lin and Wu, 2001), and by precipitation-coagulation (Cheng *et al.*, 1994; Hering *et al.*, 1997) or phytoaccumulation (Zhao *et al.*, 2002; Xie *et al.*, 2009). In addition to abiotic or microbially driven indirect oxidation of As(III) by Fe(III), direct oxidation of As(III) to As(V) depends on the activity of arsenite-oxidizing bacteria and has been described for As NA in acid mine drainages (AMDs) in Japan (Wakao *et al.*, 1988) and California (Wilkie and Hering, 1998). In the latter case, colonies on the surface of the aquatic macrophytes *Potamogeton pectinatus* (Salmassi *et al.*, 2002) contained As(III)-oxidizing bacteria, including *Agrobacterium albertimagni* AOL15 (Salmassi *et al.*, 2002) and three strains of *Hydrogenophaga* (Salmassi *et al.*, 2006).

The French site of Carnoulès (Gard, South of France), represents another good example of As NA, and has been intensely investigated using various approaches. The creek of Carnoulès drains acid mine tailings, which emerge at the surface forming the Reigous acid spring (Casiot *et al.*, 2003). The acid water (pH 2.7–3.4) is contaminated with As (100–350 mg L^{-1}), and also contains Pb and FeS$_2$. At the mine outlet, As(III) is the predominant form, and Fe occurs as

Fe(II). The concentration of As rapidly declines by about 95% between the source of the Reigous creek and its confluence with the river Amous, 1.5 km downstream. Linked to Fe(II) oxidation, As is co-precipitated as either As(III)-Fe(III) (mainly during the wet season) or As(V)-Fe(III) oxyhydroxysulfates (during the dry season) (Duquesne et al., 2003; Morin et al., 2003). Casiot et al. (2003) clearly demonstrated the active role of bacteria in this NA process. Bruneel et al. (2003) showed that two strains of Thiomonas strains from the site oxidized As(III) present in the acid mine drainage. The importance of Thiomonas strains in NA was confirmed by recent in-situ investigations (Bruneel et al., 2011). A metagenomic and proteomic study (Bertin et al., 2011) showed that the bacterial flora of the Carnoulès AMD is dominated by a group of seven bacterial strains, named CARN1 to CARN7. Only one (CARN2), a Thiomonas, had the arsenite oxidase genes aioA and aioB (Muller et al., 2003; Lett et al., 2011). The reconstruction of these strains' genomes highlighted their metabolic specificity, but also pointed towards "partnerships" among the different members of the community – several bacteria being able to fix inorganic carbon and nitrogen while others can metabolize heterotrophically. The As NA processes, including As(III)-oxidation, and As(V) immobilization by Fe and/or S co-precipitation, may be related to the As(III) oxidase activity detected in the Thiomonas strain (CARN2), coupled with Fe(II) oxidation and S oxidation by Acidithiobacillus sp. (CARN5) and the Gallionella-related strain (CARN7) respectively (Bertin et al., 2011).

Various passive treatment systems have been used, with some success, to remove or immobilize As. They are low-cost, low-maintenance options. Current technologies for their use may be further developed in the light of knowledge gained from studies of active treatments.

11.3 CONCLUSION

All the biological water treatment systems developed to remove As take advantage of As(III) oxidation as a first step, since the As(V) formed is much more easily removed by a range of methods. In most of the As-remediation processes used at present, this oxidation step is performed with strong chemical oxidants.

Since the discovery of the first As(III)-oxidizing bacteria almost a century ago, many more such bacteria have been isolated from diverse environments, whether As-contaminated or not. We have reviewed here the systems in which the chemical oxidation step has been replaced with biological oxidation. Different biological agents have been used in different systems. Some studies have involved the use of pure bacterial cultures. These unnatural systems are normally very efficient over a wide range of As concentrations. In lab experiments, the As(III) concentrations chosen were usually high (>1–$10\,mg\,L^{-1}$) to examine the maximum As(III) transformation rates. The oxidation was performed by free cells or by bacteria immobilized on various support media. Other laboratory studies have used consortia of bacteria from As-contaminated environments, rather than pure cultures. These consortia too are very efficient for As removal, and they are easier to scale up and transpose to field conditions than pure cultures. Moreover, studies of consortia have been able to use As(III) concentrations set at lower levels ($<1\,mg\,L^{-1}$) to mimic conditions in streams or groundwater. So far, no work has been done directly on drinking waters containing low As concentrations. What is clear is that no single treatment process will work in all environments, so processes must be tailored to specific environmental conditions.

Emerging "natural" technologies such as wetland passive systems are specifically designed to use natural attenuation, capitalizing on ecological and geochemical reactions. Many of these systems are new and experimental but they appear particularly appropriate for the treatment of water discharged from abandoned mines. Again, however, it is difficult to predict their efficiency with a high level of confidence, since these systems are influenced by site-specific environmental conditions – water chemistry, or seasonal variability. Only limited information is available on long-term full-scale treatment processes; in fact little is known of biotic processes at all, since in most of these systems only abiotic reactions have been studied. Indeed, no bacterial community has even been described from As-contaminated passive wetland system. Promising as it seems, the value of As(III)-oxidizing bacteria in such systems has yet to be established. In AMDs, the central role of bacteria-driven As(III) transformation has been demonstrated but only a few studies have explained the successive abiotic and biotic mechanisms resulting in the decrease of t-As. Finally, we may be sure that the As MCL of $10\,\mu g\,L^{-1}$ in drinking water will require upgrades for existing water treatment plants. The most recent As(III)-oxidizing bacteria-based remediation systems are promising at the bench scale or the pilot scale and now need to scale up to integrate existing water-supply systems. It is also important to determine the long-term performance of experimental passive treatments. Finally, these biological As remediation systems will be adopted only if we are able to demonstrate their economical viability.

ACKNOWLEDGEMENTS

FD was supported by a grant from the French Ministry of Education and Research. Financial support came from the Université de Strasbourg (Unistra), the Centre National de la Recherche Scientifique (CNRS) and the Agence Nationale de la Recherche (ANR). This work was done in the frame of the "Groupement de Recherche—Métabolisme de l'Arsenic chez les Microorganismes (GDR2909-CNRS)" (http://gdr2909.alsace.cnrs.fr/).

REFERENCES

Anderson, M.A., Ferguson, J.F. & Gavis, J.: Arsenate adsorption on amorphous aluminum hydroxide. *J. Colloid Interface Sci.* 54:3 (1976), pp. 391–399.

Andrianisa, H.A., Ito, A., Sasaki, A., Ikeda, M., Aizawa, J. & Umita, T.: Behaviour of arsenic species in batch activated sludge process: Biotransformation and removal. *Water Sci. Technol.* 54:8 (2006), pp. 121–128.

Andrianisa, H.A., Ito, A., Sasaki, A., Aizawa, J. & Umita, T.: Biotransformation of arsenic species by activated sludge and removal of bio-oxidised arsenate from wastewater by coagulation with ferric chloride. *Water Res.* 42:19 (2008), pp. 4809–4817.

Battaglia-Brunet, F., Dictor, M.C., Garrido, F., Crouzet, C., Morin, D., Dekeyser, K., Clarens, M. & Baranger, P.: An arsenic(III)-oxidizing bacterial population: Selection, characterization, and performance in reactors. *J. Appl. Microbiol.* 93:4 (2002), pp. 656–667.

Battaglia-Brunet, F., Itard, Y., Garrido, F., Delorme, F., Crouzet, C., Greffie, C. & Joulian, C.: A simple biogeochemical process removing arsenic from a mine drainage water. *Geomicrobiol. J.* 23:3–4 (2006a), pp. 201–211.

Battaglia-Brunet, F., Joulian, C., Garrido, F., Dictor, M.C., Morin, D., Coupland, K., Barrie Johnson, D., Hallberg, K.B. & Baranger, P.: Oxidation of arsenite by *Thiomonas* strains and characterization of *Thiomonas arsenivorans* sp. nov. *Antonie Leeuwenhoek* 89:1 (2006b), pp. 99–108.

Battaglia-Brunet, F., Crouzet, C., Breeze, D., Tris, H. & Morin, D.: Decreased leachability of arsenic linked to biological oxidation of As(III) in solid wastes from bioleaching liquors. *Hydrometallurgy* 107:1–2 (2011), pp. 34–39.

Bertin, P.N., Heinrich-Salmeron, A., Pelletier, E., Goulhen-Chollet, F., Arsène-Ploetze, F., Gallien, S., Lauga, B., Casiot, C., Calteau, A., Vallenet, D., Bonnefoy, V., Bruneel, O., Chane-Woon-Ming, B., Cleiss-Arnold, J., Duran, R., Elbaz-Poulichet, F., Fonknechten, N., Giloteaux, L., Halter, D., Koechler, S., Marchal, M., Mornico, D., Schaeffer, C., Smith, A.A.T., Van Dorsselaer, A., Weissenbach, J., Médigue, C. & Le Paslier, D.: Metabolic diversity among main microorganisms inside an arsenic-rich ecosystem revealed by meta- and proteo-genomics. *ISME J.* (2011).

Bhumbla, D.K. & Keefer, R.F. (1994) *Arsenic in the environment* Part I: *Cycling and characterization.* Nriagu J.O. (ed), pp. 51–82, Wiley-Interscience, New York.

Bowell, R.J.: Sorption of arsenic by iron oxides and oxyhydroxides in soils. *Appl. Geochem.* 9:3 (1994), pp. 279–286.

Brown, J.C.: Biological treatments of drinking water. *The bridge* 37:4 (2007), pp. 30–36.

Bruneel, O., Personné, J.C., Casiot, C., Leblanc, M., Elbaz-Poulichet, F., Mahler, B.J., Le Flèche, A. & Grimont, P.A.D.: Mediation of arsenic oxidation by *Thiomonas* sp. in acid-mine drainage (Carnoulès, France). *J. Appl. Microbiol.* 95:3 (2003), pp. 492–499.

Bruneel, O., Volant, A., Gallien, S., Chaumande, B., Casiot, C., Carapito, C., Bardil, A., Morin, G., Brown Jr., G.E., Personné, C.J., Le Paslier, D., Schaeffer, C., van Dorsselaer, A., Bertin, P.N., Elbaz-Poulichet, F. & Arsène-Ploetze, F.: Characterization of the active bacterial community involved in natural attenuation processes in arsenic-rich creek sediments. *Microb. Ecol.* 61:4 (2011), pp. 793–810.

Camacho, L.M., Gutiérrez, M., Alarcon-Herrera, M.T., Villalba, M.D.L. & Deng, S.: Occurrence and treatment of arsenic in groundwater and soil in northern Mexico and southwestern USA. *Chemosphere* 83:3 (2011), pp. 211–225.

Casiot, C., Morin, G., Juillot, F., Bruneel, O., Personné, J.C., Leblanc, M., Duquesne, K., Bonnefoy, V. & Elbaz-Poulichet, F.: Bacterial immobilization and oxidation of arsenic in acid mine drainage (Carnoulès creek, France). *Water Res.* 37:12 (2003), pp. 2929–2936.

Casiot, C., Pedron, V., Bruneel, O., Duran, R., Personné, J.C., Grapin, G., Drakidès, C. & Elbaz-Poulichet, F.: A new bacterial strain mediating As oxidation in the Fe-rich biofilm naturally growing in a groundwater Fe treatiment pilot unit. *Chemosphere* 64:3 (2006), pp. 492–496.

Challan Belval, S., Garnier, F., Michel, C., Chautard, S., Breeze, D. & Garrido, F.: Enhancing pozzolana colonization by As(III)-oxidizing bacteria for bioremediation purposes. *Appl. Microbiol. Biotechnol.* 84:3 (2009), pp. 565–573.

Chang, J.S., Yoon, I.H., Lee, J.H., Kim, K.R., An, J. & Kim, K.W.: Arsenic detoxification potential of aox genes in arsenite-oxidizing bacteria isolated from natural and constructed wetlands in the Republic of Korea. *Environ. Geochem. Health* 32:2 (2010), pp. 95–105.

Cheng, R.C., Liang, S., Wang, H.-C. & Beuhler, M.D.: Enhanced coagulation for arsenic removal. *J. Am. Water Works Assoc.* 86:9 (1994), pp. 79–90.

Cullen, W.R. & Reimer, K.J.: Arsenic speciation in the environment. *Chem. Rev.* 89:4 (1989), pp. 713–764.

Dastidar, A. & Wang, Y.T.: Arsenite oxidation by batch cultures of *Thiomonas arsenivorans* strain b6. *J. Environ. Eng.* 135:8 (2009), pp. 708–715.

Dastidar, A. & Wang, Y.T.: Kinetics of arsenite oxidation by chemoautotrophic *Thiomonas arsenivorans* strain b6 in a continuous stirred tank reactor. *J. Environ. Eng.* 136:10 (2010), pp. 1119–1127.

Dictor, M.C., Battaglia-Brunet, F., Garrido, F. & Baranger, P.: Arsenic oxidation capabilities of a chemo-autotrophic bacterial population: Use for the treatment of an arsenic contaminated wastewater. *J. Phys.* 107:1 (2003), pp. 377–380.

Dixit, S. & Hering, J.G.: Comparison of arsenic(V) and arsenic(III) sorption onto iron oxide minerals: Implications for arsenic mobility. *Environ. Sci. Technol.* 37:18 (2003), pp. 4182–4189.

Driehaus, W., Seith, R. & Jekel, M.: Oxidation of arsenate(III) with manganese oxides in water treatment. *Water Res.* 29:1 (1995), pp. 297–305.

Driehaus, W., Jekel, M. & Hildebrandt, U.: Granular ferric hydroxide – A new adsorbent for the removal of arsenic from natural water. *J. Water Supply Res. Technol. AQUA* 47:1 (1998), pp. 30–35.

Duquesne, K., Lebrun, S., Casiot, C., Bruneel, O., Personné, J.C., Leblanc, M., Elbaz-Poulichet, F., Morin, G. & Bonnefoy, V.: Immobilization of arsenite and ferric Iron by *Acidithiobacillus ferrooxidans* and its relevance to Acid Mine Drainage. *Appl. Environ. Microbiol.* 69:10 (2003), pp. 6165–6173.

Duquesne, K. *Rôle des bactéries dans la bioremédiation de l'arsenic dans les eaux acides de drainage de la mine de Carnoulès.* PhD Thesis, Université de la Méditerranée Aix-Marseille II, Marseille, France, 2004.

Edwards, M.: Chemistry of arsenic removal during coagulation and Fe-Mn oxidation. *J. Am. Water Works Assoc.* 86:9 (1994), pp. 64–78.

EPA (1999) *Use of monitored natural attenuation at superfund, RCRA corrective action, and underground storage tank sites.* Office of Solid Waste and Emergency Response, Directive Number 9200, pp. 4–17.

Fisher, J.C. & Hollibaugh, J.T.: Selenate-dependent anaerobic arsenite oxidation by a bacterium from Mono Lake, California. *Appl. Environ. Microbiol.* 74:9 (2008), pp. 2588–2594.

Goldberg, S.: Competitive adsorption of arsenate and arsenite on oxides and clay minerals. *Soil Sci. Soc. Am. J.* 66:2 (2002), pp. 413–421.

Grafe, M., Eick, M.J., Grossl, P.R. & Saunders, A.M.: Adsorption of arsenate and arsenite on ferrihydrite in the presence and absence of dissolved organic carbon. *J. Environ. Qual.* 31:4 (2002), pp. 1115–1123.

Haffert, L. & Craw, D.: Geochemical processes influencing arsenic mobility at Bullendale historic gold mine, Otago, New Zealand. *New Zeal. J. Geol. Geophys.* 53:2–3 (2010), pp. 129–142.

Hering, J.G., Chen, P.Y., Wilkie, J.A. & Elimelech, M.: Arsenic removal from drinking water during coagulation. *J. Environ. Eng.* 123:8 (1997), pp. 800–807.

Hoeft, S.E., Blum, J.S., Stolz, J.F., Tabita, F.R., Witte, B., King, G.M., Santini, J.M. & Oremland, R.S.: *Alkalilimnicola ehrlichii* sp. nov., a novel, arsenite-oxidizing haloalkaliphilic gammaproteobacterium capable of chemoautotrophic or heterotrophic growth with nitrate or oxygen as the electron acceptor. *Int. J. Syst. Evol. Microbiol.* 57:3 (2007), pp. 504–512.

Ike, M., Miyazaki, T., Yamamoto, N., Sei, K. & Soda, S.: Removal of arsenic from groundwater by arsenite-oxidizing bacteria. *Water Sci. Technol.* 58:5 (2008), pp. 1095–1100.

Inskeep, W.P., Macur, R.E., Hamamura, N., Warelow, T.P., Ward, S.A. & Santini, J.M.: Detection, diversity and expression of aerobic bacterial arsenite oxidase genes. *Environ. Microbiol.* 9:4 (2007), pp. 934–943.

Jiang, J.Q.: Removing arsenic from groundwater for the developing world – A review. *Water Sci. Technol.* 44:6 (2001), pp. 89–98.

Kasai, Y., Takahata, Y., Manefield, M. & Watanabe, K.: RNA-based stable isotope probing and isolation of anaerobic benzene-degrading bacteria from gasoline-contaminated groundwater. *Appl. Environ. Microbiol.* 72:5 (2006), pp. 3586–3592.

Katsoyiannis, I., Zouboulis, A., Althoff, H. & Bartel, H.: As(III) removal from groundwaters using fixed-bed upflow bioreactors. *Chemosphere* 47:3 (2002), pp. 325–332.

Katsoyiannis, I.A. & Zouboulis, A.I.: Biological treatment of Mn(II) and Fe(II) containing groundwater: kinetic considerations and product characterization. *Water Res.* 38:7 (2004), pp. 1922–1932.

Katsoyiannis, I.A., Zouboulis, A.I. & Jekel, M.: Kinetics of bacterial As(III) oxidation and subsequent As(V) removal by sorption onto biogenic manganese oxides during groundwater treatment. *Ind. Eng. Chem. Prod. Res. Dev.* 43:2 (2004), pp. 486–493.

Kim, M.J. & Nriagu, J.: Oxidation of arsenite in groundwater using ozone and oxygen. *Sci. Total Environ.* 247:1 (2000), pp. 71–79.

Lett, M.-C., Muller, D., Lièvremont, D., Silver, S. & Santini, J.M.: Unified nomenclature for genes involved in prokaryotic aerobic arsenite oxidation. *J. Bacteriol.* (2011).

Lièvremont, D., N'Negue, M.A., Behra, P. & Lett, M.C.: Biological oxidation of arsenite: Batch reactor experiments in presence of kutnahorite and chabazite. *Chemosphere* 51:5 (2003), pp. 419–428.

Lin, T.F. & Wu, J.K.: Adsorption of arsenite and arsenate within activated alumina grains: Equilibrium and kinetics. *Water Res.* 35:8 (2001), pp. 2049–2057.

Lin, Z. & Puls, R.W.: Adsorption, desorption and oxidation of arsenic affected by clay minerals and aging process. *Environ. Geol.* 39:7 (2000), pp. 753–759.

Lugtu, R.T., Choi, S.C. & Oh, Y.S.: Arsenite oxidation by a facultative chemolithotrophic bacterium SDB1 isolated from mine tailing. *J. Microbiol.* 47:6 (2010), pp. 686–692.

Lytle, D.A., Chen, A.S., Sorg, T.J., Phillips, S. & French, K.: Microbial As(III) oxidation in water treatment plant filters. *J. Am. Water Works Assoc.* 99:12 (2007), pp. 72–86.

Ma, L.Q., Komar, K.M., Tu, C., Zhang, W., Cai, Y. & Kennelley, E.D.: A fern that hyperaccumulates arsenic. *Nature* 409:6820 (2001), p. 579.

Manning, B.A. & Goldberg, S.: Modeling arsenate competitive adsorption on kaolinite, montmorillonite and illite. *Clay. Clay Miner.* 44:5 (1996), pp. 609–623.

Manning, B.A. & Goldberg, S.: Adsorption and stability of arsenic(III) at the clay mineral-water interface. *Environ. Sci. Technol.* 31:7 (1997), pp. 2005–2011.

Manning, B.A., Fendorf, S.E., Bostick, B. & Suarez, D.L.: Arsenic(III) oxidation and arsenic(V) adsorption reactions on synthetic birnessite. *Environ. Sci. Technol.* 36:5 (2002), pp. 976–981.

Mathews, S., Ma, L.Q., Rathinasabapathi, B., Natarajan, S. & Saha, U.K.: Arsenic transformation in the growth media and biomass of hyperaccumulator *Pteris vittata* L. *Bioresource Technology* 101:21 (2010), pp. 8024–8030.

Michel, C., Jean, M., Coulon, S., Dictor, M.C., Delorme, F., Morin, D. & Garrido, F.: Biofilms of As(III)-oxidising bacteria: Formation and activity studies for bioremediation process development. *Appl. Microbiol. Biotechnol.* 77:2 (2007), pp. 457–467.

Michon, J. *Etude de l'oxydation biologique de l'arsenic As(III) par le consortium bactérien CAsO1: mise au point de méthodes de détection et application à la détoxification d'effluents.* PhD Thesis, Université de Limoges, Limoges, France, 2006.

Michon, J., Dagot, C., Deluchat, V., Dictor, M.C., Battaglia-Brunet, F. & Baudu, M.: As(III) biological oxidation by CAsO1 consortium in fixed-bed reactors. *Process Biochem.* 45:2 (2010), pp. 171–178.

Mohan, D. & Pittman, J.C.U.: Arsenic removal from water/wastewater using adsorbents–A critical review. *J. Hazard. Mater.* 142:1–2 (2007), pp. 1–53.

Mokashi, S.A. & Paknikar, K.M.: Arsenic (III) oxidizing *Microbacterium lacticum* and its use in the treatment of arsenic contaminated groundwater. *Lett. Appl. Microbiol.* 34:4 (2002), pp. 258–262.

Mondal, P., Majumder, C.B. & Mohanty, B.: Treatment of arsenic contaminated water in a laboratory scale up-flow bio-column reactor. *J. Hazard. Mater.* 153:1–2 (2008), pp. 136–145.

Morin, G., Juillot, F., Casiot, C., Bruneel, O., Personné, J.C., Elbaz-Poulichet, F., Leblanc, M., Ildefonse, P. & Calas, G.: Bacterial formation of tooeleite and Mixed Arsenic(III) or Arsenic(V) – Iron(III) gels in the carnoulès acid mine drainage, France. A XANES, XRD, and SEM study. *Environ. Sci. Technol.* 37:9 (2003), pp. 1705–1712.

Muller, D., Lièvremont, D., Simeonova, D.D., Hubert, J.C. & Lett, M.C.: Arsenite oxidase *aox* genes from a metal-resistant ß-proteobacterium. *J. Bacteriol.* 185:1 (2003), pp. 135–141.

Oremland, R.S., Hoeft, S.E., Santini, J.M., Bano, N., Hollibaugh, R.A. & Hollibaugh, J.T.: Anaerobic oxidation of arsenite in Mono Lake water and by a facultative, arsenite-oxidizing chemoautotroph, strain MLHE-1. *Appl. Environ. Microbiol.* 68:10 (2002), pp. 4795–4802.

Oremland, R.S., Kulp, T.R., Blum, J.S., Hoeft, S.E., Baesman, S., Miller, L.G. & Stolz, J.F.: Microbiology: A microbial arsenic cycle in a salt-saturated, extreme environment. *Science* 308:5726 (2005), pp. 1305–1308.

Osborne, T.H., Jamieson, H.E., Hudson-Edwards, K.A., Nordstrom, D.K., Walker, S.R., Ward, S.A. & Santini, J.M.: Microbial oxidation of arsenite in a subarctic environment: Diversity of arsenite oxidase genes and identification of a psychrotolerant arsenite oxidiser. *BMC Microbiol.* 10 (2010).

Oscarson, D.W., Huang, P.M., Liaw, W.K. & Hammer, U.T.: Kinetics of oxidation of arsenite by various manganese dioxides. *Soil Sci. Soc. Am. J.* 47:4 (1983), pp. 644–648.

Ouvrard, S., Simonnot, M.O., De Donato, P. & Sardin, M.: Diffusion-controlled adsorption of arsenate on a natural manganese oxide. *Ind. Eng. Chem. Prod. Res. Dev.* 41:24 (2002), pp. 6194–6199.

Pulles, W., Rose, P., Coetser, L. & Heath, R.: Development of integrated passive water treatment systems for the treatment of mine waters. *AusIMM Bulletin* (1) (2004), pp. 58–63.

Rahman, K.Z., Wiessner, A., Kuschk, P., van Afferden, M., Mattusch, J. & Muller, R.A.: Fate and distribution of arsenic in laboratory-scale subsurface horizontal-flow constructed wetlands treating an artificial wastewater. *Ecol. Eng.* 37:8 (2011), pp. 1214–1224.

Rahman, M.A. & Hasegawa, H.: Aquatic arsenic: Phytoremediation using floating macrophytes. *Chemosphere* 83:5 (2011), pp. 633–646.

Redman, A.D., Macalady, D.L. & Ahmann, D.: Natural organic matter affects arsenic speciation and sorption onto hematite. *Environ. Sci. Technol.* 36:13 (2002), pp. 2889–2896.

Reinhold-Hurek, B. & Hurek, T.: The genera *Azoarcus*, *Azovibrio*, *Azospira* and *Azonexus*. *Prokaryotes* 5 (2006), pp. 873–891.

Rhine, E.D., Phelps, C.D. & Young, L.Y.: Anaerobic arsenite oxidation by novel denitrifying isolates. *Environ. Microbiol.* 8:5 (2006), pp. 899–908.

Ryan, M.P., Pembroke, J.T. & Adley, C.C.: *Ralstonia pickettii* in environmental biotechnology: Potential and applications. *J. Appl. Microbiol.* 103:4 (2007), pp. 754–764.

Salmassi, T.M., Venkateswaren, K., Satomi, M., Nealson, K.H., Newman, D.K. & Hering, J.G.: Oxidation of arsenite by *Agrobacterium albertimagni*, AOL15, sp. nov., isolated from Hot Creek, California. *Geomicrobiol. J.* 19:1 (2002), pp. 53–66.

Salmassi, T.M., Walker, J.J., Newman, D.K., Leadbetter, J.R., Pace, N.R. & Hering, J.G.: Community and cultivation analysis of arsenite oxidizing biofilms at Hot Creek. *Environ. Microbiol.* 8:1 (2006), pp. 50–59.

Santini, J.M., Sly, L.I., Schnagl, R.D. & Macy, J.M.: A new chemolithoautotrophic arsenite-oxidizing bacterium isolated from a gold mine: Phylogenetic, physiological, and preliminary biochemical studies. *Appl. Environ. Microbiol.* 66:1 (2000), pp. 92–97.

Simeonova, D.D., Micheva, K., Muller, D.A.E., Lagarde, F., Lett, M.C., Groudeva, V.I. & Lièvremont, D.: Arsenite oxidation in batch reactors with alginate-immobilized ULPAs1 strain. *Biotechnol. Bioeng.* 91:4 (2005), pp. 441–446.

Smedley, P.L. & Kinniburgh, D.G.: A review of the source, behaviour and distribution of arsenic in natural waters. *Appl. Geochem.* 17:5 (2002), pp. 517–568.

Stasinakis, A.S. & Thomaidis, N.S.: Fate and biotransformation of metal and metalloid species in biological wastewater treatment processes. *Crit. Rev. Environ. Sci. Tech.* 40:4 (2010), pp. 307–364.

Stolz, J.F., Basu, P., Santini, J.M. & Oremland, R.S.: Arsenic and selenium in microbial metabolism. *Annu. Rev. Microbiol.* 60 (2006), pp. 107–130.

Stoodley, P., Sauer, K., Davies, D.G. & Costerton, J.W.: Biofilms as complex differentiated communities. *Annu. Rev. Microbiol.* 56 (2002), pp. 187–209.

Sun, W., Sierra, R. & Field, J.A.: Anoxic oxidation of arsenite linked to denitrification in sludges and sediments. *Water Res.* 42:17 (2008), pp. 4569–4577.

Sun, W., Sierra-Alvarez, R., Fernandez, N., Sanz, J.L., Amils, R., Legatzki, A., Maier, R.M. & Field, J.A.: Molecular characterization and *in situ* quantification of anoxic arsenite-oxidizing denitrifying enrichment cultures. *FEMS Microbiol. Ecol.* 68:1 (2009a), pp. 72–85.

Sun, W., Sierra-Alvarez, R., Milner, L., Oremland, R. & Field, J.A.: Arsenite and ferrous iron oxidation linked to chemolithotrophic denitrification for the immobilization of arsenic in anoxic environments. *Environ. Sci. Technol.* 43:17 (2009b), pp. 6585–6591.

Sun, W., Sierra-Alvarez, R. & Field, J.A.: The role of denitrification on arsenite oxidation and arsenic mobility in an anoxic sediment column model with activated alumina. *Biotechnol. Bioeng.* 107:5 (2010a), pp. 786–794.

Sun, W., Sierra-Alvarez, R., Hsu, I., Rowlette, P. & Field, J.A.: Anoxic oxidation of arsenite linked to chemolithotrophic denitrification in continuous bioreactors. *Biotechnol. Bioeng.* 105:5 (2010b), pp. 909–917.

Sun, W., Sierra-Alvarez, R., Milner, L. & Field, J.A.: Anaerobic oxidation of arsenite linked to chlorate reduction. *Appl. Environ. Microbiol.* 76:20 (2010c), pp. 6804–6811.

Sun, W., Sierra-Alvarez, R. & Field, J.A.: Long term performance of an arsenite-oxidizing-chlorate-reducing microbial consortium in an upflow anaerobic sludge bed (UASB) bioreactor. *Bioresource Technology* 102:8 (2011), pp. 5010–5016.

Suttigarn, A. & Wang, Y.T.: Arsenite oxidation by *Alcaligenes faecalis* strain O1201. *J. Environ. Eng.* 131:9 (2005), pp. 1293–1301.

Szaleniec, M., Hagel, C., Menke, M., Nowak, P., Witko, M. & Heider, J.: Kinetics and mechanism of oxygen-independent hydrocarbon hydroxylation by ethylbenzene dehydrogenase. *Biochemistry* 46:25 (2007), pp. 7637–7646.

Teitzel, G.M. & Parsek, M.R.: Heavy metal resistance of biofilm and planktonic *Pseudomonas aeruginosa*. *Appl. Environ. Microbiol.* 69:4 (2003), pp. 2313–2320.

Wakao, N., Koyatsu, H., Komai, Y., Shimokawara, H., Sakurai, Y. & Shiota, H.: Microbial oxidation of arsenite and occurrence of arsenite-oxidizing bacteria in acid mine water from a sulfur-pyrite mine. *Geomicrobiol. J.* 6 (1988), pp. 11–24.

Wan, J., Klein, J., Simon, S., Joulian, C., Dictor, M.C., Deluchat, V. & Dagot, C.: As(III) oxidation by *Thiomonas arsenivorans* in up-flow fixed-bed reactors coupled to As sequestration onto zero-valent iron-coated sand. *Water Res.* 44:17 (2010), pp. 5098–5108.

Wang, S. & Mulligan, C.N.: Natural attenuation processes for remediation of arsenic contaminated soils and groundwater. *J. Hazard. Mater.* 138:3 (2006), pp. 459–470.

Wang, Y.T. & Suttigarn, A.: Arsenite oxidation by *Alcaligenes faecalis* strain 01201 in a continuous-flow bioreactor. *J. Environ. Eng.* 133:5 (2007), pp. 471–476.

Wang, Y.T., Suttigarn, A. & Dastidar, A.: Arsenite oxidation by immobilized cells of *Alcaligenes faecalis* strain O1201 in a fluidized-bed reactor. *Water Environ. Res.* 81:2 (2009), pp. 173–177.

Weeger, W., Lièvremont, D., Perret, M., Lagarde, F., Hubert, J.C., Leroy, M. & Lett, M.C.: Oxidation of arsenite to arsenate by a bacterium isolated from an aquatic environment. *BioMetals* 12:2 (1999), pp. 141–149.

WHO (2005) *Chlorate and chlorite in drinking water. Background document for development of WHO guidelines for drinking-water quality.* World Health Organisation, WHO/SDE/WSH/05.08/86, p. 23

Wilkie, J.A. & Hering, J.G.: Rapid oxidation of geothermal arsenic(III) in streamwaters of the eastern Sierra Nevada. *Environ. Sci. Technol.* 32:5 (1998), pp. 657–662.

Xie, Q.E., Yan, X.L., Liao, X.Y. & Li, X.: The arsenic hyperaccumulator fern *Pteris vittata* L. *Environ. Sci. Technol.* 43:22 (2009), pp. 8488–8495.

Yang, C., Xu, L., Yan, L. & Xu, Y.: Construction of a genetically engineered microorganism with high tolerance to arsenite and strong arsenite oxidative ability. *J. Environ. Sci. Health – Part A Toxic/Hazardous Substances and Environmental Engineering* 45:6 (2010), pp. 732–737.

Zhao, F.J., Dunham, S.J. & McGrath, S.P.: Arsenic hyperaccumulation by different fern species. *New Phytologist* 156:1 (2002), pp. 27–31.

CHAPTER 12

Development of biosensors for the detection of arsenic in drinking water

Christopher French, Kim de Mora, Nimisha Joshi, Alistair Elfick,
James Haseloff & James Ajioka

12.1 INTRODUCTION

12.1.1 *Overview*

Arsenic (As) is a major public health issue in Bangladesh, West Bengal and other regions. Current methods of detecting it are not altogether satisfactory. Biosensors are a potentially powerful technology for overcoming this problem, but have their own issues that must be addressed before they can be a really useful addition to the portfolio of techniques used to address the As crisis. In this chapter, we will consider the basic nature of biosensors, attempts made to apply them to As detection, and obstacles that remain to impede widespread adoption of this technology.

12.1.2 *Arsenic – a public health crisis*

During the 1970s, waterborne diseases were a major cause of sickness in Bangladesh, with hundreds of thousands dying each year (Meharg, 2005; Hossain, 2006; Chakraborti *et al.*, 2010; Akter and Ali, 2011). In 1972, UNICEF instigated a program to drill nearly one million tube wells to supply clean groundwater for drinking. This had a rapid and profound effect on the death rate, and the program was continued by other donors and by private individuals and organizations. In 2005, it was reported that approximately ten million tube wells had been sunk. Unfortunately, it had not been appreciated that the sediments through which these wells were drilled were rich in As. From 1983, doctors in Bangladesh and adjacent West Bengal noted many patients presenting with skin lesions and other conditions associated with chronic consumption of As. This was traced to groundwater containing As, mainly in the form of arsenate (AsO_4^{3-}) with As as As(V) and the much more toxic arsenite (AsO_3^{3-}) with As as As(III). It was later discovered that rice irrigated with As-containing water also contained high levels of As, and this might contribute more to the overall As burden than the drinking water itself (Zhao *et al.*, 2010). Further studies have indicated that high arsenic levels also occur in groundwater in many other parts of the world (Meharg, 2005), including Nepal (Panthi *et al.*, 2006) and Vietnam (Trang *et al.*, 2005). Worldwide, tens of millions of people are considered to be at risk from As-contaminated groundwater.

Chronic consumption of As is associated with the development of skin lesions (hyper- and hypopigmentation, followed by keratosis) and later with peripheral vascular disease and a variety of cancers, particularly of the bladder and lung. The recommended World Health Organisation (WHO) limit for long-term consumption of drinking water at the time of the original discovery of the widespread groundwater contamination was $50\,\mu g\,L^{-1}$ As. This was later reduced to $10\,\mu g\,L^{-1}$ As, though many developing countries are still working to a limit of $50\,\mu g\,L^{-1}$. Many tube wells in Bangladesh and West Bengal were found to exceed even these levels by a large margin (Meharg, 2005). Following initial surveys, unsafe wells were marked to indicate that they should not be used. However, social issues are reported to affect compliance with these recommendations. To further complicate the issue, it has also been reported that the underground plumes of As-laden groundwater can move over time, so that the As concentration in a given well can gradually change. As removal plants have been introduced, but their effectiveness has been called into question (Hossain *et al.*, 2005). In any case, it is essential that their output should be monitored.

Table 12.1.   Some commercially available field test kits for As in water.

Manufacturer	Output type	Working range	Cost per test
AAN (Japan)	semiquantitative	–	–
E-Merck (Germany)	semiquantitative	20–500 μg L^{-1}	$0.50
Global Water	semiquantitative	2–80 μg L^{-1}	–
NIPSOM (Bangladesh)	semiquantitative	0, 100, 500, 1000 μg L^{-1}	$0.30
AIIH&PH (India)	semiquantitative	–	–
ENPHO (Nepal)	semiquantitative	10–500 μg L^{-1}	Rs 114 ($1.80)
Modified AAN (Nepal)	semiquantitative	–	–
Hach-EZ (USA)	semiquantitative	10–4000 μg L^{-1}	$1–2 in USA
Wagtech Arsenator (UK)	quantitative	2–100 μg L^{-1}	$0.95 + instrument

Information from Panthi *et al.* (2006), Steinmaus *et al.* (2006), Safarzdeh-Amiri *et al.* (2011), Deshpande *et al.* (2001), and company websites. AAN: Asia Arsenic Network, Japan; NIPSOM: National Institute of Preventive and Social Medicine, Bangladesh; AIIH&PH: All India Institute of Hygiene and Public Health, Calcutta, India; ENPHO: Environmental and Public Health Organization, Nepal. Cost per test should be considered an approximate value based on information from literature and websites. A variety of similar kits are also available from other manufacturers.

It is therefore clear that a simple, inexpensive method for assaying As levels [particularly As(V) and As(III)] in drinking water is essential.

12.1.3   *Current arsenic detection technologies*

Current laboratory assays for measuring As levels in contaminated groundwater use a variety of highly sensitive techniques such as atomic absorption spectroscopy (AAS), atomic fluorescence spectroscopy (AFS), and inductively coupled plasma mass spectrometry, following reduction of As to arsine gas (AsH$_3$) (Meharg, 2005). A detailed discussion of these techniques is beyond the scope of this chapter. For our purposes, it is sufficient to note that these techniques, while sensitive and accurate, have several disadvantages. They require expensive instrumentation and highly trained operators, leading to a high cost per assay, and are fundamentally unsuited to field use, meaning that samples must be returned to a laboratory for processing. For rapid routine monitoring of wells, water treatment systems, and other samples of concern, it would be advantageous to provide simple, inexpensive assays that could be used on site by field workers or local people without requiring extensive training.

Current field test kits are based on the Gutzeit method, which involves reaction of water samples with metallic zinc and hydrochloric acid to generate arsine gas (AsH$_3$). This then reacts with an indicator paper impregnated with mercuric bromide to generate a colored spot. The intensity of the color generated indicates the concentration of As. Panthi *et al.* (2006) have summarized test kits used in Nepal; and Pande *et al.* (2001) reviewed kits used in Bangladesh and West Bengal. Some commercially available systems are listed in Table 12.1. Field test kits based on the Gutzeit method are simple and give a semi-quantitative visual output, but have a number of drawbacks. While reliably detecting wells that are heavily contaminated, they are reported to give significant numbers of false negative reactions at lower As concentrations, still within the range of concern (above 10 or even above 50 μg L^{-1}) (Rahman *et al.*, 2002), although more recent versions of the kits are reported to give good results above 15 or 20 μg L^{-1} As (Steinmaus *et al.*, 2006; Deshpande and Pande, 2005). Furthermore, arsine gas is extremely toxic, and may pose a significant hazard to operators performing large numbers of tests (Hussam *et al.*, 1999). The mercuric salts used in the assay represent a disposal issue, since mercury is also a toxic heavy metal. Owing to the visual interpretation of the assay, there is also a considerable degree of variation between operators. The Wagtech Digital Arsenator uses a colorimetric instrument to remove this variability, and is reported to give results highly correlated to those obtained by analytical laboratories

(Safarzadeh-Amiri *et al.*, 2011), but requires a greater initial investment to purchase the instrument; reportedly this comes to about $1600 for the instrument and $0.95 per assay for reagents.

Deshpande and Pande (2005) and Arora *et al.* (2009) listed the ideal features of an improved field test kit. While some of these are specific to the Gutzeit assay, such as requirements for the acids and reducing agents involved in the test, others are more generic and should apply to any useful field test. For example, it should:

- be very inexpensive: less than US$ 1 per assay
- not require special storage requirements such as cold storage
- not require separate assay equipment, but give a simple visual output
- be highly specific: it should react only with bioavailable As
- detect both arsenate and arsenite
- give reliable detection below the WHO limit of $10 \,\mu g\, L^{-1}$ As
- have simple, robust operation, be easy to set up, and operate reliably in a suitable range of ambient temperatures
- be simple to dispose of: reagents should pose no environmental threat on disposal

It seems clear that the present generation of field test kits leave some room for improvement in these areas. In the remainder of this chapter, we will consider the question of whether biosensors can form the basis of an improved field test device.

12.2 BIOSENSORS FOR DETECTION OF ENVIRONMENTAL TOXINS

12.2.1 *The basic concept of biosensors*

The term 'biosensor' is used in slightly different ways by different authors, but the common theme is that a biosensor is a device that uses a biological recognition element to detect some specific analyte, and then transduces this detection event to give a quantitative output signal. The output may be in the form of luminescence, fluorescence, or an electrical current or potential, but generally it is detected electrically and transferred to some type of electronic system for data capture and analysis.

The critical feature which distinguishes biosensors from other types of sensor is thus the presence of the biological recognition element. This is both the strength and weakness of biosensors. Molecular recognition is one of the core features of all biological systems – enzymes recognize their substrates, receptors their ligands, nucleic acids their complementary strands, and so on. These recognition events are highly specific and extremely sensitive, and biosensors aim to capture these benefits for analytical purposes. However, biological molecules are often relatively fragile, and often can be easily denatured by heat, drying, or various other stresses. The stability and storage lifetime of the biological recognition component is thus a critical design feature in the construction of a biosensor system.

12.2.2 *Biosensors based on enzymes and antibodies*

The most common recognition elements used in biosensors are enzymes, antibodies, and receptors. By far the most commercially important and widely used biosensors are enzymic biosensors for glucose, based on the fungal enzyme glucose oxidase (French and Cardosi, 2007). Enzymic biosensors are especially convenient in that the enzyme not only binds to the target analyte (the substrate), but also catalyses some sort of reaction, providing a change in the environment that can often be more or less easily transduced to generate an optical or electrical signal. The disadvantage of enzyme-based systems is that it is not yet possible to design an enzyme to act on a specific substrate (analyte); thus, if no suitable enzyme can be found in nature to act on the analyte of interest, an enzymic biosensor cannot be constructed.

Because of the nature of enzyme activities, enzymic biosensors are best suited to the detection of small molecules. The most common outputs from biosensors of these types are electrical (amperometric and potentiometric), and fluorescence. Amperometric sensors are based on enzymes that generate an electrical current owing to their activity. The glucose biosensors used widely for home blood-glucose monitoring work in this way. The enzyme glucose oxidase, immobilized at an electrode surface, oxidizes glucose to gluconolactone, reducing a flavin cofactor (flavin adenine dinucleotide, FAD) to its reduced hydroquinone form ($FADH_2$). A small molecule mediator reoxidizes the $FADH_2$ to FAD and is itself reoxidized at the electrode surface, which is held poised at a suitable potential by means of a circuit known as a potentiostat. The flow of electrons to the electrode is detected as an electrical current, which is easily measured. Potentiometric outputs, on the other hand, are based on the change in potential observed at an ion-selective electrode, which is due to the change in concentration of a particular ion. In the most common case this ion is H^+, as in the simple pH electrode, but electrodes responding to NH_3 (NH_4^+) and CO_2 (HCO_3^-) are also available. These systems are suitable for detecting the reactions of enzymes that consume or release any of these ions. Fluorescence-based biosensors are suitable for any reaction that consumes or releases a fluorescent molecule, such as the common enzyme cofactor NAD(P)H.

An alternative is to use antibodies or similar ligand-binding molecules as a recognition element. Antibodies can be raised against a wide variety of target analytes, making them potentially far more versatile than enzymes. However, the antibody simply binds to the analyte, rather than catalysing a reaction. Detecting a simple binding event is more challenging than detecting a reaction, so antibody-based sensors are often more complex than enzymic sensors, and may be less sensitive. Common examples are based on tagged competing ligands labeled with fluorescent or enzymic tags, or surface effects such as surface plasmon resonance or piezoelectric vibration, which are altered when the ligand binds to immobilized antibodies. Such techniques are best suited to relatively large analytes such as proteins, viruses or microorganisms, rather than to small molecules such as As(V) and As(III). The same considerations apply to other types of binding molecule such as aptamers (nucleic acids specifically designed or evolved to bind to a particular analyte). Surface plasmon resonance and piezoelectric biosensors are not really applicable to As detection, so will not be discussed further here; for more information, see French and Cardosi (2007) and references cited therein.

12.2.3   *Whole-cell biosensors and bioreporters*

Another class of biosensor, sometimes also referred to as a bioreporter, is based on a slightly different principle. In this case, living cells, usually bacteria, are used as the sensor element (Daunert *et al.*, 2000; Belkin, 2003; Tecon and van der Meer, 2008; van der Meer and Belkin, 2010). In the simplest case, these are bioluminescent marine bacteria such as *Photobacterium (Vibrio) fischeri*. At high population density, these organisms produce a constant blue luminescence through the action of the enzyme bacterial luciferase (LuxAB, encoded by the genes *luxAB*). This enzyme oxidizes a long-chain aldehyde such as tetradecanal to a fatty acid, in the presence of oxygen and reduced flavin mononucleotide ($FMNH_2$), and releases photons of blue light in the process. The fatty aldehyde substrate is generated by LuxCDE, requiring energy in the form of ATP, and flavin mononucleotide (FMN) is reduced to $FMNH_2$ by a flavin reductase such as LuxG, using NADPH as electron donor. The practical effect of this is that exposure to toxic substances that inhibit respiration reduces the supply of energy and reducing-power available to the luciferase system, resulting in a rapid decrease in light output. Thus, these systems are ideal for rapid screening of environmental samples (for example, from a contaminated site) for the presence of unspecified toxic substances. Samples that give a positive response in this test can then be sent for analysis using more specific (and expensive) laboratory assays. One widely used system of this type is MicroTox, supplied by Strategic Diagnostics (SDIX).

When highly sensitive detection of specific analytes is desired, a slightly different type of system is used. In this case, the cell is genetically modified so that in the presence of the target analyte an easily detectable response is generated. This relies on a genetic 'rewiring' of the cell's

Table 12.2.   Reporter genes commonly used in whole cell biosensors.

Reporter	Characteristics
lacZ (β-galactosidase)	Chromogenic and chemiluminescent substrates are available.
luxAB (bacterial luciferase)	Blue bioluminescence on addition of decanal.
luxCDABE	As above, but substrate addition is not required.
Firefly luciferase	Bioluminescence on addition of D-luciferin. Higher quantum yield than bacterial luciferase, but substrate is more expensive.
Fluorescent proteins	Fluorescence when stimulated by light of the appropriate wavelength. The original Green Fluorescent Protein (GFP) is still widely used, but many color variants are now available. Red Fluorescent Proteins (RFP) are visible by eye under normal illumination.

internal systems. Living cells have numerous built-in sensor systems which detect and respond to the presence of various types of molecule in the environment. For example, the presence of a certain sugar may cause the cell to switch on ('induce') genes that encode enzymes for the assimilation of that sugar. Likewise, the presence of toxic substances may cause the cell to switch on a detoxification system. In bacteria, these systems most commonly rely on genetic switches known as inducible promoters. The promoter is simply a sequence of DNA that acts as a switch to control the activity of a set of genes for a particular purpose. Also present in the cell is an effector protein, which binds to the promoter and either causes it to become active (in which case it is known as an activator protein) or, more commonly, causes it to become inactive (in which case it is known as a repressor protein). The activator or repressor protein can also bind to the molecule to which the promoter responds, which in this context is called the inducer. Binding of the inducer to a repressor protein causes it to release the promoter, allowing it to become active; binding of the inducer to an activator protein causes it to bind the associated promoter, causing it to become active (see Fig. 12.1 below for an illustration of this with respect to the *ars* promoter.)

Typically, cells contain hundreds or thousands of inducible promoters, responding to a wide variety of different molecules. The art in constructing a bioreporter is thus to find or introduce a promoter that responds to the analyte of interest, and then cause this promoter to activate a gene that encodes some easily detected protein. Such a gene is known as a reporter gene. The most widely used reporter genes produce chromogenic responses (i.e. color change), luminescence, or fluorescence. Some of the most commonly used are shown in Table 12.2. Chromogenic reporters such as LacZ can be read by spectrophotometers or plate-readers, but are also used for rapid visual assessment of the response. Luminescent and fluorescent responses require a luminometer or fluorimeter to detect and quantify the response.

Whole cell biosensor systems and their applications have been reviewed by (among others) Daunert *et al.* (2000), Belkin (2003), Tecon and van der Meer (2008), and van der Meer and Belkin (2010). The recent development of the new discipline of synthetic biology, which allows more extensive and rational re-engineering of living systems, raises some exciting possibilities for further development of a new generation of whole-cell biosensor systems, incorporating features such as tunable response characteristics, multi-stage outputs, novel output characteristics optimized for visual or electrical detection systems, and desirable host features such as robustness for storage and distribution. Some of these possibilities will be discussed below. For a further discussion of the possibilities raised, see van der Meer and Belkin (2010).

12.3   BIOSENSORS FOR ARSENIC

12.3.1   *Enzymic biosensors*

A number of enzymes are known to act on As. These fall into three categories (Stolz *et al.*, 2006; Silver and Phung, 2005; Mukhopadyay *et al.*, 2002): the widely distributed cytoplasmic As(V) reductases involved in As(V) detoxification, which reduce As(V) to As(III) prior to its expulsion from the cell by an As(III) efflux pump (discussed further below); the periplasmic/membrane-bound As(V) reductases involved in reduction of As(V) as a terminal electron acceptor (see Chapter 4); and the As(III) oxidases responsible for the use of As(III) as an electron donor (described in Chapter 7). As these are redox enzymes, they are potentially suitable for construction of specific electrochemical (amperometric) enzymic biosensors. We have found only one such system described in the literature (Male *et al.*, 2007). This involved immobilization of the As(III) oxidase from a chemolithoautotrophic bacterium, *Rhizobium* sp. strain NT-26, on the surface of a carbon electrode modified by deposition of multi-walled carbon nanotubes. The enzyme oxidized As(III) to As(V), and then passed the electrons to the electrode, held poised at a suitable potential, generating a detectable current. As(III) could be detected within 10 seconds, at levels of $1 \, \mu g \, L^{-1}$ As, with a linear response up to $500 \, \mu g \, L^{-1}$. The sensor was reported to work well with spiked water samples. This system presumably would not detect As(V) unless a pre-reduction step were included; however, As(III) is considerably more toxic than arsenate (Meharg, 2005), so detection of arsenite alone might be useful.

A number of reports have described electrochemical biosensor systems that, rather than using enzymes specifically active against As(V) or As(III), rely upon the inhibition of various enzyme activities by As(V) or As(III). An obvious limitation of this approach is that it lacks specificity for As; any substance that inhibits the activity of the analytical enzyme will give a response. For example, Pal *et al.* (2009) reported an electrochemical biosensor based on urease. The activity of urease was quantified by electrochemical oxidation of ammonium, released by urease activity, at an electrode modified with rhodium, which catalyses ammonia oxidation. Toxic heavy metals including mercury, cadmium and As could be detected by their inhibition of urease activity. The As detection limit was reported as $1.5 \, \mu g \, L^{-1}$. This type of device clearly is not specific for As, but might be useful if As were expected to be the only toxic heavy metal present in a sample. Sanllorente-Mendez *et al.* (2010) reported an amperometric biosensor device based on inhibition of the activity of acetylcholinesterase immobilized on a screen-printed carbon electrode. In this system, acetylcholinesterase acts on acetylthiocholine iodide to release the electroactive molecule thiocholine iodide, which is oxidized at the electrode surface to generate an amperometric signal (electric current). As(III) inhibits the activity of acetylcholinesterase, leading to a reduced signal. A detection limit of $1.1 \times 10^{-8} \, M$, corresponding to $0.83 \, \mu g \, L^{-1}$ As, was reported. However, As(V) does not inhibit acetylcholinesterase, so is not detected.

Sarkar *et al.* (2010) reported a slightly different amperometric sensor system. Rather than an enzyme, this system uses the amino acid L-cysteine to reduce As(V) to As(III) using screen-printed carbon electrodes. Nitrate, another potential oxidizing species, was found not to interfere under the conditions tested. As concentrations as low as $1.2 \, \mu g \, L^{-1}$ could be detected, and it was stated that the system could be used for field testing using a hand-held potentiostat. Since this system does not use a specific biological recognition element, it is not really a biosensor in the strict sense, and may lack the specificity of a true biosensor, but under the relatively predictable conditions of contaminated groundwater, such a system could potentially be a simple and low-cost device for field use. As with other electrochemical sensors based on As(V) reduction, this system would presumably detect only As(V), and not As(III), unless As(III) were first oxidized to As(V) before the assay (adding a sample pre-processing step), or the system were modified to measure As(III) oxidation as well as As(V) reduction. Perhaps an As(III)-oxidizing electrochemical biosensor could be combined with an As(V)-reducing sensor to produce a simple system that could give simultaneous measurements of both As(V) and As(III).

Numerous non-biological electrochemical assay systems for As have been reported in the literature, based on a variety of techniques. Since these are not biosensors, a detailed description is beyond the scope of this review; the reader is referred to Mays and Hussam (2009) for further details.

In addition to these devices, which are designed for the quantification of As, electrochemical sensors have also been used to study the damage to DNA caused by As. In devices of this type, DNA is immobilized at an electrode surface, and a signal due to its direct or indirect oxidation is

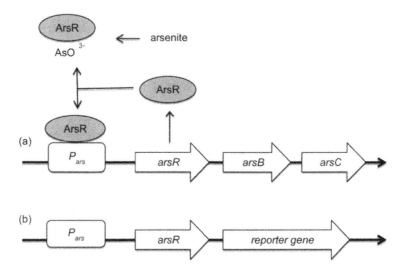

Figure 12.1. (a) Normal regulation of the *ars* operon. The gene *arsR* encodes repressor protein ArsR, which normally binds the promoter, P_{ars}, and inactivates it. Arsenite binds ArsR and prevents it from associating with the promoter, allowing transcription of the operon and production of proteins ArsR, ArsB and ArsC. (b) Typical arrangement of a bioreporter for As. The reporter gene is inserted following P_{ars} and *arsR*, usually on a multi-copy plasmid. Figure based on information collated from references cited in the main text.

measured. For example, Ozsoz *et al.* (2003) studied the interaction of As trioxide with DNA based on alterations to the guanine oxidation signal, and Labuda *et al.* (2005) studied DNA damage caused by As(III) and a variety of other As-containing compounds based on alteration to the signal caused by DNA oxidation catalysed by a rubidium or cobalt complex.

The majority of As biosensors reported in the literature are whole-cell biosensors, and in the remainder of this chapter, we will focus on the characteristics and potential of these devices.

12.3.2 *Whole-cell biosensors and bioreporters responding to arsenic*

Many reports of As biosensors in the literature describe whole-cell biosensors (bioreporters); for a recent review, see Diesel *et al.* (2009). As noted above, development of a whole-cell biosensor for a given analyte relies on finding a promoter that specifically responds to that analyte. Fortunately, many bacteria possess an As detoxification system that responds in a rapid and sensitive way to the presence of As(V) or As(III). This detoxification system is known as the *ars* operon. In its simplest form it consists of the inducible *ars* promoter, controlling the expression of three genes: *arsR*, encoding the *ars* repressor protein ArsR; *arsB*, encoding an As(III) efflux pump, which pumps As(III) out of the cell; and *arsC*, encoding the As(V) reductase (ArsC), which reduces As(V) to As(III), as described above. The repressor ArsR binds to the *ars* promoter and causes it to be inactive. As(V) or As(III) bind ArsR and prevent it from binding the promoter, leaving the promoter active to generate ArsB and ArsC (Fig. 12.1a). In the process, more ArsR is also produced, so that the promoter tends to switch itself off once it has produced enough ArsB and ArsC to deal with the problem (negative autoregulation). The *Escherichia coli* chromosomal *ars* operon is of this type (Diorio *et al.*, 1995), as is that found on *Staphylococcus* plasmid pI258 (Ji and Silver, 1992). The chromosomal *ars* operon of *Bacillus subtilis* includes an extra gene of unknown function, which does not appear to be required for response to As (Sato and Kobayashi, 1998). More complex systems also exist. For example, one of the best studied systems is that of the *E. coli* plasmid R773. This has a more complex arsenite efflux pump encoded by *arsAB*, as well as a second repressor, ArsD, which also acts as a 'chaperone' to carry As(III) to the efflux pump (Lin *et al.*, 2006).

Generation of a whole-cell biosensor for As(V) and As(III), then, simply requires insertion of a reporter gene adjacent to the *ars* promoter (Fig. 12.1b). This is a relatively trivial undertaking using modern genetic manipulation techniques. Many such bioreporter systems have been constructed to study the regulation of the *ars* operon. For example, Ji and Silver (1992) fused the promoter and *arsR* gene of *Staphylococcus* plasmid pI258 to β-galactosidase (*lacZ*) and bacterial luciferase (*luxAB*) to study its expression in *Staphylococcus aureus* and *E. coli*. In *E. coli*, expression of the reporter gene was induced by As(V) and As(III); in *S. aureus*, induction also occurred in the presence of antimonite and bismuth(III). Clear induction with As(III) was seen in both hosts at concentrations as low as 0.1 μM (equivalent to 7.5 μg L^{-1} As). Corbisier *et al.* (1993) described similar experiments with fusions of *luxAB* to the As and cadmium promoter elements of pI258, though induction seemed considerably less sensitive in this case. Factors affecting the sensitivity of whole-cell biosensors are discussed further below. Wu and Rosen (1993) characterized regulation of the *ars* promoter from *E. coli* plasmid R773 using fusions to a different reporter gene, β-lactamase. Diorio *et al.* (1995) discovered the *E. coli* chromosomal *ars* operon (as distinct from the previously known *ars* operon of plasmid R773) by randomly inserting *lacZ* (encoding β-galactosidase) onto the chromosome and screening for clones in which β-galactosidase activity was induced in the presence of As(III). Weaker induction was seen with As(V), and the promoter also responded to antimonite. Cai and DuBow (1996) constructed *luxAB* fusions to the promoter-*arsR* region of this operon, and demonstrated rapid (30-minute) induction by As(III). Sato and Kobayashi (1998) constructed *lacZ* fusions to the *ars* operon of *Bacillus subtilis*, and similarly found that As(III) and antimonite were strong inducers, with As(V) giving weaker induction. Again, As(III) gave good induction at 0.1 μM (equivalent to 7.5 μg L^{-1} As).

During the late 1990s, the As crisis in Bangladesh and West Bengal was becoming widely known, and it was rapidly appreciated that such systems could be used as biosensors for the detection of As in groundwater. For example, Tauriainen *et al.* (1997) reported fusion of the firefly luciferase reporter gene to the As promoter from pI258, and tested the resulting plasmids for response in three different hosts: *S. aureus, B. subtilis* and *E. coli*. As previously noted, induction was seen in the presence of As(V), As(III), and antimonite, with response to As(V) being least sensitive. Unexpectedly, the cells also responded to cadmium. The most sensitive response was seen in *S. aureus*, with detectable induction seen with 33 nM As(III) (2.5 μg L^{-1} As) following 2 hours induction. Freeze-dried cells gave a similar response, with somewhat weaker induction. Further *E. coli*-based sensors were also reported in a later paper (Tauriainen *et al.*, 1999) which also investigated the differences between direct measurement of luminescence in living cells, as compared to assay of luciferase activity after cell lysis. Freeze-dried forms of these sensors, as well as others responding to mercury, cadmium and lead, were tested against natural water samples spiked with heavy metals, and were found to give good responses with induction times as low as 30 minutes, with optimal induction occurring after 1 to 4 hours of incubation depending on the system (Tauriainen *et al.*, 2000).

Scott *et al.* (1997) described the construction of *E. coli*-based As biosensor cells based on the *ars* promoter of plasmid R773, using β-galactosidase as a reporter gene. Rather than the more conventional chromogenic assays for detecting β-galactosidase activity, they opted for more sensitive alternatives. Initially, detection was electrochemical, with *p*-aminophenyl-β-D-galactopyranoside as substrate. In this system, *p*-aminophenol released by enzyme activity is oxidized at an electrode surface to generate a detectable current. With this system, the detection limit for both As(III) and antimonite seemed to be around 0.1 μM (7.5 μg L^{-1} As). Ramanathan *et al.* (1998) tested the same sensor system using an alternative assay method, in which β-galactosidase acts on a chemiluminescent substrate, Galacto-Light Plus (Applied Biosystems), to generate a luminescent signal. As with the electrochemical system, cell lysis was necessary for this assay to be performed. Polymyxin B sulfate was found to give better lysis than Triton X-100 for this application. This system showed remarkable sensitivity, with antimonite detectable at 10^{-15} M. Ramanathan *et al.* (1997) also described a bioluminescent biosensor system based on expression of bacterial luciferase, LuxAB. Again, sensitivity to As(III) and antimonite at around 10^{-15} M was reported with 30 minutes induction. Interestingly, in both the chemiluminescent and bio-luminescent systems, a biphasic response was seen, with an initial increase in luminescence at around 10^{-15} M, followed by a leveling off, then a further increase at around 10^{-7} M. When a host strain lacking the chromosomal *ars* operon was used, the initial increase disappeared, giving a less sensitive response curve, suggesting that the chromosomally encoded ArsR repressor might be binding to the plasmid-borne *ars* operator with higher affinity than the plasmid-encoded ArsR (Ramanathan *et al.*, 1998).

Stocker *et al.* (2003) presented a set of As biosensor organisms based on *E. coli* using the promoter from plasmid R773 fused to one of three different reporter genes: β-galactosidase, *luxAB*, or Green Fluorescent Protein (GFP). To reduce basal activity of the promoter in the absence of As, a second copy of the repressor binding site was introduced between the *arsR* gene and the reporter gene (discussed further below). The β-galactosidase-based system was used to develop a semiquantitative assay using cells vacuum-dried on paper strips, with a chromogenic substrate (X-gal) leading to production of a blue spot in the presence of As. Such a system would have obvious advantages for field use, being readable by eye rather than requiring complex and expensive laboratory equipment. The luciferase-based assay, conducted in 96-well plates, gave a good response with samples containing between 8 and 78 µg L^{-1} As as As(III) following 1 hour incubation. The GFP-based sensor showed somewhat lower sensitivity but could be used to examine the response in single cells by epifluorescence microscopy, a useful tool in studying the stochastic nature of the response. The visually-read β-galactosidase system showed a clear difference from the As-free control at 0.2 µM As(III) (equivalent to 16 µg L^{-1} As) with 30 minutes incubation, a level suitable for screening of water samples in the field. The luciferase-based system was later field-tested in Vietnam for examination of genuine As contaminated water samples (Trang *et al.*, 2005), a test that surprisingly few systems in the literature have undergone. The same system was later used to test levels of As in rice (Baumann and van der Meer, 2007). Rice takes up As from water, so if irrigated with contaminated groundwater it may pose another, under-appreciated, route of exposure (Zhao *et al.*, 2010). Another system to be field-tested was an *E. coli* based system using the pI258 *ars* promoter with a red-shifted variant of GFP as reporter gene (Liao *et al.*, 2005). This was tested against As-contaminated water from Yun-Lin County, Taiwan. Results were very similar to those previously determined by chemical analysis (discussed further below).

12.3.3   *Speciation and bioavailability*

There are a number of potential advantages associated with the use of living microorganisms as biosensor elements. For example, they are 'self-manufacturing', and need merely be grown and harvested to be ready for use, whereas enzymes for enzymic biosensors must be purified and immobilized on the electrode or other sensor surface. Thus, whole-cell biosensors are potentially cheaper and easier to manufacture. Also, in the specific context of As, most cells will naturally provide As(V) reductase activity, or can easily be modified to do so; thus they will respond to both As(V) and As(III), whereas systems that rely on chemical or enzymic redox reactions will generally detect only one or the other form, unless a pre-processing interconversion step is added to the assay. Having said this, it is generally reported that the response to As(III) is considerably more sensitive than that to As(V). In some cases, a simple modification to the growth medium may help; for example, Turphinen *et al.* (2003) reported that a luminescent bacterial As biosensor grown in a minimal growth medium responded to As(III) at concentrations 100 times lower than those required for detection of As(V), whereas cells grown in a rich medium showed a 20-fold improvement in sensitivity to As(V) [that is, the response was approximately 5 times less sensitive than that to As(III)]. Possibly even better results might be achieved by constitutive expression of an increased level of As(V) reductase. A reasonably similar response to As(V) and As(III) would be highly desirable, as otherwise results would vary according to the ratio of these species in the sample.

Another advantage often cited is that such devices detect only 'bioavailable' As; that is, they respond only to that fraction of As available to interact with living organisms. The implicit assumption made in this case is that bioavailability to humans is the same as bioavailability to bacteria. For example, Turphinen et al. (2003) examined the differences between total water-soluble As (as determined by atomic absorption spectroscopy) and bioavailable As (as determined by use of a luminescent bacterial As biosensor) in As-contaminated soil samples from wood-processing sites in Finland. There was considerable variation in the results, with bioavailable As from two long-disused sites being around 15% of total As (t-As), whereas in a site still in use, the bioavailable As was around 35% of t-As measured, suggesting that As may become less bioavailable over time, at least under these conditions. It seems clear that bioavailability is a significant issue and that whole-cell biosensors are a suitable tool to study it. Interestingly, the field tests of groundwater in Taiwan reported by Liao et al. (2005), using a different bioreporter, showed a much closer correlation between AAS and bioreporter values; one well gave $165 +/- 50\,\mu g\,L^{-1}$ As by biosensor analysis, compared to $218\,\mu g\,L^{-1}$ by chemical analysis; another, $336 +/- 14\,\mu g\,L^{-1}$ by biosensor as compared to $362\,\mu g\,L^{-1}$ by chemical analysis. A similarly close correlation was reported for Vietnamese groundwater by Trang et al. (2005). This is perhaps related to the nature of the sample (groundwater as opposed to soil). It was noted that the presence of iron in groundwater samples greatly reduced biosensor response, possibly owing to the adsorption of arsenite to iron oxyhydroxide precipitates. Harms et al. (2005) examined this effect in more detail, and reported that 0.1 mM EDTA was able to release As from the precipitate, removing iron interference. Joshi et al. (2009) and de Mora et al. (2011) reported that the presence of bicarbonate ions in samples led to an apparent increase in the sensitivity of a bacterial bioreporter to As(V), an effect that does not seem to have been reported elsewhere. However, the response to As(III) was not investigated in these studies. Clearly any biosensor system intended for field use will have to be thoroughly tested for such effects.

12.3.4   Storage and distribution of whole cell biosensors

Clearly the use of whole cells, rather than isolated enzymes, as a recognition element, requires consideration of a number of factors. In particular, the biosensor organism must be stored and distributed in a way that preserves its viability; if the cells die, no response will be seen. Most bacteria can be preserved for long periods by freeze-drying (lyophilization), and this is an obvious choice for preservation of biosensor organisms; for example, Tauriainen et al. (1997) reported freeze drying of As biosensor organisms based on Staphylococcus aureus expressing firefly luciferase under the control of the pI258 ars promoter. Other methods may also be possible; for example, Kuppardt et al. (2009) tested various preservation methods for Escherichia coli-based As biosensor cells, and reported that the best results were achieved by vacuum drying at ambient temperature in the presence of polyvinylpyrrolidone and trehalose, with good results obtained after 12 weeks storage. Stocker et al. (2003) vacuum-dried cells from a raffinose-containing protectant solution on paper strips, as described above. These strips could be stored for 2 months at $-20°C$, $4°C$ or $30°C$, and then used with no detectable reduction in sensitivity. De Mora et al. (2010) obtained reasonable survival levels over several weeks using cells air dried from a 10% w/v lactose solution; lactose was used because trehalose, a more commonly used protective agent, was incompatible with the detection system (described further below).

An alternative option for storage and distribution is to use a host organism that naturally forms a dormant resting state that is stable under long term storage. Bacillus subtilis, a well characterized Gram positive bacterium, naturally forms a dormant resting state known as endospores. When B. subtilis cells detect that conditions are becoming unfavorable for growth, each cell undergoes an asymmetrical cell division to produce a large mother cell and a smaller cell known as a forespore. The mother cell grows to engulf the forespore, and secretes layers of protein coats around it, while the forespore produces a protective gel of calcium dipicolinate and a set of small acid-soluble proteins to protect its DNA and ribosomes. When the process is complete, the mother cell lyses to release the mature endospore. Endospores are extremely resistant to heat, drying and other stresses. There are well attested reports of endospores surviving in dry form for decades or even centuries (Nicholson *et al.*, 2000) and more controversial reports of survival for millions of years in unusual conditions, such as the intestines of insects preserved in amber (Cano and Borucki, 1995). When the spore detects that conditions are favorable, it germinates rapidly to regenerate an actively growing vegetative cell within a period of 30 minutes or so. The potential advantages of endospores for storage and distribution of biosensor organisms are obvious.

As noted above, *B. subtilis* possesses an As detoxification system similar to that found on the *E. coli* chromosome (Sato and Kobayashi, 1998). Biosensors can be prepared by introducing a suitable reporter gene adjacent to the *ars* promoter. A number of systems of this kind have been reported in the literature. As described above, Tauriainen *et al.* (1997, 1998) described biosensors based on *B. subtilis* cells expressing firefly luciferase under the control of the *ars* promoter, but did not report the use of endospores as a biosensor format. However, Date *et al.* (2007) reported the construction and testing of As biosensors based on endospores of *B. subtilis* and the related organism *B. megaterium*, using β-galactosidase as a reporter gene, with a chemiluminescent substrate or, alternatively, Enhanced Green Fluorescent Protein (EGFP). It was reported that the spores could be stored for more than 6 months in dry form prior to use. Similar biosensors for zinc have also been constructed. Fantino *et al.* (2009) reported the development of a device they referred to as a 'Sposensor', using *B. subtilis* endospores for detection of zinc and bacitracin, though not for As. In this system, β-galactosidase was used as a reporter gene, together with a chromogenic substrate, and spores were dried on filter paper discs. We ourselves developed a demonstration 'Bacillosensor', based on *B. subtilis* endospores expressing catechol-2,3-dioxygenase (XylE) under the control of the *ars* promoter. XylE acts on catechol, an inexpensive aromatic compound, to produce a bright yellow derivative, 2-hydroxy-*cis,cis*-muconic semialdehyde. Our initial system showed rather low sensitivity (Joshi *et al.*, 2009), but a refined version was able to detect As [as As(V)] at levels below the WHO limit of 10 μg L^{-1} (L. Montgomery and C. French, unpublished). Furthermore, spores could be boiled for two minutes prior to performing the assay. This step serves two purposes: it activates the spores for rapid germination, while also killing other bacteria in the water sample that might interfere with the assay. The potential advantages of this for a field assay are obvious.

12.3.5 *Outputs for visual detection*

For a field test assay, it might well be advantageous to develop a system that can easily be read visually without requiring the use of expensive laboratory equipment. One simple way to achieve this is to use a reporter gene encoding an enzyme that acts on a chromogenic substrate to generate a colored product. As described above (Table 12.2), the most common reporter gene used in this way is β-galactosidase (LacZ), which acts on the chromogenic substrate X-gal (5'-bromo-4'-chloro-3'-indolyl-β-D-galactopyranoside) to generate an insoluble blue pigment related to indigo, well suited for visual assessment. As described above, Stocker *et al.* (2003) developed a system of this type, with *E. coli* cells immobilized on a paper strip, producing a blue pigment on exposure to As. It was sensitive down to 16 μg L^{-1} As [as As(III)]. Alternatively, ONPG (*o*-nitrophenyl-β-D-galactopyranoside) is acted on by β-galactosidase to release a soluble yellow pigment, *o*-nitrophenol, which is better suited to spectrophotometric quantitation.

An alternative reporter suitable for visual detection is the catechol-2,3-dioxygenase (XylE) of *Pseudomonas putida*, as described above; this acts on catechol to produce the bright yellow product 2-hydroxy-*cis,cis*-muconic semialdehyde. The gene *xylE* is smaller than the complete form of *lacZ*, making it more convenient to work with, and catechol is very cheap, but unfortunately rather unstable in solution, rapidly oxidizing to produce brown polymeric substances, meaning that catechol solutions must be made up fresh and can not be stored for long periods (Joshi *et al.*, 2009; L. Montgomery and C. French, unpublished).

Some of the various relatives and derivatives of Green Fluorescent Protein, particularly the various forms of Red Fluorescent Protein (RFP), such as DsRed and mCherry, are also clearly visible by eye under normal illumination. Red Fluorescent Proteins do not seem to have been widely applied in As bioreporters, but Hakkila *et al.* (2002) investigated DsRed as a potential reporter, along with firefly luciferase, bacterial luciferase, and GFP, for Hg and As bioreporters. Bacterial and firefly luciferases gave faster and more sensitive responses than fluorescent proteins. Results were described for a DsRed-based Hg biosensor, but not for the equivalent As biosensor; DsRed gave a considerably lower and slower response than GFP, consistent with its longer folding time.

Another possibility is to use a pH-based response, which can easily be detected visually using a pH indicator solution. This gives a very bright and clear response and does not depend on the concentration of a chromogenic substrate or its product. The first such design, to the best of our knowledge, was proposed by a team of students from the University of Edinburgh as their entry for the 2006 International Genetically Engineered Machine competition (iGEM), an annual competition in synthetic biology organized by researchers at the Massachusetts Institute of Technology (MIT). The originally proposed form of the Edinburgh As biosensor (Aleksic *et al.*, 2007) involved a multi-stage output, with the growth medium becoming alkaline at very low levels of As, neutral at moderate levels, and acidic at dangerous levels. The *in vivo* signal processing system required to achieve this multi-stage output is discussed further below. The alkaline response was to be generated by the action of urease on urea, which was shown to raise the pH to around 10. The acidic response was produced by inducing β-galactosidase activity, allowing the fermentation of lactose by the mixed acid fermentation pathway naturally present in *E. coli*, which lowers the pH to around 4.5. These changes were to be made visible using a universal pH indicator solution (Riedel-de Häen, 36828), which is blue in alkaline solutions, green in neutral and red in acidic conditions. In practice, though the individual components were tested in isolation, only the acid-generating portion was assembled, using the *E. coli* chromosomal *ars* promoter; however, this was found to give a highly sensitive and visually obvious response to As(V) concentrations as low as $2.5\,\mu g\,L^{-1}$ (Aleksic *et al.*, 2007). This is an encouraging result: such bioreporters are usually much less sensitive to As(V) than to As(III) (see above), but As(III) sensitivity was not investigated in this study. Initially it was planned that an acidic (positive) response would be indicated as a red color using methyl red as a pH indicator; but methyl red was unstable in the presence of live cells in liquid cultures, with the red color rapidly fading. Bromothymol blue was therefore adopted as an alternative, with a positive response indicated by a change from blue to yellow (Joshi *et al.*, 2009). This gives an extremely bright and clear visual response. An inexpensive automated monitoring system was demonstrated by de Mora *et al.* (2011), consisting of a web-cam that simultaneously monitors an array of assays. Color information is extracted using open-source software. This is analyzed to determine the time at which color changes from blue to yellow, which is strongly correlated with the As concentration of the water sample. Such a system would seem well suited to local and regional testing laboratories. Effective As detection was demonstrated using genuine As-contaminated groundwater samples from Hungary. As with all iGEM projects, the biological components of the Edinburgh As biosensor are freely available in modular BioBrick™ form from the Registry of Standard Biological Parts at MIT.

Finally, it may be possible to use an endogenous pigment. Several systems of this type have been reported in the literature, based on organisms that naturally produce carotenoid pigments. Fujimoto *et al.* (2006) engineered the photosynthetic bacterium *Rhodovulum sulfidophilum* so that the *crtA* gene, encoding the enzyme responsible for the conversion of yellow spheroidene to red spheroidenone, was placed under the control of the *E. coli ars* promoter (introduced on a plasmid along with its regulatory *arsR* gene). In the presence of As(III), cell pigmentation changed from yellow to red. This color change was clearly visible by eye at $5\,\mu g\,L^{-1}$ As(III) after incubation in the light for 24 hours. However, the CrtA reaction is oxygen-dependent, so it is highly sensitive to the level of oxygen in the medium, which can be difficult to control. To overcome this, Yoshida *et al.* (2008) developed a similar system based on the related organism *Rhodopseudomonas palustris*. Here the gene *crtI*, encoding phytoene dehydrogenase, was placed under the control of the *E. coli ars* promoter in a *crtI* mutant host strain, which was blue-green in color owing to the production of non-carotenoid pigments. As(III) induced CrtI activity, allowing production of the red pigment lycopene. After 24 hours of incubation, a panel of observers judged that $10 \mu g L^{-1}$ As(III) could be detected on the basis of the red color produced. Sensitivity to As(V) was much lower, and the authors speculated that this might be due to low endogenous As(V) reductase activity in this species, as the *ars* promoter is known to respond more strongly to As(III) than to As(V).

To facilitate the construction of chromogenic biosensors, it would be useful to have a library of genetic cassettes resulting in the production of differently colored pigments. To this end, the University of Cambridge iGEM team (2009) have produced a set of genetic cassettes in BioBrick™ format, based on different combinations of genes from the carotenoid and violacein biosynthetic pathways, resulting in the formation of a variety of pigments including red, orange, yellow, green, blue and violet. All of these cassettes are freely available from the Registry of Standard Biological Parts.

12.3.6 *Outputs for direct electrical detection of reporter gene induction*

For quantitative output, and for data storage and analysis, it is advantageous if biosensors produce an output that can be easily converted to an electrical signal without the use of expensive equipment. The standard chromogenic, fluorescent and luminescent responses are usually detected and quantified using relatively expensive laboratory equipment, but some steps have been taken towards integrating such biosensor organisms into instruments (described further below).

One potential advantage of the pH-based Edinburgh As biosensor is that in addition to the obvious visual response obtained from the use of a pH-indicator solution, pH is easily quantified using a cheap glass pH electrode or a solid-state pH sensor (ion-selective field effect transistor, ISFET) (Aleksic *et al.*, 2007). This is an example of a potentiometric biosensor system in which the response is generated in the form of a voltage change.

The other major category of electrical output systems for whole cell biosensors is amperometric, where an electrical current is generated and quantified. Scott *et al.* (1997) described an amperometric system for detection of As(III), in which the reporter gene β-galactosidase acts on the substrate *p*-aminophenyl-β-D-galactopyranoside to release *p*-aminophenol, which is oxidized at an electrode held at a suitable potential. This system had the disadvantage of requiring cell lysis for the assay to be performed. In principle, a real time assay with living cells can be achieved by connection of the bacterial respiratory chain to an electrode. Bacterial respiration naturally generates a flow of electrical current, which can be coupled to an electrode by means of redox-active small molecules known as mediators. Some bacteria, such as *Shewanella* and *Geobacter*, can even transfer electrons to an electrode directly without the requirement for such mediators (Lovley, 2006). Relatively few amperometric whole-cell biosensor systems seem to have been reported in the literature. Two examples arose from student entries in iGEM. The 2007 iGEM team from the University of Glasgow reported a biosensor system in which the analyte of interest controlled a biosynthetic pathway producing a mediator, pyocyanin; the 2008 iGEM team from Harvard University reported a system in which one of the critical *Shewanella* proteins for transfer of electrons to an electrode, MtrB, was placed under the control of an analyte-responsive promoter. While neither system was designed to detect As, the modular nature of whole-cell biosensor systems means that it would be easy to modify them for this purpose, simply by using an As-responsive promoter.

12.3.7 *Integration of whole-cell biosensors into instruments*

In most reports, whole-cell biosensors are analyzed using laboratory instruments such as spectrophotometers, fluorimeters, and luminometers. Some are also designed for visual interpretation, as described above. However, a number of reports have also described the incorporation of living whole-cell biosensor organisms into instruments to make a dedicated biosensor system, as is commonly done with enzymic biosensors. One well-known example is the Bioluminescent Bioreporter Integrated Circuit (BBIC) system produced at the Centre for Environmental Biotechnology, University of Tennessee (Nivens *et al.*, 2004; Vijayaraghavan *et al.*, 2007). This incorporates bioluminescent biosensor organisms, expressing bacterial luciferase, into a chip containing a complementary metal oxide semiconductor (CMOS) microluminometer together with the circuitry required for signal processing and output. As far as we know, these devices have not been applied to As detection, but they should be fully compatible with the bioluminescent As sensors described above. A device of this type could provide a simple and inexpensive hand-held instrument that could be used for field testing for As and other pollutants.

Ivask et al. (2007) described an optical biosensor with biosensor organisms responsive to As and mercury, consisting of E. coli cells expressing bacterial luciferase under the control of the relevant promoters. The biosensor cells were immobilized in a layer of alginate gel on the surface of a fiber optic probe. Close contact between the cells and the probe surface eliminates possible interference due to turbidity in the water sample; but sensitivity was reduced in comparison with systems using non-immobilized bacteria, possibly indicating a diffusion limitation.

Several recent papers have also described the integration of whole-cell biosensor organisms into microfluidic devices. Rothert et al. (2005) generated an As biosensor consisting of E. coli expressing Green Fluorescent Protein, and incorporated this into a centrifugal microfluidic device in which pressure for liquid pumping was provided by rotation of the device. Fluorescence was quantified using a separate laboratory instrument, with a fiber-optic probe held above the detection chamber. As(III) could be detected at concentrations above 1×10^{-6} M (equivalent to 75 µg L^{-1} As) with a meaningful response generated over a range of two orders of magnitude. A later report (Date et al., 2010) described a similar device incorporating Bacillus endospore-based biosensor cells (see section 12.3.4 above), responsive to As and zinc, with both luminescence and fluorescence outputs. As(III) was detected at 10^{-7} M (7.5 µg L^{-1} As) using a lacZ-based detection system in which β-galactosidase cleaves 6-O-β-galactopyranosyl luciferin to release D-luciferin, which is then used a substrate by firefly luciferase. The time required for germination of spores and induction of the reporter gene was 2 hours. Diesel et al. (2009) reported the incorporation of an As biosensor based on E. coli expressing Green Fluorescent Protein into a microfluidic system. The luminescence from approximately 200 cells could be visualized using a separate camera. Subsequently Buffi et al. (2011) described a system using E. coli cells encapsulated in 50 µm agarose beads. As could be reliably detected at 1.6 µg L^{-1}, and the chips, with cells incorporated, could be stored at $-20°$C for at least one month. Again, a separate instrument was used to quantify fluorescence.

Thus, several reports have described the use of As bioreporters in microfluidic systems. Incorporation of miniaturized detectors, such as the BBIC, into such devices could be a useful step towards generation of simple handheld biosensor devices for use in the field.

12.3.8   Tuning, sensitivity modulation and in vivo signal processing

Critical parameters in the design of a biosensor, or any sensor, include the sensitivity (the lowest analyte concentration that can be detected) and the dynamic range (the range of analyte concentrations over which the analyte concentration can be determined from the sensor response). In enzymic biosensors, with an enzyme immobilized on a sensor surface, the kinetics of the response are controlled either by the affinity of the biological recognition element for the analyte, or by the rate of diffusion of the analyte to the sensor surface (French and Cardosi, 2007). The situation is more complex in the case of whole-cell biosensors (for a detailed discussion, see van der Meer et al., 2004). The analyte must first enter the cells. In the case of As(V) and As(III), this probably occurs via the phosphate uptake system for the former [Mukhopadyay et al., 2002; Silver and Phung, 2005]. It must then interact with the intracellular repressor protein, moderating the interaction of that protein with the promoter region of the DNA. This then alters the affinity with which RNA polymerase interacts with the promoter, which in turn alters the rate at which messenger RNA is produced. Finally, translation of the mRNA must occur to yield the reporter protein. The rate at which this occurs is dependent on the affinity of the ribosomes for the ribosome binding site preceding the coding sequence. Additionally, the level of reporter protein will depend on the rate at which messenger RNA and protein are degraded by the cell machinery, which depend on many factors.

In practice, the complexity of this system means that the sensitivity and dynamic range of the sensor can be 'tuned' by making a wide variety of straightforward modifications to the system, either rationally or using a random combinatorial approach. Importantly, the system can be tuned without any requirement to modify the repressor protein itself so as to change its affinity for the ligand. To give one trivial example, the pH-based Edinburgh biosensor was experimentally reassembled using different ribosome binding sites and altering the order of the genes, so that the reporter gene came between the promoter and *arsR*, rather than after *arsR*. This resulted in decreased sensitivity but an expanded dynamic range under the assay conditions tested (X. Wang and C. French, unpublished). In the original version of the sensor, under the conditions tested, the response saturated at less than $5 \mu g L^{-1}$ As, whereas in the reassembled version, saturation occurred at between 20 and $100 \mu g L^{-1}$ As. While the mechanism of this was not investigated, one plausible explanation would be a higher concentration of the repressor protein due to the stronger ribosome binding site used.

A number of other authors have described more targeted modifications to the biosensor system, to achieve particular aims. For example, Stocker *et al.* (2003) reported the addition of a second copy of the repressor binding site between *arsR* and the reporter gene, with the aim of reducing background expression in the absence of As. This led to considerably improved induction characteristics. Wackwitz *et al.* (2008) stated that modification of the activity or synthesis rate of the reporter protein, in this case cytochrome *c* peroxidase or β-galactosidase, led to strong changes to the response of the system to As at different concentrations. Tani *et al.* (2009) reported the generation of a GFP-based *E. coli* As biosensor bearing three tandem copies of the promoter-*gfp* cassette, with *arsR* expression controlled by a separate promoter. This doubled the signal to noise ratio, and reduced the detection limit from $20 \mu g L^{-1}$ to $7.5 \mu g L^{-1}$ As [as As(III)], as compared with the system with a single promoter-reporter cassette. A more ambitious and systematic exploration of this concept was undertaken by the iGEM team of Peking University in 2010 (iGEM 2010, Peking University). In this case, the biosensor used was for Hg, rather than for As, but the systems are very similar in nature. By placing the regulatory gene *merR*, encoding the regulatory protein for the Hg-responsive promoter, under the control of ribosome-binding sites of different strengths, a range of different sigmoidal response curves were obtained. A similar result was achieved by making a range of point mutations in the MerR binding site of the promoter.

By these methods, it is possible to generate a range of biosensors with different response curves. Given a suitably chosen array of such biosensors, it is possible to generate a type of system designated as a 'traffic-light' biosensor (Wackwitz *et al.*, 2008; van der Meer and Belkin, 2010). In this type of system, several different biosensor organisms are present in separate compartments. A given analyte (As) concentration will activate the more sensitive of the biosensors to give a visible response, but not the less sensitive ones. Thus, a type of semi-quantitative 'bar graph'-like response will be obtained, in which the analyte concentration can be estimated from the number of wells which show a positive response. This could be extremely useful for a field test system.

More complex and rational tuning systems have also been proposed. Rather than having the analyte-responsive promoter control the reporter gene directly, the analyte-responsive promoter can be used to control expression of a separate activator or repressor protein, which controls the expression of a second promoter, which in turn controls expression of the reporter gene. For tuning of biosensors using indirect reporter gene systems, the University of Cambridge iGEM team in 2009 presented a set of genetic cassettes, each consisting of the gene encoding a bacteriophage activator protein, followed by a transcription termination sequence, followed by a promoter that responds to the activator protein in question. These cassettes can be placed between an analyte-responsive promoter and a reporter gene to modulate the expression. A combinatorial approach was taken, using promoters from bacteriophages P2 (promoters P_F, P_O, P_P and P_V) and P4 (promoters P_{sid} and P_{LL}), and the activators from bacteriophages P2 (Ogr), P4 (δ protein), PSP3 (Pag) and φR73 (δ protein) (Julien and Calendar, 1996). Fifteen promoter-activator combinations were characterized, so that the designer of a biosensor can choose a cassette that will give the desired response characteristics. These genetic cassettes are available from the Registry of Standard Biological Parts.

Systems of this indirect type, while considerably more complex, offer greatly increased flexibility. For example, the response can be 'inverted': if the analyte-responsive promoter causes expression of a repressor protein that reduces expression of the second promoter, reporter gene expression will be reduced rather than increased by the presence of analyte. This type of arrangement can be used to generate a system with a kind of *in vivo* signal processing, for example, to generate multiple output signals depending on the analyte concentration. The original As biosensor design proposed by the University of Edinburgh iGEM team (Aleksic *et al.*, 2007), which was designed and mathematically modeled but never actually constructed, was of this type. The design called for a three-stage output. Without As, an alkaline pH signal would be generated through the action of urease. In the presence of moderate amounts of As, a highly sensitive As-responsive promoter would cause the expression of a repressor protein, which would prevent the expression of urease, leading to a neutral pH. In the presence of higher As concentrations a second, less sensitive, As-responsive promoter would cause the expression of β-galactosidase, as described above, leading to an acidic pH. Thus, depending on the As concentration, three qualitatively different responses would be seen, which could easily be distinguished by eye using a universal pH indicator solution: in As-free water, alkaline (blue); in water with borderline As concentrations, neutral (green or yellow); and in water with high As concentrations, acid (red). Such a qualitative response could be extremely useful in a field test to be used by local people with minimal training.

12.3.9   *Regulatory considerations relating to field use of genetically modified biosensor organisms*

Whole-cell biosensors of the type described in the preceding sections are, by their nature, genetically modified microorganisms and, as such, are regulated under various regimes depending on the context in which they are to be used. Use within a laboratory is relatively straightforward, with clear and well-defined rules. In the United Kingdom, such use is regulated by the Health and Safety Executive under the Genetically Modified Organisms (Contained Use) Regulations, 2000. Generally there is a requirement for a safety assessment, considering possible harm to employees or to the environment, a need for proper disposal of genetically modified organisms after use, and consideration of procedures to be employed in the case of accidents.

Some types of whole-cell biosensor, such as the SOS Chromotest (a mutagenicity-testing system based on recombinant *E. coli* expressing *lacZ* under the control of a promoter which responds to DNA damage), are designed to be used solely within a laboratory context, and it is clear that many proposed As biosensor formats could also be used in this way, for example in local or regional testing laboratories, for regular screening of samples. However, given the widely distributed nature and rural locations of many tube wells, it would be advantageous if devices could be prepared and taken into the field for rapid on-site screening, or to allow local people to monitor their own water supply. Many reports in the literature have described biosensor formats that are aimed at such use. For example, as described above, Stocker *et al.* (2003) described a format with *E. coli* cells immobilized in a strip of gel, generating a colored spot *via* a chromogenic reporter system, and Lumin Sensors (www.luminsensors.com) has proposed a disposable plastic device as a format for field use of the pH-based Edinburgh As biosensor. Some authors have also considered the possibility of implementing As-biosensor systems in bacteria able to colonize contaminated soil. For example, Petänen *et al.* (2001) reported construction of a bioluminescent As biosensor system in the soil bacterium *Pseudomonas fluorescens*, with the stated aim of constructing strains that could be used to study the distribution of bioavailable As in microcosms or soil core samples. Potentially, such an organism could also be used for direct *in situ* studies.

The regulations regarding genetically modified organisms in out-of-laboratory contexts seem to be less clear, since once the organism is outside the laboratory, containment can not be guaranteed, especially if the device is intended to be used by field workers or local people without extensive training. In the UK, the most relevant legislation would seem to be the Genetically Modified Organisms (Deliberate Release) Regulations, 2002 (implementing EU Directive 2001/18/EC). However, this legislation is designed to regulate organisms such as genetically modified plants and microorganisms designed to be able to survive and persist in the environment (e.g., plant-interacting microorganisms for agricultural use, or microorganisms to be used in the bioremediation of contaminated land). The application procedure is therefore complex and expensive, and requires designated release sites and extensive post-release monitoring. These stringent regulations may be inappropriate for biosensor organisms, which can be designed specifically not to survive in the environment, and not to compete effectively with wild-type microorganisms. In fact, special design for this purpose may not even be necessary, since most reported biosensor systems are based on laboratory host strains, which possess multiple disabling mutations and show reduced growth rates, and are therefore intrinsically incapable of effectively competing with wild-type microorganisms (see, for example, Chart *et al.*, 2000). In the USA, the relevant legislation at the federal level would appear to be the Toxic Substances Control Act (Sayre and Seidler, 2005). Other countries may not have well developed legislation regarding the use of genetically modified microorganisms, but it would seem ethically questionable at best to promote the use of a system that could not legally be used in the country where it was developed.

The current U.K. legislation on the use of genetically modified microorganisms outside a laboratory context appears to limit the attractiveness of genetically modified whole-cell biosensors for field use. It would be useful if future legislation could specifically consider the case of genetically modified organisms that are specifically designed not to be able to survive in the environment. A test application is required to demonstrate proof of principle and to force consideration of these issues. Once one such system has been approved, it will set a precedent and open the way for further applications. Since the As crisis is a major public health issue, and several As biosensor systems are at a development stage where field testing would be a realistic next step, it would seem to provide an excellent test case for clarification of these issues.

12.4 CONCLUSIONS

A wide variety of biosensor systems for detection of As(V) and As(III) have been reported in the literature. Generally these show excellent sensitivity, responding to As(III) concentrations well below $10 \, \mu g \, L^{-1}$. There are also a wide range of available output modalities. However, to the best of our knowledge, no devices have yet been adopted for use, despite apparent advantages over the existing screening technologies. Very few systems have been tested under field conditions; few, if any, are commercially available at the time of writing. In terms of whole-cell biosensors more generally, as far as we know, no such device is in general use apart from the non-specific toxicity sensors, such as MicroTox, and mutagen-testing kits such as SOS Chromotest. However, improved formats are appearing in the literature with increasing frequency. Since the detection of As in drinking water is a very major application with a clear public good, and since it is relatively straightforward to implement whole-cell biosensors, it may be that this will become a flagship application for such devices. Widespread adoption of one such device might resolve regulatory issues and allow a range of other applications to enter the field. Several groups are working towards bringing their own biosensor concepts to the market. It will be interesting to watch developments over the next few years.

ACKNOWLEDGEMENTS

The authors would like to acknowledge valuable input from Dr. David Grimshaw (Practical Action) and Mr. Matthew Owens (Engineers Without Borders). Some unpublished experiments described in this chapter were performed in the laboratory of C. French by Lucy Montgomery and Xiaonan Wang. The authors have received no funding from any commercial enterprise for work described in this chapter.

REFERENCES

Akter, A. & Ali, M.H.: Arsenic contamination in groundwater and its proposed remedial measures. *Int. J. Environ. Sci. Tech.* 8:2 (2011), pp. 433–443.

Aleksic, J., Bizzari, F., Cai, Y., Davidson, B., de Mora, K., Ivakhno, S., Seshasayee, S.L., Nicholson, J., Wilson, J., Elfick, A., French, C.E., Kozma-Bognar, L., Ma, H. & Millar, A.: Development of a novel biosensor for the detection of arsenic in drinking water. *IET Synthetic Biology* 1:1–2 (2007), pp. 87–90.

Arora, M., Megharaj, M. & Naidu, R.: Arsenic testing field kits: some considerations and recommendations. *Environ. Geochem. Health* 31 (2009), pp. 45–48.

Baumann, B. & van der Meer, J.R.: Analysis of bioavailable arsenic in rice with whole cell living bioreporter bacteria. *J. Agr. Food Chem.* 55 (2007), pp. 2115–2120.

Belkin, S.: Microbial whole-cell sensing systems of environmental pollutants. *Curr. Opin. Microbiol.* 6 (2003), pp. 206–212.

Buffi, N., Merulla, D., Beutier, J., Barbaud, F., Beggah, S., van Lintel, H., Renaud, P. & van der Meer, J.R.: Development of a microfluidics biosensor for agarose-bead immobilized bioreporter cells for arsenite detection in aqueous samples. *Lab Chip* 11 (2011), pp. 2369–2377.

Cai, J. & DuBow, M.S.: Expression of the *Escherichia coli* chromosomal *ars* operon. *Can. J. Microbiol.* 42:7 (1996), pp. 662–671.

Cano, R.J. & Borucki, M.K.: Revival and identification of bacterial spores in 25 million year old to 40 million year old Dominican amber. *Science* 268:5213 (1995), pp. 1060–1064.

Chakraborti, D., Rahman, M.M., Das, B., Murrill, M., Dey, S., Mukherjee, S.C., Dhar, R.K., Biswas, B.K., Chowdhury, U.K., Roy, S., Sorif, S., Selim, M., Rahman, M. & Quamruzzaman, Q.: Status of groundwater arsenic contamination in Bangladesh: a 14 year study report. *Water Res.* 44 (2010), pp. 5789–5802.

Chart, H., Smith, H.R., La Ragione, R.M. & Woodward, M.J.: An investigation into the pathogenic properties of *Escherichia coli* strains BLR, BL21, DH5α and EQ1. *J. Appl. Microbiol.* 89 (2000), pp. 1048–1058.

Corbisier, P., Ji, G., Nuyts, G., Mergeay, M. & Silver, S.: *luxAB* gene fusions with the arsenic and cadmium resistance operons of *Staphylococcus aureus* plasmid pI258. *FEMS Microbiol. Lett.* 110 (1993), pp. 231–238.

Date, A., Pasini, P. & Daunert, S.: Construction of spores for portable bacterial whole-cell biosensing systems. *Anal. Chem.* 79 (2007), pp. 9391–9397.

Date, A., Pasini, P. & Daunert, S.: Integration of spore-based genetically engineered whole-cell sensing systems into portable centrifugal microfluidic platforms. *Anal. Bioanal. Chem.* 398 (2010), pp. 349–356.

Daunert S., Barret, G., Feliciano, J.S., Shetty, R.S., Shrestha, S. & Smith-Spencer, W.: Genetically engineered whole-cell sensing systems: coupling biological recognition with reporter genes. *Chem. Rev.* 100 (2000), pp. 2705–2738.

de Mora, K., Joshi, N., Balint, B.L., Ward, F.B., Elfick, A. & French, C.E.: A pH-based biosensor for detection of arsenic in drinking water. *Anal. Bioanal. Chem.* 400 (2011), pp. 1031–1039.

Deshpande, L.S. & Pande, S.P.: Development of arsenic testing field kit – a tool for rapid on-site screening of arsenic contaminated water sources. *Environ. Monit. Assess.* 101 (2005), pp. 93–101.

Diesel, E., Schreiber, M. & van der Meer, J.R.: Development of bacteria-based bioassays for arsenic detection in natural waters. *Anal. Bioanal. Chem.* 394 (2009), pp. 687–693.

Diorio, C., Cai, J., Marmor, J., Shinder, R. & DuBow, M.S.: An *Escherichia coli* chromosomal *ars* operon homolog is functional in arsenic detoxification and is conserved in Gram negative bacteria. *J. Bacteriol.* 177:8 (1995), pp. 2050–2056.

Erickson, B.E.: Field kits fail to provide accurate measure of arsenic in groundwater. *Environ. Sci. Technol.* 37:1 (2003), pp. 35A–38A.

Fantino, J.-R., Barras, F. & Denizot, F.: Sposensor: a whole-cell bacterial biosensor that uses immobilized *Bacillus subtilis* spores and a one-step incubation/detection process. *J. Mol. Microbiol. Biotechnol.* 17 (2009), pp. 90–95.

French, C.E. & Cardosi, M.F.: Biosensors in bioprocess monitoring and control. In: E.M.T. El-Mansi, C.F.A. Bryce, A.L. Demain & A.R. Allman (eds): *Fermentation microbiology and biotechnology.* 2nd edition, CRC Press, Boca Raton, FL, USA, 2007, pp. 363–406.

Fujimoto, H., Wakabayashi, M., Yamashiro, H., Maeda, I., Isoda, K., Kondoh, M., Kawase, M., Miyasaka, H. & Yagi, K.: Whole cell arsenite biosensor using photosynthetic bacterium *Rhodovulum sulfidophilum.* *Appl. Microbiol. Biot.* 73:2 (2006), pp. 332–338.

Hakkila, K., Maksimow, M., Karp, M. & Virta, M.: Reporter genes *lucFF, luxCDABE, gfp,* and *dsred* have different characteristics in whole-cell bacterial sensors. *Anal. Biochem.* 301 (2002), pp. 235–242.

Harms, H., Rime, J., Leupin, O., Hug, S.J. & van der Meer, J.R.: Effect of groundwater composition on arsenic detection by bacterial biosensors. *Microchim. Acta* 151 (2005), pp. 217–222.

Hossain, M.F.: Arsenic contamination in Bangladesh – an overview. *Agric. Ecosystems Environ.* 113 (2006), pp. 1–16.

Hossain, M.A., Sengupta, M.K., Ahamed, S., Rahman, M.M., Mondal, D., Lodh, D., Das, B., Nayak, B., Roy, B., Mukherjee, A. & Chakraborti, D.: Ineffectiveness and por reliability of arsenic removal plants in West Bengal, India. *Environ. Sci. Technol.* 39 (2005), pp. 4300–4306.

Hussam, A., Alauddin, M., Khan, A.H., Rasul, S.B. & Munir, A.K.M.: Evaluation of arsine generation in arsenic field kit. *Environ. Sci. Technol.* 33 (1999), 3686–3688.

International Genetically Engineered Machine Competition (iGEM) 2006, University of Edinburgh: http://parts.mit.edu/wiki/index.php/University_of_Edinburgh_2006 (accessed 29 July 2011).

International Genetically Engineered Machine Competition (iGEM) 2007, University of Glasgow: http://parts.mit.edu/igem07/index.php/Glasgow (accessed 29 July 2011).

International Genetically Engineered Machine Competition (iGEM) 2008, Harvard University: http://2008.igem.org/Team:Harvard (accessed 29 July 2011).

International Genetically Engineered Machine Competition (iGEM) 2009, University of Cambridge: http://2009.igem.org/Team:Cambridge (accessed 29 July 2011).

International Genetically Engineered Machine Competition (iGEM) 2010, Peking University: http://2010.igem.org/Team:Peking (accessed 29 July 2011).

Ivask, A., Green, T., Polyak, B., Mor, A., Kahru, A., Virta, M. & Marks, R.: Fibre-optic bacterial biosensors and their application for the analysis of bioavailable Hg and As in soils and sediments from Aznalcollar mining area in Spain. *Biosens. Bioelectron.* 22 (2007), pp. 1396–1402.

Ji, G. & Silver, S.: Regulation and expression of the arsenic resistance operon from *Staphylococcus aureus* plasmid pI258. *J. Bacteriol.* 174 (1992), pp. 3684–3694.

Joshi, N., Wang, X., Montgomery, L., Elfick, A. & French, C.E.: Novel approaches to biosensors for detection of arsenic in drinking water. *Desalination* 248:1–3 (2009), pp. 517–523

Julien, B. & Calendar, R.: Bacteriophage PSP3 and φR73 activator proteins: analysis of promoter specificities. *J. Bacteriol.* 178:19 (1996), pp. 5668–5675.

Kuppardt, A., Chatzinotas, A., Breuer, U., van der Meer, J.R. & Harms, H.: Optimization of preservation conditions of arsenic (III) bioreporter bacteria. *Appl. Microbiol. Biot.* 82 (2009), pp. 785–792.

Labuda, J., Bubnicova, K., Kovalova, L., Vanickova, M., Mattusch, J. & Wennrich, R.: Voltammetric detection of damage to DNA by arsenic compounds at a DNA biosensor. *Sensors* 5 (2005), pp. 411–423.

Liao, V.H.-C. & Pu, K.-L.: Development and testing of a green fluorescent protein-based bacterial biosensor for measuring bioavailable arsenic in contaminated groundwater samples. *Environ. Toxicol. Chem.* 24:7 (2005), pp. 1624–1631.

Lin, Y.-F., Walmsley, A.R. & Rosen, B.P.: An arsenic metallochaperone for an arsenic detoxification pump. *P. Natl. Acad. Sci. USA* 103:42 (2006), pp. 15617–15622.

Lovley, D.: Bug Juice: harvesting electricity with microorganisms. *Nat. Rev. Microbiol.* 4:7 (2006), pp. 497–508.

Male, K.B., Hrapovic, S., Santini, J.M. & Luong, J.H.T.: Biosensor for arsenite using arsenite oxidase and multiwalled carbon nanotube modified electrodes. *Anal. Chem.* 79 (2007), pp. 7831–7837.

Mays, D.E. & Hussam, A.: Voltammetric methods for determination and speciation of inorganic arsenic in the environment – a review. *Anal. Chim. Acta* 646 (2009), pp. 6–16.

Meharg, A.: *Venomous earth: how arsenic caused the world's worst mass poisoning.* Macmillan, Basingstoke, U.K. 2005.

Mukhopadyay, R., Rosen, B.P., Phung, L.T. & Silver, S.: Microbial arsenic: from geocycles to genes and enzymes. *FEMS Microbiol. Rev.* 26 (2002), pp. 311–325.

Nicholson, W.L., Munakata, N., Horneck, G., Melosh, H.J. & Setlow, P.: Resistance of *Bacillus* endospores to extreme terrestrial and extraterrestrial environments. *Microbiol. Mol. Biol. Rev.* 64:3 (2000), pp. 548–572.

Nivens, D.E., McKnight, T.E., Moser, S.A., Osbourn, S.J., Simpson, M.L. & Sayler, G.S.: Bioluminescent bioreporter integrated circuits: potentially small, rugged and inexpensive whole-cell biosensors for remote environmental monitoring. *J. Appl. Microbiol.* 96 (2004), pp. 33–46.

Oremland, R.S. & Stolz, J.F.: The ecology of arsenic. *Science* 300:5621 (2003), pp. 939–944.

Ozsoz, M., Erdem, A., Kara, P., Kagan, K. & Ozkan, D.: Electrochemical biosensor for the detection of interaction between arsenic trioxide and DNA based on guanine signal. *Electroanal.* 15 (2003), pp. 613–619.

Pal, P., Bhattacharyay, D., Mukhopadyay, A. & Sarkar, P.: The detection of mercury, cadmium, and arsenic by the deactivation of urease on rhodinized carbon. *Environ. Eng. Sci.* 26:1 (2009), pp. 25–32.

Pande, S.P., Deshpande, L.S. & Kaul, S.N.: Laboratory and field assessment of arsenic testing field kits in Bangladesh and West Bengal, India. *Environ. Monit. Assess.* 68 (2001), pp. 1–18.

Panthi, S.R., Sharma, S. & Mishra, A.K.: Recent status of arsenic contamination in groundwater of Nepal – a review. *Kathmandu Univ. J. Sci. Eng. Technol.* 2 (2006), pp. 1–11.

Petänen, T., Virta, M., Karp, M. & Romantschuk, M.: Construction and use of broad host range mercury and arsenite sensor plasmids in the soil bacterium *Pseudomonas fluorescens* OS8. *Microbial Ecol.* 41 (2001), pp. 360–368.

Rahman, M., Mukherjee, D., Sengupta, M.K., Chowdhury, U.K., Lodh, D., Chanda, C.R., Roy, S., Selim, M., Quamrussaman, Q., Milton, A.H., Shahidullah, S.M., Rahman, M.T. & Chakraborti, D.: Effectiveness and reliability of arsenic field testing kits: are the million dollar screening projects effective or not? *Environ. Sci. Technol.* 36 (2002), pp. 5385–5394.

Ramanathan, S., Shi, W., Rosen, B.P. & Daunert, S.: Sensing antimonite and arsenite at the subattomole level with genetically engineered bioluminescent bacteria. *Anal. Chem.* 69:16 (1997), pp. 3380–3384.

Ramanathan, S., Shi, W., Rosen, B.P. & Daunert, S.: Bacteria-based chemiluminescence sensing system using β-galactosidase under the control of the ArsR regulatory protein of the *ars* operon. *Anal. Chim. Acta* 369 (1998), pp. 189–195.

Registry of Standard Biological Parts: http://partsregistry.org/Main_Page (accessed 29 July 2011).

Rothert, A., Deo, S.K., Millner, L., Puckett, L.G., Madou, M.J. & Daunert, S.: Whole-cell-reporter-gene-based biosensing systems on a compact disk microfluidics platform. *Anal. Biochem.* 342 (2005), pp. 11–19.

Safarzdeh-Amiri, A., Fowlie, P., Kazi, A.I., Siraj, S., Ahmed, S. & Akbor, A.: Validation of analysis of arsenic in water samples using Wagtech Digital Arsenator. *Sci. Total Environ.* 409 (2011), pp. 2662–2667.

Sanllorente-Méndez, S., Dominguez-Renedo, O. & Martinez, M.J.: Immobilization of acetylcholinesterase on screen-printed electrodes: application to the determination of arsenic(III). *Sensors* 10 (2010), pp. 2119–2128.

Sarkar, P., Banerjee, S., Bhattacharyay, D. & Turner, A.P.F.: Electrochemical sensing systems for arsenate estimation by oxidation of L-cysteine. *Ecotox. Environ. Safe.* 73 (2010), pp. 1495–1501.

Sato, T. & Kobayashi, Y.: The *ars* operon in the skin element of *Bacillus subtilis* confers resistance to arsenate and arsenite. *J. Bacteriol.* 180:7 (1998), pp. 1655–1661.

Sayre, P. & Seidler, R.J.: Application of GMOs in the US: EPA research and regulatory considerations related to soil systems. *Plant Soil* 275:1/2 (2005), pp. 77–91.

Scott, D., Ramanathan, S., Shi, W., Rosen, B.P. & Daunert, S.: Genetically engineered bacteria: electrochemical sensing system for antimonite and arsenite. *Anal. Chem.* 69 (1997), pp. 16–20.

Silver, S. & Phung, L.T.: A bacterial view of the periodic table: genes and proteins for toxic inorganic ions. *J. Ind. Microbiol. Biot.* 32 (2005), pp. 587–605.

Stocker, J., Balluch, D., Gsell, M., Harms, H., Feliciano, J., Daunert, S., Malik, K.A. & van der Meer, J.R.: Development of a set of simple bacterial biosensors for quantitative and rapid measurements of arsenite and arsenate in potable water. *Environ. Sci. Technol.* 37 (2003), 4743–4750.

Stolz, J.F., Basu, P., Santini, J.M. & Oremland, R.S.: Arsenic and selenium in microbial metabolism. *Annu. Rev. Microbiol.* 60 (2006), pp. 107–130.

Steinmaus, C.M., George, C.M., Kalman, D.A. & Smith, A.H.: Evaluation of two new arsenic field test kits capable of detecting arsenic water concentrations close to $10\,\mu g\,L^{-1}$. *Environ. Sci. Technol.* 40 (2006), pp. 3362–3366.

Tani, C., Inoue, K., Tani, Y., Harun-ur-Rashid, M., Azuma, N., Ueda, S., Yoshida, K. & Maeda, S.: Sensitive fluorescent microplate bioassay using recombinant *Escherichia coli* with multiple promoter-reporter units in tandem for detection of arsenic. *J. Biosci. Bioeng.* 108:5 (2009), pp. 414–420.

Tauriainen, S., Karp, M., Chang, W. & Virta, M.: Recombinant luminescent bacteria for measuring bioavailable arsenite and antimonite. *Appl. Environ. Microbiol.* 63:11 (1997), pp. 4456–4461.

Tauriainen, S., Virta, M., Chang, W. & Karp, M. Measurement of firefly luciferase reporter gene activity from cells and lysates using *Escherichia coli* arsenite and mercury sensors. *Anal. Biochem.* 272 (1999) 191–198.

Tauriainen, S.M., Virta, M.P.J. & Karp, M.T.: Detecting bioavailable toxic metals and metalloids from natural water samples using luminescent sensor bacteria. *Water Res.* 3:10 (2000), pp. 2661–2666.

Trang, P.T.K., Berg, M., Viet, P.H., Mui, N.V. & van der Meer, J.R.: Bacterial bioassay for rapid and accurate analysis of arsenic in highly variable groundwater samples. *Environ. Sci. Technol.* 39 (2005), pp. 7625–7630.

Turphinen, R., Virta, M. & Häggblom, M.H.: Analysis of arsenic bioavailability in contaminated soils. *Environ. Chem.* 22:1 (2003), pp. 1–6.

van der Meer, J.R., Tropel, D. & Jaspers, M.: Illuminating the detection chain of bacterial bioreporters. *Environ. Microbiol.* 6:10 (2004), pp. 1005–1020.

van der Meer, J.R. & Belkin, S.: Where microbiology meets microengineering: design and applications of reporter bacteria. *Nat. Rev. Microbiol.* 8 (2010), pp. 511–522.

Vijayaraghavan, R., Islam, S.K., Zhang, M., Ripp, S., Caylor, S., Bull, N.D., Moser, S., Terry, S.C., Blalock, B.J. & Sayler, G.S.: Bioreporter bioluminescent integrated circuit for very low-level chemical sensing in both gas and liquid environments. *Sensor Actuator* B 123 (2007), pp. 922–928.

Wackwitz, A., Harms, H., Chatzinotas, A., Breuer, U., Vogne, C. & van der Meer, J.R.: Internal arsenite bioassay calibration using multiple reporter cell lines. *Microb. Biot.* 1 (2008), pp. 149–157.

Wu, J. & Rosen, B.P.: Metalloregulated expression of the ars operon. *J. Biol. Chem.* 268 (1993), pp. 52–58.

Yoshida, K., Inoue, K., Takahashi, Y., Ueda, S., Isoda, K., Yagi, K. & Maeda, I.: Novel carotenoid-based biosensor for simple visual detection of arsenite: characterization and preliminary evaluation for environmental application. *Appl. Environ. Microbiol.* 74 (2008), pp. 6730–6738.

Zhao, F.-J., McGrath, S.P. & Meharg, A.A.: Arsenic as a food chain contaminant: mechanisms of plant uptake and metabolism and mitigation strategies. *Annu. Rev. Plant Biol.* 61 (2010), pp. 535–559.

Author index

Subject index

Printed and bound by CPI Group (UK) Ltd, Croydon, CR0 4YY

18/10/2024

01776253-0007